De Gruyter Graduate
Darvas, Dormán, Hessel • Flow Chemistry

Also of Interest

Sustainable Process Engineering – Prospects and Opportunities
Koltuniewicz, 2014
ISBN 978-3-11-030875-4, e-ISBN 978-3-11-030876-1

Chemical Reaction Technology
Murzin, 2015
ISBN 978-3-11-033643-6, e-ISBN 978-3-11-033644-3

Process Integration and Intensification – Saving Energy, Water and Resources
Klemeš, Varbanov, Wan Alwi, Manan, 2014
ISBN 978-3-11-030664-4, e-ISBN 978-3-11-030685-9

Chemical Photocatalysis
König, 2013
ISBN 978-3-11-026916-1, e-ISBN 978-3-11-026924-6,
Set-ISBN 978-3-11-026925-3

Green Processing and Synthesis
Hessel (Editor-in-Chief)
ISSN 2191-9550

Flow Chemistry

Volume 2: Applications

Edited by
Ferenc Darvas, György Dormán, Volker Hessel

DE GRUYTER

Editors
Ferenc Darvas
Florida International University
College of Medicine
University Park, 495
11 200 S.W. 8th St.
Miami 33 199
USA
e-mail: ferenc.darvas@darholding.com

György Dormán
ThalesNano Nanotechnology Inc
Graphisoft Park
Zahony u. 7
Budapest 1031
Hungary
e-mail: gyorgy.dorman@thalesnano.com

Volker Hessel
Eindhoven Univ. of Technology
Micro Flow Chem. & Proc. Techn. Group
Dept. of Chemistry & Chemical Eng.
Den Dolech 2
5600 MB Eindhoven
The Netherlands
e-mail: v.hessel@tue.nl

ISBN 978-3-11-036707-2
e-ISBN 978-3-11-036750-8

Library of Congress Cataloging-in-Publication Data
A CIP catalog record for this book has been applied for at the Library of Congress.

Bibliographic information published by the Deutsche Nationalbibliothek
The Deutsche Nationalbibliothek lists this publication in the Deutsche Nationalbibliografie;
detailed bibliographic data are available on the Internet at http://dnb.dnb.de.

© 2014 Walter de Gruyter GmbH, Berlin/Boston
Typesetting: le-tex publishing services GmbH, Leipzig
Printing and binding: CPI books GmbH, Leck
Cover image: Book cover design by Reka Darvas (www.zenurdel.com)
♾ Printed on acid-free paper
Printed in Germany

www.degruyter.com

Preface

Flow chemistry – the use of small flow reactors to perform chemical synthesis – has matured over the past two decades from early demonstrations of simple chemical transformations in microstructured reactors (microreactors) to complex, multistep synthesis relevant to fine chemistry and pharmaceuticals in commercial systems. This evolution in synthetic methods and equipment has been motivated by advantages inherent to continuous synthesis in small scale, specifically enhanced rates from improved heat and mass transfer along with an expanded space of reactions and process conditions. Continuous operation also eliminates headspace issues and avoids accumulation of reactive or toxic intermediates offering opportunities for telescoping of reactions. Synthesis applications are further enhanced by automated optimization as well as mechanistic and kinetic information gained from integrating reaction components with sensors, actuators, and automated fluid handling. Moreover, the steady state operation inherent in continuous operation provides robustness, stability, and scalability.

The expansion in flow chemistry applications and equipment has been detailed in numerous review papers and monographs, but there has been a longstanding need for a comprehensive coverage of the many concepts underlying flow chemistry for graduate students in chemistry and chemical engineering. The present *Graduate Textbook on Flow Chemistry* fills the gap in graduate education by covering chemistry and reaction principles along with current practice, including examples of relevant commercial reaction, separation, automation, and analytical equipment. It motivates the reasons for flow chemistry and importantly when flow chemistry will and will *not* be advantageous compared to batch processing. Basic theory and practical considerations are summarized to enable the reader to appreciate the difference between conventional batch chemistry and flow chemistry as well as to implement flow chemistry in the laboratory. A very useful feature is the inclusion of validate reactions that can serve as laboratory test experiments. The subsequent treatment of theoretical foundations for flow chemistry, also know as reaction engineering, provides useful in depth understanding of continuous reactions.

The second portion of the *Graduate Textbook on Flow Chemistry* covers specific reaction classes, concepts, and experimental methods. Homogeneous and heterogeneous catalysis, supercritical processes, photochemistry, green chemistry, and radiolabelled chemistry applications are described in individual chapters along with examples of flow chemistry for nanotechnology and materials science. Practical oriented chapters address (i) analytical techniques, specifically in-line monitoring methods, (ii) examples of automation, (iii) how to build your own flow chemistry set-up as well an overview of commercially available units, and (iv) importantly, safety aspects of flow chemistry systems and processes.

The Editors of this *Graduate Textbook on Flow Chemistry*, Drs. Ferenc Darvas, Volker Hessel and György Dormán are commended for having taken the initiative to bring together experts from the field to provide a comprehensive treatment of fundamental and practical considerations underlying flow chemistry. It promises to become a useful study text and as well as reference for the graduate students and practitioners of flow chemistry.

June 2014

Klavs Jensen
Department Head Chemical Engineering,
Massachusetts Institute of Technology, USA

The Editors would like to express their gratitude to the many people who helped to complete this textbook. They are indebted to all the authors for their outstanding contribution and the valuable and constructive suggestions during the planning. They are very grateful to Prof. Dr. Jan van Hest (POAC Committee, Radboud University Nijmegen, The Netherlands); to Prof. Floris Rutjes (Radboud University Nijmegen); to Dr. Varsha Kapoerchan (Organisation for Scientific Research NWO, Advanced Chemical, Technologies for Sustainability (ACTS), The Netherlands) and to Darholding Inc. (Hungary) for their financial support. Prof. Volker Hessel kindly acknowledges the funding provided by the Advanced European Research Council Grant "Novel Process Windows – Boosted Micro Process Technology" (no 267 443). Special thanks should be given to all the instrument suppliers for their contributions to the Microreactor Chapter (Chemtrix, FutureChemistry, Invenios, Microinnova, Syrris, ThalesNano, Uniqsis). The Editors' thanks is extended to Ms. Szilvia Gilmore (Flow Chemistry Society) for the coordination and monitoring duties during the preparation of the textbook, to Ms. Karin Sora, Editorial Director Chemistry/Materials Science and Ms. Julia Lauterbach, Project Editor STM, DeGruyter Publishing House for their enthusiasm, continuing motivation and technical support as well as to Reka Darvas for the great cover design.

Contents

Preface —— v
About the editors —— xiii
Contributing authors —— xv
Abbreviations —— xvii

Part I   Catalysis and activation

Clemens Brechtelsbauer and King Kuok (Mimi) Hii
1   Catalysis in flow —— 3
1.1   Introduction —— 3
1.1.1   Flow versus batch chemistry —— 3
1.1.2   Development of catalytic reactions and flow for organic synthesis —— 3
1.2   Reactor types, catalytic reactions and productivity —— 5
1.2.1   Solid-liquid reactors —— 6
1.2.2   Solid-liquid-gas systems —— 18
1.3   Conclusion —— 25

Claude de Bellefon
2   Catalytic engineering aspects of flow chemistry —— 31
2.1   Introduction —— 31
2.2   Basis of (catalytic) reactor engineering —— 33
2.2.1   Flow motion in reactors —— 33
2.2.2   Relevant physics —— 36
2.2.3   Characteristic times —— 36
2.2.4   Characteristic lengths —— 38
2.2.5   Surface area —— 40
2.2.6   Mixing —— 41
2.2.7   Heat issues —— 42
2.3   Describing the chemistry —— 43
2.3.1   Kinetic rate laws —— 43
2.3.2   Rate measurement and reaction time —— 45
2.3.3   Catalyst deactivation —— 47
2.4   Methodology for Flow reactor dimensioning —— 51
2.4.1   Batch versus Flow reactor comparison —— 51
2.4.2   Checking for mass and heat transfer limitations —— 54
2.4.3   Basis for reactor scale-up —— 59
2.5   Conclusion —— 61

Thomas H. Rehm
3 Continuous-flow photochemistry in microstructured environment —— 63
3.1 Environmental impact in view of *Green Chemistry* —— 63
3.2 Physical considerations – reasons why microstructured equipment is preferred for flow photochemistry —— 64
3.2.1 Absorption of light by molecules in solution —— 64
3.2.2 Role of solvent —— 66
3.2.3 Micrometer-sized structures as key elements of reactor equipment for flow photochemistry —— 66
3.3 Technological considerations for flow photochemistry —— 68
3.3.1 Light sources —— 68
3.3.2 Reactor concepts for flow photochemistry —— 73
3.4 Chemical considerations for flow photochemistry —— 78
3.4.1 Photochemical reactions without catalyst material —— 78
3.4.2 Heterogeneous flow photocatalysis —— 80
3.4.3 Flow photocatalysis with organic dyes or noble metal complexes —— 84
3.5 Summary and outlook —— 91

Julian Schuelein and Holger Loewe
4 Electrochemistry in flow —— 99
4.1 Introduction —— 99
4.2 Electrochemistry in flow —— 100
4.3 Microreactor design —— 103
4.3.1 Thin gap cells —— 104
4.3.2 ELMI – microstructured high pressure single pass thin gap flow cell —— 111
4.3.3 Segmented thin gap flow cells —— 114
4.4 Electrochemistry in microreactors —— 116
4.4.1 Direct product synthesis —— 116
4.4.2 Electrolyte free synthesis —— 117
4.4.3 Activation of chemicals —— 119
4.5 Ionic liquids in electrochemistry —— 122

Part II Cutting-edge applications in advanced and functional materials

L. Zane Miller, Jeremy L. Steinbacher, and D. Tyler McQuade
5 Synthesis of materials in flow – principles and practice —— 133
5.1 Introduction —— 133
5.2 Unique properties of microreactors —— 133
5.2.1 Mixing —— 133
5.2.2 Thermal and pressure control —— 134
5.2.3 Fluid behavior —— 134

5.3	Synthesis of materials in flow —— 140
5.3.1	Linear polymers —— 140
5.3.2	Beads, disks, and other solid polymeric materials —— 144
5.3.3	Janus materials —— 149
5.3.4	Capsules —— 150
5.3.5	Membranes and fibers —— 152
5.3.6	Nanoparticles and inorganic nonpolymeric materials —— 154
5.4	Conclusions —— 156

Genoveva Filipcsei, Zsolt Otvos, Reka Angi, and Ferenc Darvas

6	**Flow chemistry for nanotechnology** —— 161
6.1	Introduction to nanotechnology and graphene technology —— 161
6.1.1	Introduction —— 161
6.1.2	Definition and concepts —— 161
6.1.3	Brief history of nanotechnology —— 162
6.1.4	Why nanotechnology? —— 163
6.1.5	Batch and flow-chemistry based nanonization technologies —— 164
6.1.6	Overview and principles of microfluidic reactors —— 165
6.2	Nanomaterials —— 166
6.2.1	Structure and properties: is the smaller better? —— 166
6.2.2	Organic nanoparticles: biologically active small molecules —— 169
6.2.3	Inorganic nanoparticles: metallic, bimetallic and semiconductor particles —— 171
6.2.4	Hybrid nanoparticles —— 172
6.3	Theoretical background of nanoparticle synthesis using flow-chemistry based approaches —— 173
6.3.1	Principles of nanoparticle stabilization —— 173
6.3.2	Classical nucleation theory —— 174
6.4	Application of flow technology in nanoparticle synthesis —— 176
6.4.1	Synthesis of metal nanoparticles —— 176
6.4.2	Synthesis of semiconductor nanoparticles —— 177
6.4.3	Synthesis of biologically active organic nanoparticles —— 178
6.5	Impact of nanotechnology: an outlook —— 182

Samar Haroun, Paul C. H. Li

7	**Continuous-flow synthesis of carbon-11 radiotracers on a microfluidic chip** —— 189
7.1	Introduction to continuous-flow microreactors and carbon-11 radiolabeling —— 189
7.2	Microfluidic synthesis of raclopride —— 192
7.2.1	Microfluidic nonradioactive synthesis of raclopride —— 194
7.2.2	Microchip radioactive synthesis of [^{11}C]raclopride —— 196

7.3	Computational fluid dynamics (CFD) —— **200**
7.3.1	Reaction engineering lab®(REL) module – "ideal" flow-reactor model —— **201**
7.3.2	Microelectromechanical system (MEMS) module – "geometry-dependent" flow study —— **203**
7.4	Conclusion —— **207**

Part III Additional features of the Flow Process: in-line analytics, safety and green principles

Ferenc Darvas, György Dormán, and Melinda Fekete

8	**Lab environment: in-line separation, analytics, automation & self optimization —— 213**
8.1	The role of analytics in flow applications —— **213**
8.1.1	Applications of mass spectroscopy —— **214**
8.1.2	ReactIR flow cell —— **218**
8.1.3	Nuclear magnetic resonance (NMR) —— **224**
8.2	Automation and self optimization —— **228**
8.2.1	General description of the self-optimization methods —— **228**
8.2.2	Automation and feedback control systems —— **230**
8.2.3	Nelder–Mead Simplex method —— **234**
8.2.4	Multidimensional optimization —— **235**
8.2.5	Optimization and scale-up —— **236**
8.2.6	Flow reactors with built-in optimization —— **238**
8.3	In-line separation —— **239**
8.3.1	Liquid-liquid separators —— **239**
8.3.2	Scavenger and chromatography columns —— **241**
8.3.3	Simulated moving Bed Chromatography —— **243**

Jean-Christophe Monbaliu, Ana Cukalovic, and Christian V. Stevens

9	**Safety aspects related to microreactor technology —— 253**
9.1	Introduction —— **253**
9.1.1	Chemical processes —— **253**
9.1.2	Safety in chemical processes —— **254**
9.2	Inherently safer processes using microreaction technology —— **254**
9.2.1	Advantages of microreaction technology to safety —— **254**
9.2.2	Recent examples of processes involving dangerous reagents/reactions under MRT conditions —— **258**
9.2.3	MRT processes involving harsh conditions (elevated temperatures and pressures) —— **274**
9.3	Conclusions —— **275**

Volker Hessel, Qi Wang, and Dana Kralisch
10 From green chemistry principles in flow chemistry towards green flow process design in the holistic viewpoint —— 283
10.1 Introduction of Green Chemistry principles —— 283
10.1.1 Green principles —— 283
10.1.2 Green flow chemistry —— 285
10.2 Flow process design and relation to green chemistry/engineering —— 285
10.2.1 Flow processing – major means in process intensification —— 285
10.2.2 Transport intensification – the flow-scale —— 286
10.2.3 Chemical intensification – the reactor scale —— 286
10.2.4 Process-design intensification – the full-process scale —— 287
10.2.5 Elemental green criteria with proven impact of flow process design —— 287
10.2.6 Elemental green criteria with suspected impact of flow process design —— 288
10.2.7 Elemental green criteria with uncertainty over impact of flow process design —— 288
10.3 Holistic methodology introduction for systematic green flow process design —— 289
10.4 Green flow process design for fine chemicals/pharmaceuticals —— 293
10.4.1 Technology comparison for green pharmaceutical process design —— 293
10.4.2 Flow process design of a green biphasic fine chemical synthesis —— 295
10.4.3 Exergetic LCA for improvement of an existing pharmaceutical production process —— 297
10.5 Green flow process design for bulk chemicals and benchmark to conventional process —— 298
10.5.1 Process simulation —— 299
10.5.2 LCA for continuous flow synthesis of ADA —— 302
10.5.3 LCA for two-step conventional synthesis of ADA —— 303
10.5.4 Complete LCA picture —— 303
10.5.5 Enlightment —— 305
10.6 Outlook for green flow process design —— 306

Answers to the study questions —— 313
Index —— 327

About the editors

Prof. Ferenc Darvas acquired his degrees in Budapest, Hungary (medical chemistry MS, computer sciences BS, degree in patent law, PhD in experimental biology). He has been teaching in Hungary, Spain, Austria, and in the United States of America at different universities, presently serves as associate professor at the Florida International University in Miami. He is author of 140 pre-reviewed papers and 5 books. Dr. Darvas has been involved in introducing microfluidics/flow chemistry methodologies for synthetizing drug candidates since the late 90's, which led him to found ThalesNano. One of his team's inventions, the desktop high pressure/high temperature flow hydrogenator H-Cube won several innovation awards in the United States of America and also in Europe, and has been used in more than 60 countries. Dr. Darvas is also the founder and active President of the Flow Chemistry Association located in Switzerland.

Prof. György Dormán obtained his Ph.D. in organic chemistry from the Technical University of Budapest in 1986. Between 1986–1988 and 1996–1999 he worked at Sanofi–Chinoin in Budapest. In 1988–1989 he spent a post-doctoral year in the UK (University of Salford). Between 1992 and 1996 he was a Visiting Scientist at the State University of New York, Stony Brook. Between 1999 and 2008 he served ComGenex/AMRI as Chief Scientific Officer. Since 2008 he is responsible for the scientific innovation of ThalesNano. Dr. Dormán is involved in many training courses in the area of (bio)organic and flow chemistry. In 2011 he became Professor at University of Szeged. He is an author of 85 scientific papers and book chapters. He is a member of the editorial board of Molecular Diversity and the advisory board of J. Flow Chemistry.

Prof. Volker Hessel studied chemistry at Mainz University (PhD in organic chemistry, 1993). In 1994 he entered the Institut für Mikrotechnik Mainz GmbH (1996: group leader microreaction technology). In 2002, Prof. Hessel was appointed Vice Director R&D at IMM and in 2007 as Director R&D. In 2005 and 2011, he was appointed as part-time and full professor at Eindhoven University of Technology, respectively, for the chair of "Micro Flow Chemistry and Process Technology". He is (co-)author of more than 270 peer-reviewed publications, with 18 book chapters and 5 books. He received the AIChE award "Excellence in Process Development Research" in 2007 and in 2010 the ERC Advanced Grant "Novel Process Windows". Prof. Hessel is in the scientific advisory board of the "International Conference on Microreaction Technology". He is Editor-in-Chief of the journal "Green Processing and Synthesis".

Contributing authors

Reka Angi
NanGenex
Budapest, Hungary
e-mail: reka.angi@nangenex.com
Chapter 6

Claude de Bellefon
CPE Lyon
University of Lyon
Lyon, France
e-mail: claude.debellefon@lgpc.cpe.fr
Chapter 2

Clemens Brechtelsbauer
Department of Chemical Engineering
Imperial College London
London, UK
e-mail: c.brechtelsbauer@imperial.ac.uk
Chapter 1

Ana Cukalovic
SynBioC
Department of Sustainable Organic Chemistry
and Technology
Ghent University
Gent, Belgium
e-mail: ana.cukalovic@ugent.be
Chapter 9

Ferenc Darvas
College of Medicine
Florida International University
Miami, USA
e-mail: ferenc.darvas@darholding.com
Chapter 6, 8

György Dormán
ThalesNano Nanotechnology Inc
Budapest, Hungary
e-mail: gyorgy.dorman@thalesnano.com
Chapter 8

Melinda Fekete
ThalesNano
Budapest, Hungary
e-mail: fekete.mela@gmail.com
Chapter 8

Genoveva Filipcsei
NanGenex
Budapest, Hungary
e-mail: filipcsei.genoveva@nangenex.com
Chapter 6

Samar Haroun
Department of Chemistry
Simon Fraser University
Burnaby, Canada
e-mail: sharoun2@gmail.com
Chapter 7

Volker Hessel
Dept. of Chemistry & Chemical Eng.
Eindhoven Univ. of Technology
Eindhoven, The Netherlands
e-mail: v.hessel@tue.nl
Chapter 10

King Kuok (Mimi) Hii
Department of Chemistry
Department of Chemical Engineering
Imperial College London
London, UK
e-mail: mimi.hii@imperial.ac.uk
Chapter 1

Dana Kralisch
Institute for Technical Chemistry and
Environmental Chemistry
Friedrich-Schiller-University
Jena, Germany
e-mail: dana.kralisch@uni-jena.de
Chapter 10

Paul C.H. Li
Department of Chemistry
Simon Fraser University
Burnaby, Canada
e-mail: paulli@sfu.ca
Chapter 7

Holger Loewe
Fraunhofer ICT-IMM
and
Institute for Organic Chemistry
Johannes Gutenberg-University
Mainz, Germany
e-mail: Loewe@imm-mainz.de
Chapter 4

David Tyler McQuade
Department of Chemistry and Biochemistry
Florida State University
Tallahassee, FL, USA
e-mail: mcquade@chem.fsu.edu
Chapter 5

L. Zane Miller
Department of Chemistry and Biochemistry
Florida State University
Tallahassee, FL, USA
e-mail: levimiller@chem.fsu.edu
Chapter 5

Jean-Christophe Monbaliu
Department of Chemistry
Center for Integrated Technology and Organic Synthesis
University of Liège, Sart-Tilman
Liège, Belgium
e-mail: jc.monbaliu@ulg.ac.be
Chapter 9

Zsolt Otvos
NanGenex
Budapest, Hungary
e-mail: zsolt.otvos@nangenex.com
Chapter 6

Thomas H. Rehm
Fraunhofer ICT-IMM
Mainz, Germany
e-mail: Rehm@imm-mainz.de
Chapter 3

Julian Schuelein
Fraunhofer ICT-IMM
and
Institute for Organic Chemistry
Johannes Gutenberg-University
Mainz, Germany
e-mail: julschue@students.uni-mainz.de
Chapter 4

Jeremy L. Steinbacher
Department of Chemistry and Biochemistry
Canisius College
Buffalo, NY, USA
e-mail: steinbaj@canisius.edu
Chapter 5

Christian V. Stevens
SynBioC
Department of Sustainable Organic Chemistry and Technology
Ghent University
Gent, Belgium
e-mail: Chris.Stevens@UGent.be
Chapter 9

Qi Wang
Micro Flow Chemistry & Process Technology
Department of Chemical Engineering and Chemistry
Eindhoven University of Technology
Eindhoven, The Netherlands
e-mail: Q.Wang1@tue.nl
Chapter 10

Abbreviations

Ad	adamantyl
ADA	adipic acid
ADP	abiotic resource depletion
AE	atom economy
AO	anthraquinone oxidation
AP	acidification
API	active pharmaceutical ingredient
ATR	attenuated total reflectance
BEMP	2-tert-2-diethylamino-1,3-dimethyl-perhydro-1,2,3-diazophosphorine
C-11	carbon-11
Ca	capillary number
CEENE	extraction from the natural environment
CFD	computational fluid dynamics
cGMP	current good manufacturing practice
CQD	colloidal quantum dots
CRTR	continuous recycled tube reactor
CSB	chemical safety and hazard investigation board
CSTR	continuous stirred-tank reactor
CV	coefficients of variation
DBU	1,8-diazabicyclo-[5.4.0]undec-7-ene
DEDAM	diethyl(diallyl)malonate
DFT	density functional theory
DHA	dihydroacetone
DIBAL-H	diisobutyl-aluminum hydrid
DIPEA	N,N-diisopropylethylamine
DLS	dynamic light scattering
DLVO theory	Deryaguin and Landau and Verwey and Overbeek
DMF	N,N-dimethyl formamide
DMIT	dimethyl itaconate
DMR	desmethyl raclopride
DMSO	dimethyl sulfoxide
ee	enantiomeric excess
EGDMA	ethyleneglycol dimethacrylate
ELMI	electrochemical microreactor
EMIM	[CF_3SO_3] 1-ethyl-3-methilimidazolium trifluoromethanesulfonate
EP	eutrophication
ESI	electrospray ionisation

ETP	eco-toxicity
FDG	fluorodeoxyglucose
FEP	fluorinated ethylenepropylene
FT-IR	Fourier transform-infrared spectroscopy
GC	gas chromatography
GNRs	gold nanorods
GWP	global warming
HOMO	highest occupied molecular orbital
HPLC	high performance liquid chromatography
HSV	hourly space velocity
HTP	human toxicity
ILs	ionic liquids
IR	infrared
IS	internal standard
ISO	the International Standard Organisation
KFT	Karl–Fischer titration
LC	liquid chromatography
LC/MS	liquid chromatography/mass spectrometry
LCA	life cycle assessment
LCC	life cycle costing
LED	light emitting diode
LH	Langmuir Hinshelwood
LHSV	liquid hourly space velocity
LIGA	lithography galvanic molding
LOC	lab-on-a-chip
LU	land use
LUMO	lowest unoccupied molecular orbital
MCT	mercury cadmium telluride
MD	molecular dynamics
MEMS	microelectromechanical system module
MI	mass intensity
MMA	alpha-acetamidoacrylic acid methyl ester
MRT	micro reaction technology
NCA	lysine, alanine, leucine, or glutamic acid
NMO	N-methylmorpholine-N-oxide
NMR	nuclear magnetic resonance
NPV	net present value
NSAIDs	non-steroidal anti-inflammatory drugs
NTU	number of transfer units
ODP	ozone depletion
OLEDs	organic light emitting diodes
OSN	organic solvent nanofiltration

PBRs	packed-bed reactors
PDI	polydispersity index
PDMS	poly(dimethylsiloxane)
Pe	peclet number
PEEK	polyether ether ketone
PET	positron emission tomography
PF	plug flow
PFA	perfluoroalkoxy
PFR	plug-flow reactor
PLGA	poly(d,l-lactic acid-co-glycolic acid)
PMI	process mass intensity
POCP	photochemical ozone creation
PPi	pyrophosphate
PS-TBD	polystyrene-supported 1,5,7-triazabicyclo[4.4.0]dec-5-ene
PTFE	polytetrafluoroethylene
PVA	poly(vinylalcohol)
PVC	polyvinyl chloride
PVP	poly(vinyl)pyridine
RAD	radioactivity detector
RAFT	reversible addition-fragmentation chain transfer
RCM	ring closing metathesis
RCY	radiochemical yield
Re	Reynolds Number
REL	reaction engineering laboratory module
REO	robust, efficient and orthogonal
Rf	radiofrequency
RME	reaction mass efficiency
ROMP	ring-opening polymerization
RTD	resistive thermal device
RTILs	room temperature ionic liquids
RU	repeating units
S/C ratio	substrate/catalyst
SCFs	supercritical fluids
SET	single electron transfer
SFT	staggered fed tube
SILP	supported ionic-liquid phase
SLCA	simplified LCA
SM	Suzuki–Miyaura
SMB	simulated moving-bed
SNR	signal-to-noise ratio
SSRE	solid-state-reference electrodes
STBE	solketal t-butyl ether

TEM	transmission electron microscopy
TFSI	trifluoromethylsulfonyl)imide
TMAOH	tetramethylammonium hydroxide
TOF	turnover frequency
TON	turnover number
TPGDA	tripropyleneglycol diacrylate
UV	ultraviolet
We	Weber number
WHSV	weight hourly space velocity
µSSRE	miniaturized solid-state-reference electrodes

Part I: Catalysis and activation

Clemens Brechtelsbauer and King Kuok (Mimi) Hii
1 Catalysis in flow

1.1 Introduction

The practice of catalysis on the industrial scale is closely associated with flow. Indeed, the most extensive and important industrial processes, namely, the Haber–Borsch process, water-gas shift reaction and methanol production, are all conducted under continuous flow (CF) over (heterogeneous) catalysts.

1.1.1 Flow versus batch chemistry

Catalytic flow chemistry embraces some of the most important principles of Green Chemistry and Engineering. The basic function of a catalyst is to lower the activation energy, and therefore the overall energy consumption, of a given reaction, while side-products can be eliminated or reduced by designing more atom-economical and selective processes. Concurrently, engineering tools can be applied to shift an unfavorable thermodynamic equilibrium towards product formation via the operation of Le Chatelier's principle (as is the case with the three industrial processes mentioned above).

Traditionally, fine and pharmaceutical industrial sectors have largely favored the use of batch reactors, due to the long sequences of unit operations required for the assembly of complex molecules, their low production volume, and last but by no means least, the inherent similarity between a chemist's round bottom flask and an engineer's stirred tank reactor. In recent years, however, there is increasing recognition of the potential value of flow processes [1], particularly in combination with catalytic reactions [2], as a way of mitigating the high E-factors associated with the production of complex molecules [3]. Even more importantly, the use of flow reactors enables highly reactive intermediates and gases to be handled safely, thus enabling otherwise hazardous reactions to be performed at scale with no risks to the operator.

In this subchapter, the application of immobilized organo- and metal catalysts in combination with flow chemistry will be discussed.

1.1.2 Development of catalytic reactions and flow for organic synthesis

For the development of catalytic flow processes, it is almost always preferable to work with multiphase reactions, where the catalyst is immobilized to allow it to be easily separated and/or recovered from the product. There are many different ways of keeping catalysts separated from the reaction mixture and/or the product stream, which have been described in recent reviews [4–6]. As organic reactants are almost always dissolved in a solvent, the immobilization of catalyst invariably generates biphasic solid-liquid or liquid-liquid, or sometimes, gas-liquid-solid tri-phasic reaction mixtures. In turn, catalytic activity can be limited by mass transport efficiency or through catalyst deactivation. Almost all pharmaceutically active substances are multifunctional molecules, and it is not uncommon that the reaction partners themselves can poison the catalyst. Furthermore, catalyst leaching is also a significant issue, particularly for the manufacturing of pharmaceutical products where the amount of impurities in the final product (including residual metal) is subject to strict quality control. Catalyst leaching is dependent on several factors, including the nature of the support, the choice of solvent, reaction temperature and the nature of the reactant/products.

> **Fixed-bed (or packed-bed) reactor:** Usually a tubular reactor in which the heterogeneous catalyst is fixed in a packed-bed between filters. The packed-bed can consist solely of the catalyst, or can be diluted through the use of inert material of similar or larger particle size. This can be used to, for example, dissipate heat, or reduce the amount of pump energy that is required to generate the flow, that is, lower back-pressure.
>
> **Mass transfer:** The movement of chemical species from one physical phase (e.g., a gas) into a different phase (e.g., a liquid or a solid) which is governed by diffusion and convection. The rate at which mass transfer occurs has a direct impact on the rate of catalytic reactions, as only molecules that have been transported to the active centers of the catalyst can react.

Most commercially-available catalysts consist of metallic particles supported on inorganic supports (e.g., alumina, silica, titania, ceria, and zeolites), or incarcerated within organic matrixes (e.g., polyurea, PVP). While these catalysts are highly effective for certain processes (such as hydrogenation reactions), they are not generally sufficiently active or selective enough for organic synthesis. In contrast, catalysts based on discreet organometallic complexes, organocatalysts, or enzymes can offer exquisite (stereo-)selectivity, but will require some form of immobilization, most commonly by attachment onto an insoluble solid support [7], or be retained in a distinct liquid phase, for example an ionic liquid [8].

The ultimate design of the flow reactor is very much determined by the nature of the catalyst and their support, particularly physical properties such as catalyst loading, density, pore and particle sizes. When a fixed-bed reactor is deployed, a compromise often has to be struck between residence time for maximum conversion, mass transport and unacceptable back-pressure; certain polymeric supports, for example polystyrene, swell upon exposure to organic solvents. The size and configuration of a flow reactor are also determined by catalyst activity and stability, which will determine the optimal residence time (rate and selectivity), and heat management (heat of reaction).

The unique physical properties of ionic liquids (IL's) and supercritical CO_2 (sc-CO_2) have been widely exploited in the development of continuous flow processes involving organometallic [9] and bio-catalysts [10]. Ionic liquids may be used to dissolve the catalyst, generating a solution which will remain immiscible with the organic solvent during the reaction [11]. To circumvent the mass-transport problem associated with the mixing between two liquid phases, a thin film of the catalytic solution can be dispersed onto a solid support with a high surface area, such as silica, for deployment in a fixed-bed reactor. The method is known as Supported Ionic-Liquid Phase (SILP) catalysis. On the other hand, sc-CO_2 has been substituted for organic solvents in several flow systems. This is particularly useful as the physical property of CO_2 can be varied between a gas and a liquid at different pressures. Therefore, leaching is less problematic with such systems, facilitating recovery and reuse of the catalyst. However, solubility of polar molecules in sc-CO_2 is a major limitation against its wider use in organic chemistry. The potential for CO_2 to react with nucleophilic reagents, for example amines, is another major consideration.

1.2 Reactor types, catalytic reactions and productivity

The productivity of a catalytic flow process is largely dependent upon catalytic activity, defined by the electronic and steric nature of the catalyst as well as the reactant(s). Application of an appropriate flow reactor enhances mass transport (e.g., mixing, adsorption/desorption of substrates onto a catalyst surface), heat management (e.g., effective dissipation of a reaction exotherm) and precise control of residence time (flow rate). For multiphasic systems, the design and implementation of the flow reactor is equally important as the nature of the catalyst. In this subchapter, the turnover frequency (TOF), as mmol(product)/mmol(cat)/h, will be adopted as the indication of productivity. However, it has to be noted that these numbers are often extrapolated and may not correspond to catalyst longevity, that is, the extent of catalyst deactivation and leaching, which can only be revealed by time-on-stream studies. These are not always performed in the laboratory, but are essential before a process can be scaled up into production.

Turnover frequency: A measurement of how often the catalytic cycle occurs per unit of time and, therefore, a quantitative description of the activity and effectiveness of a catalyst. Although strictly speaking associated with the number of catalytically active sites, in practice it gets more conveniently quantified by amount of product generated per amount of catalyst and unit of time.

Catalyst longevity: Time that a catalyst is able to maintain a certain level of activity or productivity when exposed to the reaction system, that is, the lifetime of a catalyst. All catalysts lose activity with time, which influences their useful *time-on-stream*.

Time-on-stream: The length of time that a catalyst can remain in a reactor at reaction conditions without considerable loss of activity or selectivity.

The aim of this subchapter is to provide an overview of different immobilized catalysts developed for flow chemistry. Two main types of multiphasic system (solid-liquid or solid-liquid-gas) will be discussed, and examples of C-C and C-X bond forming reactions will be selected to demonstrate some of the key issues of these systems.

1.2.1 Solid-liquid reactors

Solid-liquid packed-bed reactors (PBRs) can be easily assembled, even in a modestly-resourced laboratory: the simplest arrangement composes of a chromatographic column packed with the heterogeneous/immobilized catalyst, connected to high-performance liquid chromatography (HPLC) or syringe-pumps for the delivery of liquid reactants under ambient conditions. There are two general arrangements for PBRs – single-pass or recirculation mode (Figure 1.1). The single-pass arrangement is the traditional mode of operation for a tubular reactor under continuous flow conditions, whereas complete recirculation flow mode is essentially the equivalent of a batch process, performed in a perfectly mixed stirred vessel. Through the application of suitable regulation valves, the degree of re-circulation can be varied by substituting a percentage of product draw-off with fresh feed thus converting the batch reaction into a truly continuous process. This gives the option to approach the mixing mode of a continuous stirred tank, which is inherently different to that of a tubular reactor in single pass mode. The different mixing environments can give rise to different catalytic selectivity. This is particularly important in the case of competitive reaction schemes, where by-product formation may determine the feasibility of a process.

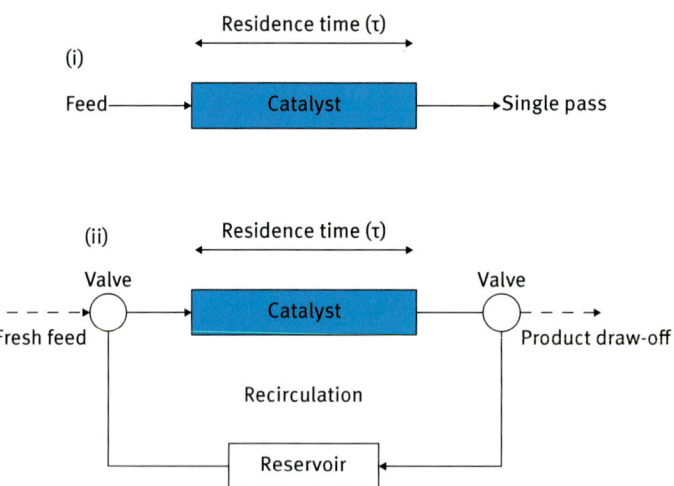

Fig. 1.1: Single-pass/recirculation flow mode arrangements for PBRs.

> **Residence time:** Also called space-time, is the length of time calculated by the ratio of reactor volume and volumetric flow rate. It defines the average time a molecule spends in a chemical reactor, and is roughly the equivalent of a *batch time* for continuous flow situations.
>
> **Batch time:** The length of time a reaction mixture spends in a chemical laboratory reactor, that is, the reaction time. Distinct from the plant batch time, which includes times needed for charging, discharging and cleaning operations.

In later production, catalyst deactivation can be managed in both modes through using two or more reactors in parallel, with one reactor on stream while the other is on standby. The quality of the product is monitored through a suitable on-line analytical method – as soon as it drops below an acceptable value, the feed stream is switched to the fresh reactor. The reactor with deactivated catalyst can now be cleaned or replaced until another switch is necessary. Due to the (normally) short residence times in these systems the product loss during non-steady state operation is relatively small, but it can also be factored into the design of the control algorithms for these systems.

The section on solid-liquid reactors will be largely focused on the application of flow to two different C-C bond forming processes, arguably the most important catalytic reactions discovered in the twentieth century, acknowledged by the award of Nobel Prizes in Chemistry to Yves Chauvin, Robert H. Grubbs and Richard R. Schrock *"for the development of the metathesis method in organic synthesis"* (2005), followed by Richard F. Heck, Ei-ichi Negishi and Akira Suzuki *"for palladium-catalyzed cross-couplings in organic synthesis"* (2010) [12]. This will be followed by a section on the application of enzymes in kinetic-resolution processes, which is industrially relevant, and conclude with the related area of metal-free organocatalysts for C-C bond forming reactions.

1.2.1.1 Olefin (ring closing) metathesis reactions

Olefin metathesis reactions constitute one of the most important classes of C=C bond forming processes in industry, ranging from the production of oleochemicals, to polymers and pharmaceutical products [13–16]. For organic synthesis, ring closing metathesis (RCM) reactions are widely used for the formation of macrocyclic compounds [17]. The value of the RCM reaction has been demonstrated with the commercial synthesis of Ciluprevir **1** (Figure 1.2), an NS3 serine protease inhibitor marketed by Boehringer Ingelheim for the treatment of hepatitis C [18].

Although RCM reactions are entropically favored, they are liable to thermodynamic equilibria depending on the size of the (macro-)cyclic ring, so the removal of a volatile by-product during the reaction, most frequently ethylene, is necessary to facilitate the formation of the desired product. At the same time, dilute solutions are necessary to suppress the formation of oligomers. This requires a highly active catalyst that does not deactivate easily, so that meaningful turnovers may be obtained within

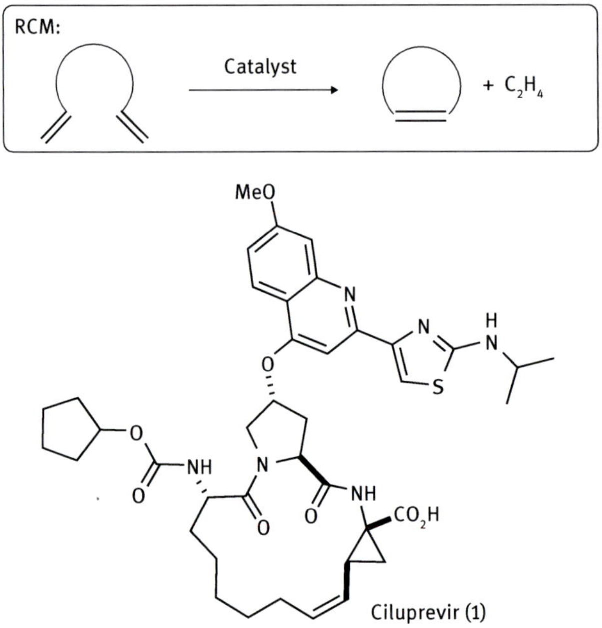

Fig. 1.2: General reaction scheme of olefin ring-closing metathesis (RCM) reaction and the structure of Ciluprevir.

a reasonable timescale [19]. A myriad of catalysts have been developed for the olefin metathesis reaction, with the air- and moisture- stable ruthenium-based (Grubbs') catalysts being particularly popular. The removal of Ru residues from the metathesis product is also a significant problem in these reactions [20].

Immobilized catalysts in packed-bed reactors (Figure 1.3, Table 1.1*).* Buchmeiser et al. reported the preparation of a variant of Grubbs' 2nd generation catalyst **2** (Ad = adamantyl); modified at the backbone of the imidazolidine group so that it can be grafted onto a monolithic support of well-defined microglobules (1.5 µm) via a ring-opening polymerization (ROMP) reaction [21]. For the RCM of diethyl(diallyl)malonate (DEDAM), a TOF of 25 min^{-1} was recorded with a residence time of 3 minutes, corresponding to a productivity of 500 (Table 1.1, entry 1). However, leaching of the Ru was observed in the product ($\leq$ 70 ppm), and catalyst stability over a longer timescale was not assessed. Subsequently, a modified Grubbs'-Hoveyda (G-H) catalyst **3** was prepared, where a Ru-acetate linkage was used to attach the catalyst to the monolithic structure [22]. The productivity of the resultant catalyst **3** appears to be lower than that of **2** (Table 1.1, entry 2). Significantly, less than 0.2% of the catalyst was lost to the product. However, significant catalyst deactivation was observed – starting from 70% single-pass conversion initially, decreasing to just 15% after 2 h.

Catalyst **4** was prepared by a modification of the benzylidene ligand backbone, in order to graft the G-H catalyst onto a mesoporous silica support, via a 1,3-dipolar

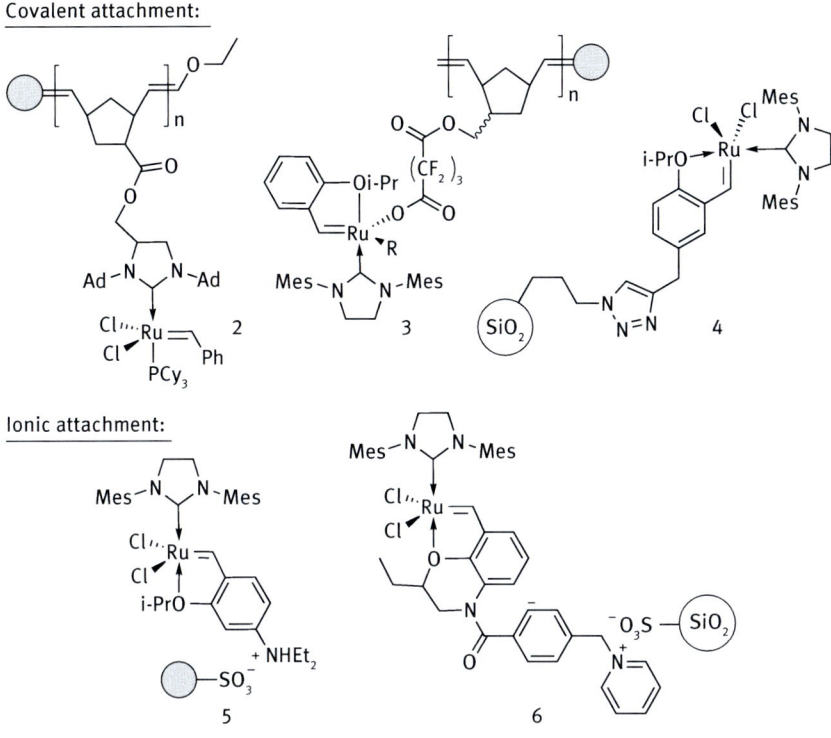

Fig. 1.3: Different ways of immobilizing RCM catalysts for use in packed-bed reactors (see also Table 1.1; Mes = 1,3,5-trimethylphenyl).

Huisgen cycloaddition [23]. The solution containing the substrate (DEDAM) was recirculated through the catalyst packed-bed with a high flow rate, to expel ethylene which has an inhibitory effect on catalysis. The productivity was low compared to the previous systems reported by Buchmeiser, and catalyst deactivation and/or leaching remained significant problems.

A collaborative project between the research groups of Kirshning and Grela explored the use of noncovalent interactions as a means of attaching the catalyst to a support. In this work, catalyst **5** was prepared, whereby a diethylammonium moiety used as an 'ionic tag' to attach the G-H catalyst onto a sulfonated polystyrene support via electrostatic interactions. The polymeric material was subsequently incorporated into a monolith within a microreactor (PassFlow) [24]. The cyclization of *bis*(allyl)phenylsulfonate can be achieved under (complete) recirculation flow, but the productivity of the system was quite low (Table 1.1, entry 4). Subsequent reuse of the reactor resulted in lower yields, losing catalytic activity on the third run. Catalytic activity can be regenerated by reloading the monolith with more Ru catalyst.

Similarly, a pyridinium group was used to immobilize a modified G-H catalyst to a sulfonated silica (**6**) for use in a borosilicate glass column for the cyclization of DEDAM

Table 1.1: Solid-immobilized RCM catalysts in packed-bed reactors.

Entry	Catalyst	Reaction conditions	Reactor	Productivity[a]	Ref
1	2 (monolith)	X = C(CO$_2$Et)$_2$, 43 °C	Borosilicate column, τ = 3 min	500	[21]
2	3 (monolith)	X = C(CO$_2$Et)$_2$, 45 °C	Peek column, 0.1 mL/min.	214	[22]
3	4	X = C(CO$_2$Et)$_2$, 50 °C, CH$_2$Cl$_2$	Packed-bed reactor, recirculation flow, 5 mL/min	91	[23]
4	5	X = NTs, CH$_2$CL$_2$, 40 °C	PassFlow reactor, recirculation, 2 mL/min, 5 h	4	[24]
5	6	X = C(CO$_2$Et)$_2$, toluene, 40 °C	Glass column (100 × 2.5 mm), 2.5 mL/min, recirculation flow	4.5	[25]

[a] mmol(product)/mmol(cat)/h.

(Table 1.1, entry 5) [25]. Once again, catalyst leaching was found to be pronounced under flow conditions.

In summary, solid-supported RCM catalysts generally suffer from leaching issues, which requires the use of a separate scavenger column to remove the heavy metal residue from the product. Although a low level of catalyst leaching was observed with catalyst **3**, it deactivates quickly over 2 h.

Supported Ionic-Liquid Phase (SILP) catalysis. Ionic liquids were investigated as a means of support for suitably functionalized G-H catalysts. Ionically-tagged G-H catalyst **7** (Figure 1.4) can be immobilized in [bmim][PF$_6$], which is then dispersed on an organic solvent nanofiltration (OSN) membrane for use in a catalytic membrane reactor [26]. The system was operated in a semicontinuous process for the cyclization of diallyltosylamine, but loses activity upon the 3rd cycle of use. Shortly after this report, a very similar SILP catalyst **8** was supported in [bmim][NTf$_2$] and loaded onto silica [27]. This was then used in a packed-bed reactor, in combination with sc-CO$_2$ as a transport vector, a very respectable productivity of 91 mol(product)/mol(cat)/h can be achieved for the RCM of the DEDAM at 50 °C, but the catalyst was observed to deactivate over 6 h. Nevertheless, by controlling the temperature and pressure, leaching of catalyst into the product can be suppressed by manipulating the phase behavior between CO$_2$ and starting materials, such that the catalyst remained largely dissolved in the IL.

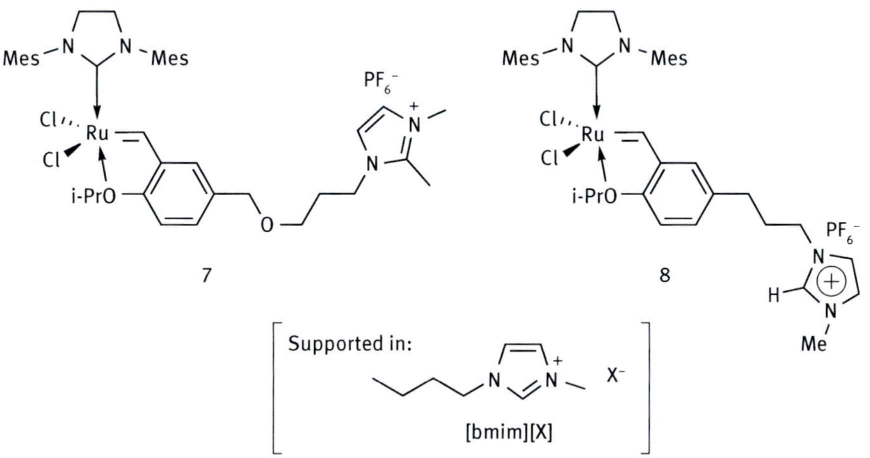

Fig. 1.4: Modified Grubbs'–Hoveyda catalysts used in SILP catalysis.

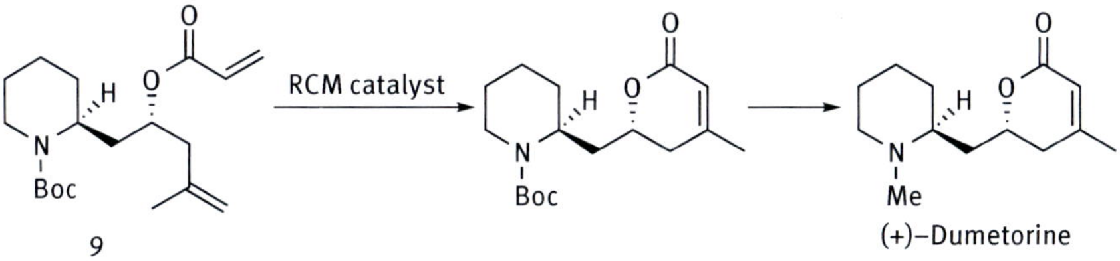

Fig. 1.5: Synthesis of (+)-dumetorine using the RCM reaction.

It is important to note that in almost all of the cases above, DEDAM is often used as the model substrate for the RCM reaction. But it does not possess the level of complexity required for 'real' applications. For example, the diene **9** was employed in a late-stage synthesis of (+)-dumetorine by a RCM reaction (Figure 1.4) [28]. The reaction can be achieved with complete conversion in the homogeneous solution under flow conditions using either 1st or 2nd generation Grubbs' catalyst. Several heterogeneous systems were also assessed, including Kischining's and Buchmeiser's systems, but were deemed impractical due to too much leaching, or insufficient productivity (a PEG-supported G-H catalyst was used to conduct the reaction in a batch reactor eventually, and the catalyst was recovered by precipitation afterwards). This serves to highlight the general need for the academic community to include more challenging substrates in the development of catalytic flow methodologies.

1.2.1.2 Pd-catalyzed cross-coupling reactions

Suzuki–Miyaura cross-coupling. There are also many reports of immobilized Pd catalysts used in Suzuki–Miyaura (SM) reactions (Figure 1.7) in packed-bed reactors. However, unlike the RCM reaction, where only volatile alkene by-product is generated, the cross-coupling reactions generate a stoichiometric amount of salt by-product, which can be problematic for flow reactors, as they can cause clogging of the reaction channels.

$$Ar^1-X + Ar^2-B(OH)_2 \xrightarrow[\text{Solvent, base}]{\text{Pd catalyst}} Ar^1-Ar^2$$

Fig. 1.6: The Suzuki–Miyaura cross-coupling reaction.

SM reactions with aryl iodides, bromides, and certain aryl chlorides can be effected by colloidal or nanoparticulate Pd. As a result, many 'simple' heterogeneous palladium catalysts, for example Pd/C, can be utilized for these reactions, even at room temperature [29–31]. In such cases, many studies have been dedicated to different ways of minimizing leaching of Pd into the solution phase. Ley and co-workers were the first to describe a microencapsulated Pd(OAc)$_2$ catalyst (Pd EnCat^{TM}) for the SM reaction [32]. The system was subsequently employed in a packed-bed arrangement, where the use of organic solvents or sc-CO$_2$ as reaction media was compared (Table 1.2, entry 1 versus entry 2) [33]. While SM reaction of aryl iodides can be achieved, the corresponding reaction with aryl bromide was far less effective in the absence of ligands. The limited solubility of the inorganic base in the reaction media is also an important limiting factor, which led to low productivity figures. Deposition of by-products (salts and boronic ester) causes blockage problems in such systems.

Kirschning *et al.* reported a way of incorporating Pd(0) particles onto an ion-exchange monolithic support within a microreactor for Suzuki–Miyaura cross-coupling reactions of aryl iodides (Table 1.2, entry 3) [34]. The system was implemented with recirculation flow using the PASS*Flow*^{TM} reactor, and the productivity figure is comparable to Pd EnCat (entry 2 versus entry 3). Catalyst leaching was found to be dependent upon the solvent polarity and reaction temperature. In the best cases, the microreactor can be reused up to 20 times.

The use of aryl bromides and aryl chlorides are generally cheaper than aryl iodides and are also preferred in terms of reducing the E-factor [35] of the process. The use of less activated substrates (particularly electron-rich and/or sterically bulky chlorides), however, requires more active organometallic precatalysts. Kirschning and co-workers investigated the use of poly(vinyl)pyridine (PVP) as a functionalized solid support to immobilize palladium complexes containing the oxime [36] and NHC (PEPPSI) [37] ligands, **10** and **11** (Figure 1.7), respectively, into monolithic supports in the Pass*Flow*^{TM} reactor (Table 1.2, entries 4 and 5). The NHC-ligated catalyst **11** is clearly the more ac-

Table 1.2: Immobilized Pd catalysts for SM cross-coupling reactions in flow reactors.

Entry	Catalyst and packing mode	Reaction conditions	Flow reactor type	Productivity[a]	Ref
1	Pd-Encat 40^{TM} in a packed-bed	ArX = PhI, 4-MeC$_6$H$_4$B(OH)$_2$, n-Bu$_4$NOMe, toluene/methanol (9:1), 100 °C.	HPLC column + HPLC pump, 6.4 mL/min, $\tau \approx$ 4 min	ca. 1.1[b]	[33]
2	Pd-Encat 40^{TM} in a packed-bed	ArX = PhI, 4-MeC$_6$H$_4$B(OH)$_2$, sc-CO$_2$, MeOH, 100 °C, 166 bar	High-pressure ModCol column (10 cm×25.4 mm), 6.4 mL/min, $\tau \approx$ 32 min, single pass	ca. 0.63[b]	[33]
3	Pd(0) supported on OH-functionalized monolithic support	X = I, Base = hydroxide ions, DMF/H$_2$O (10:1), 120 °C.	PASS*Flow*^{TM} microreactor, flow rate = 3 mL/min, 24 h, recirculation flow	0.11–0.4	[34]
4	**10** Monolithic support	X = Br, Base = CsF, DMF/H$_2$O (10:1), 100 °C.	PASS*Flow*^{TM} microreactor, flow rate = 2.5 mL/min, 24 h, recirculation flow	ca. 2.7×10^{-2} [b]	[36]
5	**11** Monolithic support	X = Cl, Base = potassium $tert$-pentoxide, i-PrOH, 80 °C.	PASS*Flow*^{TM} microreactor, flow rate = 2 mL/min, recirculation flow	2.05–3.07	[37]
6	**12** in a packed-bed	X = Cl, Br, I, OTf; KOH, THF/H$_2$O (1:1), 60 °C.	Vapourtec (R2+R4), τ = 5 min, single pass	1.8–3.75	[38]
7	5 wt% Pd/Al$_2$O$_3$ in a packed-bed	ArX = 4-BrC$_6$H$_4$CN, PhB(OH)$_2$, K$_2$CO$_3$, DMF, 80 °C.	Capillary flow reactor, microwave, 90 W, 55–63 °C, τ = 60 s	1.5×10^{-2}	[39]
8	Pd-EnCat 30^{TM} in a packed-bed	X = Br, n-Bu$_4$NOAc, EtOH, 70 °C.	Glass U-tube flow reactor, 0.2 mL/min, μwave, 50 W with cooling, τ = 32.5 s	Up to 17.8	[40]
9	**13**, in a packed-bed	ArX = 2-BrC$_6$H$_4$CN, PhB(OH)$_2$, CsF, DMF/H$_2$O (10:1), inductively heated at 25 kHz, 100 °C	Glass reactor (9 mm×14 cm), 100 °C, τ = 20 min, recirculation flow.	4	[41]

[a] mmol (product) mmol(cat)$^{-1}$ h^{-1}. [b] Estimated (amount of catalyst unspecified).

tive, offering a much greater productivity over the other systems (up to ten fold), even when aryl chlorides were used as substrates.

More recently, a silica-supported phosphine-ligated catalyst **12** (SiliaCat® DPP-Pd, Figure 1.7) has been shown to be a good catalyst for SM reactions, displaying a wide scope for a variety of aryl halides and triflates [38]. In this case, a Vapourtec system (column reactor) was employed to give good productivity in a single pass (Table 1.2, entry 6). Similar results were also obtained using a simple setup comprising of a column (4.6 × 50 mm), syringe pumps and a column heater. Even more remarkably, continuous flow can be maintained for 8 h without any loss of activity. Significantly, less than 10 ppb of Pd was found in the organic fraction collected.

Microwave irradiation has been widely applied for Pd catalyzed cross-coupling reaction under continuous flow conditions, and the subject has been reviewed [42]. The system was first described in 2004 in a capillary tube (Haswell *et al.*) using Pd supported on silica or alumina, utilizing a gold patch to enhance the heating in the catalyst bed (Figure 1.8 and Table 1.2, entry 7) [33]. The productivity of this system can be improved by using a U-shaped glass tube in combination with the Pd-EnCat 30 catalyst, as reported by Bexandale *et al.* (Table 1.2, entry 8) [40].

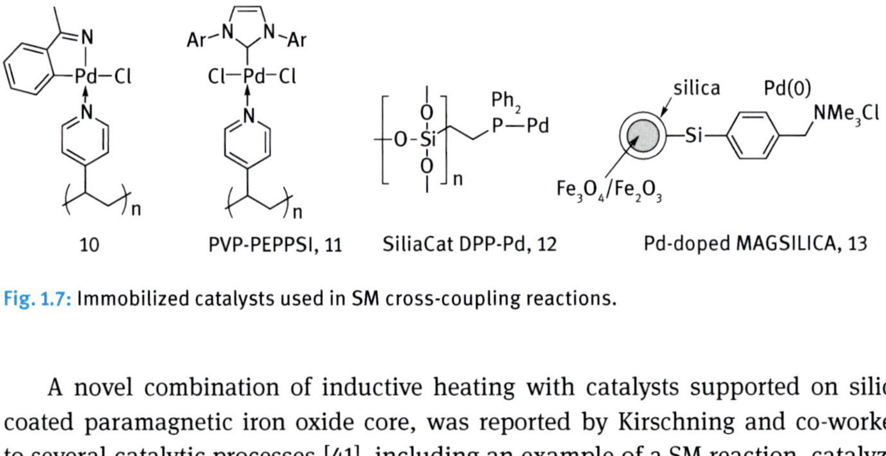

Fig. 1.7: Immobilized catalysts used in SM cross-coupling reactions.

A novel combination of inductive heating with catalysts supported on silica-coated paramagnetic iron oxide core, was reported by Kirschning and co-workers to several catalytic processes [41], including an example of a SM reaction, catalyzed Pd-doped MAGSILICA **13** (Figure 1.7, Table 1.2, entry 9).

Sonogashira cross-coupling. In some cases, the surface of the reactor can be rendered catalytically active. This was demonstrated for Sonogashira cross-coupling by electroplating a copper tubular reactor with Pd, which is placed in tandem with an uncoated Cu tubing (Figure 1.9) [43]. Leached Pd from the first reactor is carried into the copper reactor, where the reactive organocopper reagent is generated and consumed immediately in the sp^2-sp coupling. Productivities of such systems are difficult to define as the number of catalyst active sites cannot be precisely determined.

In summary, a wide variety of flow systems have been reported for Pd-catalyzed cross-coupling reactions. In almost all cases, catalyst leaching is a significant prob-

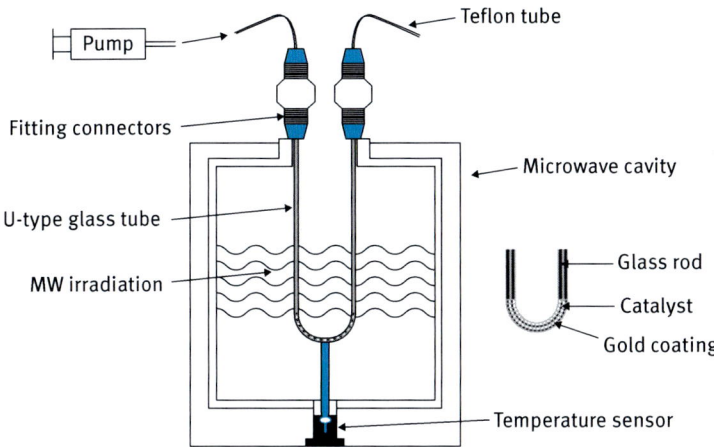

Fig. 1.8: The arrangement of a CF microwave reactor using a capillary tube [33].

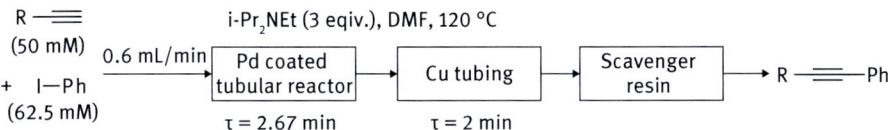

Fig. 1.9: A dual reactor for the Sonogashira coupling.

lem. In some cases, additional scavenger modules were needed for post-reaction treatment. Even so, column chromatography for the purification of the product is still necessary to remove the salt by-product and any excess reactant. This highlights the need for the development of catalytic methodologies that are either 100% atom economical, or only by-products that can be easily removed and preferably benign (e.g., water).

1.2.1.3 (Chemo)enzymatic reactions

As demonstrated in the previous sections, leaching of metal catalysts from heterogeneous supports is a considerable problem, as the dative/ionic bonding between the metal and functionalized supports is inherently weak. In light of this, the use of enzymes and organocatalysts in continuous flow systems can be very attractive as they can be immobilized through a covalent bond.

In addition to packed-bed reactors, membrane reactors are also widely employed in enzymatic processes (Figure 1.10). The inert membrane reactor contains a membrane that simply acts as a barrier to separate feed from permeate. As immobilized enzymes are invariably very much larger than substrates and products, they can be mechanically retained effectively using an ultrafiltration membrane, where the catalyst and the substrates are dissolved or suspended in the liquid phase. In contrast, the cat-

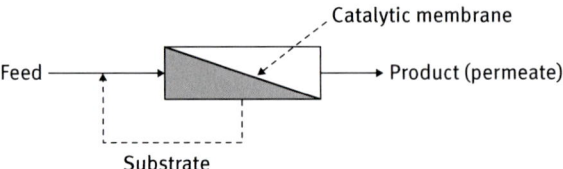

Fig. 1.10: Membrane reactor utilized in enzymatic flow processes.

alytic membrane reactor contains the catalyst either as a coating or embedded within the membrane, such that catalysis and separation may be achieved simultaneously.

Biocatalysis. Enzymes are often immobilized for use in chemical processes [44]. Very often, some loss of enzyme activity is observed, which may be caused by the immobilization procedure, but also reduced mass transport due to reduced diffusion of the substrates to the active site. On the other hand, the immobilization of enzymes can prevent them from denaturing in certain solutions and/or at high temperatures.

Immobilized enzymes are very often employed in continuous flow reactors, and the subject has been reviewed [4]. It is beyond the scope of this chapter to provide a comprehensive review of chemo(enzymatic) flow reactions. Instead, selected examples will be presented in this section to highlight the advantages of this approach, namely: (i) to eliminate product inhibition; (ii) achieve continuous kinetic resolution; and (iii) effect tandem processes in one continuous flow.

The use of packed-bed reactors is particularly effective for enzyme catalysis, as by-product(s) that could inhibit the active site can be continuously purged from the system by the flow stream. This is demonstrated by the production of oleamide from oleic acid and ammonium carbamate as a source of ammonia, catalyzed by Novozyme 435 (Figure 1.11) [45]. The reaction exists in thermodynamic equilibrium with the product. Removal of water and ammonia was achieved by using 2-methyl-2-butanol (2M2B) as the solvent, and controlling the flow rate. Under the optimal conditions, a productivity of 4.5 tonnes of oleamide/kg enzyme (ca. 100 g/L/h) can be achieved; at least a four-fold improvement on previously reported processes.

Lipases are extensively used in kinetic resolution of chiral acids and alcohols by catalyzing the esterification of one particular enantiomer. The kinetic resolution of two non-steroidal anti-inflammatory drugs (NSAIDs) **14** and **15** (Figure 1.12) into their ethyl esters was accomplished by using the lipase *Pseudomonas* sp. immobilized into a capillary polyamide membrane [46], which is used in a reactor in a batch recycle mode. The result obtained for the resolution of ibuprofen (**15**) compared very favorably to an earlier report using a batch stirred tank reactor, where only 85% enantiomeric excess (ee) was achieved [47]. Notably, the activity of the catalytic membrane can be maintained for at least 500 h.

Many enzymatic processes proceed in water. This allows biochemical processes to be coupled to effect consecutive transformations in a single flow-through process. This

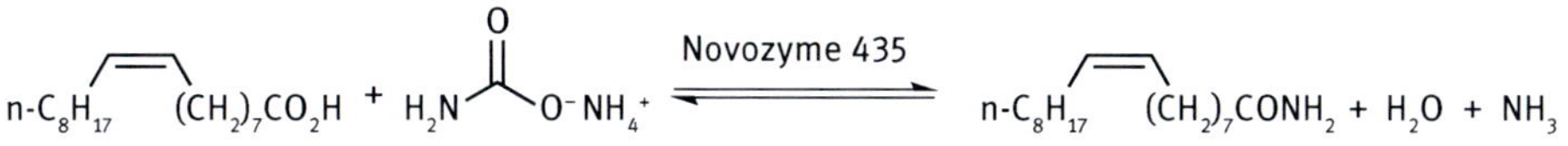

Fig. 1.11: Enzymatic transformation of oleic acid to oleamide by immobilized Novozyme 435 in a packed-bed reactor.

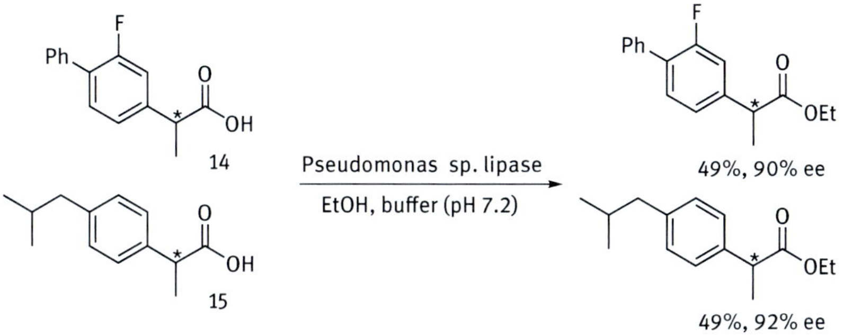

Fig. 1.12: Kinetic resolution of pharmaceutically relevant molecules using a catalytic membrane reactor under batch recycled.

was demonstrated elegantly by Babich *et al.*, using immobilized acid phosphatase (PhoN-Sf) and two aldolases (RAMA and RhuA) in a sequential packed-bed arrangement, to produce carbohydrate analogues from dihydroacetone (DHA) and aldehydes (Figure 1.13) [48, 49]. In the first reactor, DHA is phosphorylated at one of the primary alcohols to form the phosphate derivative **16**. This effectively activates the DHA to undergo an aldol condensation with an aldehyde in the presence of the aldolase enzyme RAMA in the second reactor, producing a single diastereomer ($3S, 4R$)-**17**. Finally, hydrolysis in the third reactor gave optically active compound **18**, a precursor for carbohydrate synthesis. Using a different aldolase in the second reactor produces the opposite enantiomer ($3R, 4S$)-**17** as part of a 3 : 2 diastereomeric mixture. The system was utilized for the continuous synthesis of various aldol products in 0.2–0.8 g scale.

The success of the process depends on the ability to perform the entire operation in a single-pass without the need to change the reaction medium (water). Each transformation step is highly selective, atom-economical, and does not produce any byproducts that are insoluble or can inhibit the activity of the enzyme in the subsequent reactor. Provided that these criteria can be met, a single-pass tandem continuous flow strategy is immensely powerful for constructing multiple bonds.

Organocatalysis. Up to two adjacent stereogenic centers can be constructed simultaneously with 100% atom-economy by using aldol and Michael-type reactions, making them highly valuable for organic synthesis. Asymmetric variants of these reactions have benefited greatly from advances in organocatalysis in recent years. Organocatalysts often have to be employed in high catalytic loading (typically 10 mol%), so their

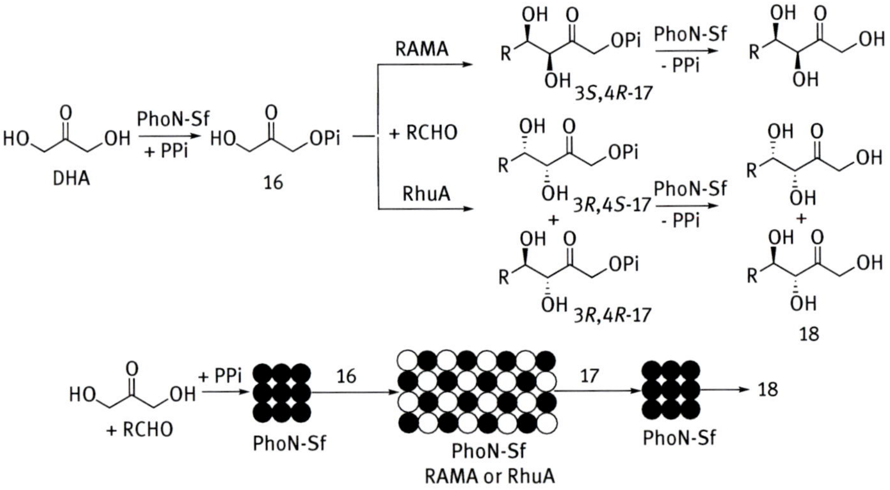

Fig. 1.13: Sequential packed-bed reactors to enable a single-pass tandem process (PPi = pyrophosphate).

immobilization will enable their recovery and reuse. Peptidic catalysts, such as **20–22** (Figure 1.14) can be easily constructed on polymer resins via well-established procedures. As the catalysts are attached covalently to the solid supports, they are much more resilient against leaching. In some cases, reuse of these catalysts is possible without any loss in activity. Nevertheless, the productivities of such systems are quite low. Despite using large excess of one reactant, long residence time is required to achieve good conversions, even for activated substrates. As such, they are currently limited to lab-scale demonstrations. In some cases, a compromise has to be struck between achieving optimal yield and selectivity.

1.2.2 Solid-liquid-gas systems

Catalytic reactions utilizing gaseous reactants (such as H_2, CO and O_2) are almost always atom-efficient and clean, as any excess can simply be vented from the reaction mixture. However, the use of gaseous reagents adds considerable complexity to the flow system, generally associated with the need to deploy pressurized systems. This is the case for most microreactors, which typically have an upper working limit of ca. 20 bar; limiting their utility to small-scale production of molecules, for example, heterogeneous-catalyzed hydrogenation using segmented flow in capillary columns [54] was used to introduce deuterium into pharmaceutically relevant N-heterocycles [55].

Redox catalysis constitutes the largest class of reactions under study. For this reason, this section will discuss the application of multiphase flow reactors for catalytic reactions involving H_2 and O_2. While catalytic hydrogenation in the liquid phase may

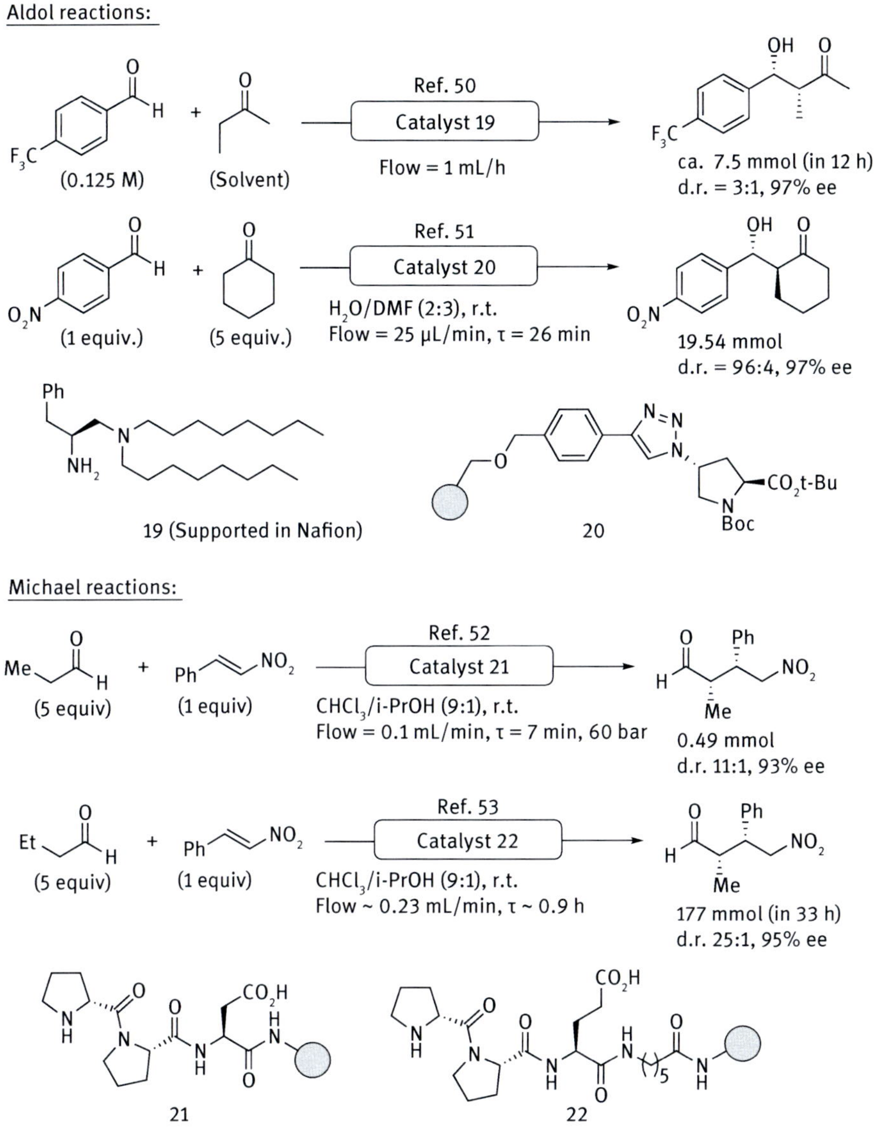

Aldol reactions:

Ref. 50 — Catalyst 19, Flow = 1 mL/h
F_3C-C$_6H_4$-CHO (0.125 M) + ethyl methyl ketone (Solvent) → aldol product, ca. 7.5 mmol (in 12 h), d.r. = 3:1, 97% ee

Ref. 51 — Catalyst 20, H_2O/DMF (2:3), r.t., Flow = 25 μL/min, τ = 26 min
O_2N-C$_6H_4$-CHO (1 equiv.) + cyclohexanone (5 equiv.) → aldol product, 19.54 mmol, d.r. = 96:4, 97% ee

19 (Supported in Nafion): Ph-CH$_2$-CH(NH$_2$)-CH$_2$-N(octyl)$_2$

20: polymer-O-CH$_2$-C$_6H_4$-triazole-pyrrolidine(Boc)-CO$_2$t-Bu

Michael reactions:

Ref. 52 — Catalyst 21, CHCl$_3$/i-PrOH (9:1), r.t., Flow = 0.1 mL/min, τ = 7 min, 60 bar
Me-CHO (5 equiv) + Ph-CH=CH-NO$_2$ (1 equiv) → Michael adduct, 0.49 mmol, d.r. 11:1, 93% ee

Ref. 53 — Catalyst 22, CHCl$_3$/i-PrOH (9:1), r.t., Flow ~ 0.23 mL/min, τ ~ 0.9 h
Et-CHO (5 equiv) + Ph-CH=CH-NO$_2$ (1 equiv) → Michael adduct, 177 mmol (in 33 h), d.r. 25:1, 95% ee

21: Pro-Pro-Asp-supported peptide catalyst

22: Pro-Pro-Glu-(CH$_2$)$_5$-supported peptide catalyst

Fig. 1.14: Asymmetric aldol- and Michael-related reactions mediated by organocatalysts in flow.

be considered as a mature technology (particularly in the food industry for the hydrogenation of unsaturated fats) [56], there are currently no aerobic oxidation reactions performed industrially in the liquid phase. The challenges associated with these reactions will be highlighted in the following sections.

1.2.2.1 Chemoselective catalytic hydrogenation

Hydrogen is a highly flammable gas that poses an immediate fire and explosive hazard with concentrations > 4% in air. For these reasons, pressurized reactions had to be run using specialized equipment (typically autoclaves) fed from gas cylinders in designated areas. In this regard, the H-cube marketed by ThalesNano provided an innovative solution [57], by using a high-pressure electrochemical cell for the on-demand delivery of H_2, thus effectively reducing the pressurized hydrogen inventory to the comparatively small reaction volume. Applying such hydrogen generators, these pressurized reactions are becoming routinely accessible in the synthetic laboratory. Nowadays, other meso-scale continuous flow reactors are also commercially available, and a comprehensive review of the subject area was carried out by Kappe and co-workers in 2011, including a list of common catalysts and substrates [58].

Despite the myriad of heterogeneous catalysts reported for the reduction of C=C bonds, there are still considerable challenges in attaining good chemoselectivity for molecules containing more than one unit of unsaturation. An example of such a substrate is citral. Containing conjugated C=C and C=O and a trisubstituted olefin group, the selective hydrogenation of any of these moieties can lead to valuable products for the fine chemical industry for the production of flavors, fragrances and insect repellents (Figure 1.15). A recent review surveyed the use of heterogeneous and homogeneous catalysts for these reactions [59]. In general, selectivity of the first reduction to either citronellal or geraniol/neral is largely dictated by the nature of the catalyst: selective C=C reduction is favored by later transition metal catalysts such as Pd, Ni and Rh, whereas the reduction of C=O is achieved by using ruthenium or iron catalysts. In either case, further reduction to citronellol and/or over-reduction to the 3,7-dimethyl-1-octanol can be controlled by mass transport, reactor design and residence time.

Most often than not, the rate of the C=C reduction process is limited by the efficiency of the mass transport (diffusion) of H_2 to the catalyst surface. This can be enhanced either by increasing the mass transport coefficient itself (e.g., varying the degree of turbulence) or enhancing the contact area between the gas and the liquid phase (e.g., gas sparging) to boost the solution concentration of hydrogen. In a patent published in 1999 by BASF, the process was accelerated by conducting the reaction in a (batch liquid phase) bubble column reactor over 5 wt% Pd/C at 70 °C, whereby very high conversion (99.5%) and selectivity (94%) to citronellal can be achieved in under 3 hours [60]. Bubble columns are ideal for comparatively slow processes where the majority of the reaction takes place in the bulk liquid. On the other hand, fast

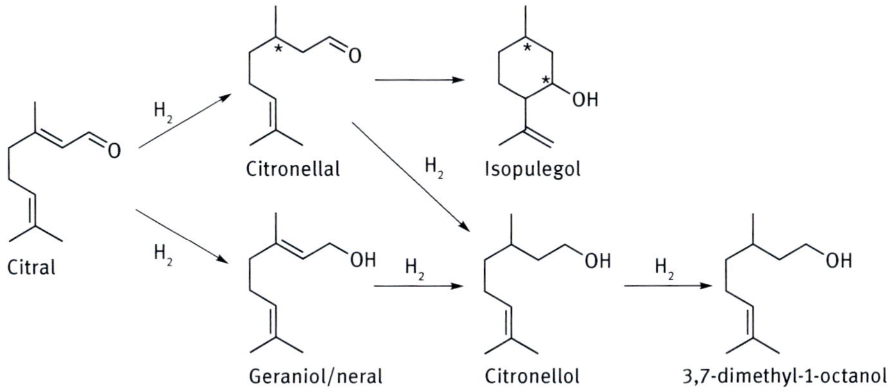

Fig. 1.15: Possible products resulting from the catalytic hydrogenation of citral.

reactions are best performed in reactors which maximize the thin liquid boundary layer between gas phase and solid catalyst. For example, the selective reduction of the carbonyl group was achieved by using a trickle bed reactor containing magnesia-supported Sn-Pt catalyst, where quantitative conversion and 97% selectivity for geraniol can be obtained with a residence time of 79.5 s [61].

1.2.2.2 Asymmetric hydrogenation

Asymmetric hydrogenation is a key process for the production of optically active compounds in the fine and pharmaceutical industries, and is the preferred technology for the reduction of C=C, C=O and C=N moieties. By far, the immobilization of organometallics containing chiral mono- and di-phosphine ligands is the most successful strategy for conducting asymmetric hydrogenation reactions in continuous reactors.

Reduction of C=C bonds (Figure 1.16). Phosphotungstic acid on alumina (PTA/Al_2O_3) was used to support [Rh(MonoPhos)$_2$(COD)] for the asymmetric hydrogenation using a H-Cube [62]. Very high selectivity (96% ee) can be achieved in the hydrogenation of α-acetamidoacrylic acid methyl ester (MMA). However, without any modification of the catalyst structure, the productivity was extremely low (36 mg/h). Leaching of the catalyst was also a substantial issue, dependent on the temperature and choice of solvent. At room temperature and using ethyl acetate as solvent, the loss of Rh can be reduced to 0.3 mg/L. The same support was used to immobilize a range of chiral diphosphine ligands for the hydrogenation of dimethyl itaconate (DMIT) using sc-CO_2 as a reaction medium [63]: JosiPhos was found to afford a respectable 83% ee, but the conversion per pass was low. On the other hand, Leitner and co-workers have shown that by using a molecular Rh catalyst ligated by 1-napthylQUINAPHOS (**23**) can be utilized under conditions of SILP catalysis in sc-CO_2 [64]. This was used to reduce

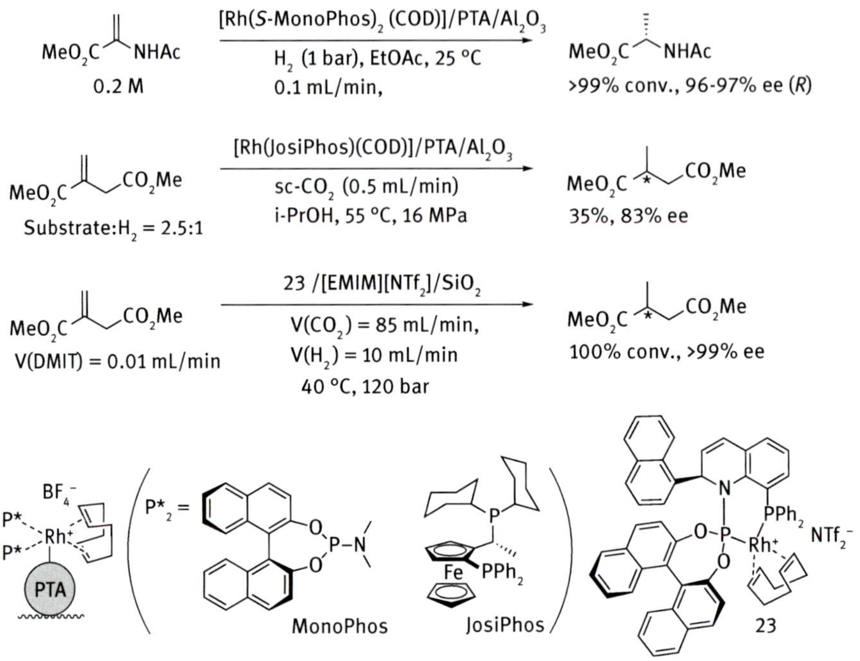

Fig. 1.16: Asymmetric hydrogenation of MMA and DMIT using supported mono- and bi-dentate phosphorus ligands.

dimethyl itaconate with > 99% ee, with extremely high TOF of > 2000 h^{-1}. Although catalyst leaching can be kept to a minimum, the enantioselectivity of the reaction stream can only be maintained for 10 h, after which it slowly erodes, attributed to partial decomposition of the catalyst.

In a different approach, the need for a heterogeneous support is eliminated by using the chiral MonoPhos ligand to assemble an extensive network of Rh metal centres through multi-topic linkers (Figure 1.17). These self-supported catalysts can be used in a packed-bed reactor for asymmetric hydrogenation of α-dehydroamino acid methyl esters, with > 90% ee across a range of different substrates [65]. The excellent activity of the catalyst allows the H$_2$ pressure to be lowered to 2 bar without detrimental effect on activity and selectivity, and catalyst activity can be maintained for 144 h.

Reduction of C=O bonds (Figure 1.18). The reduction of unsymmetrical ketones to chiral alcohols is most often achieved using isopropanol both as the solvent and hydride source, which is oxidized to acetone during the process. The development of an immobilized Ru catalyst on modified silica (**24**) for continuous transfer asymmetric hydrogenation of ketones was reported by Reek and van Leeuwen [66]. The catalyst **24** was employed in a packed-bed reactor, converting acetophenone to 1-phenylethanol in 95% yield and 90% ee under optimized conditions (Figure 1.18, Equation 1). The catalyst was reported to be more stable when it is employed in the continuous flow

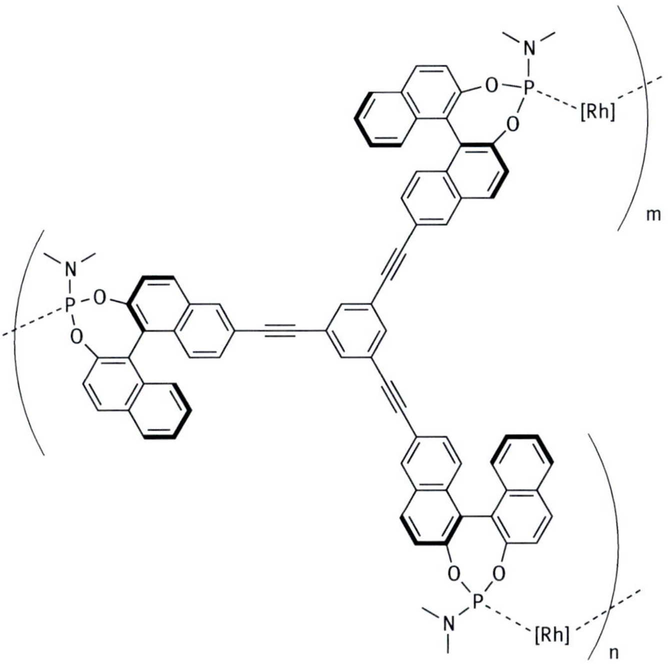

Fig. 1.17: Self-supporting catalyst.

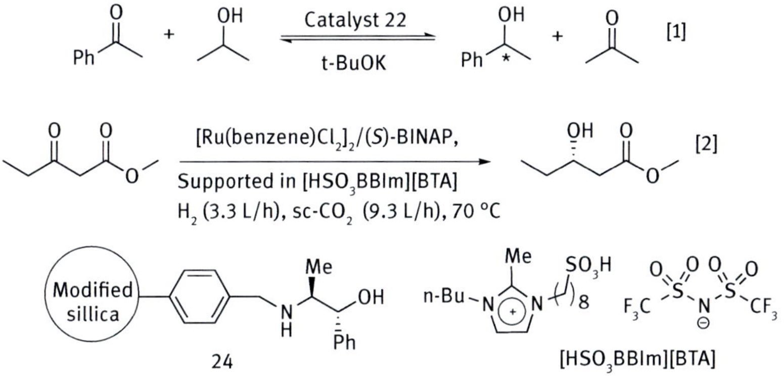

Fig. 1.18: Asymmetric hydrogenation of ketones under continuous flow conditions.

reactor; no loss of conversion or selectivity occurred over a period of one week. As the system operates under equilibrium, catalyst performance is highly dependent upon the flow rate (residence time); in this case, high enantioselectivity was achieved at a high flow rate.

Racemization of the product may be avoided by using H_2 as a reductant. A SILP catalytic system was reported recently by Leitner and co-workers [67], whereby the asymmetric hydrogenation of a β-keto ester (methyl propionylacetate) can be achieved using IL-supported BINAP-Ru catalyst and sc-CO_2 as a mobile phase (Figure 1.18, Equation 2). The reaction was accelerated by an acid additive, hence an acid moiety was incorporated into the IL to simplify the purification of the product. Using the system, single-pass conversion of 95% was observed with 80% ee can be maintained over 100 h time-on-stream.

1.2.2.3 Oxidation

Liquid-phase oxidation of alcohols to their corresponding carbonyl compounds (Figure 1.19) is an enormously important transformation in many industrial processes. Catalytic oxidation of alcohols to carbonyl groups in the liquid phase can be effected by a number of heterogeneous catalysts using O_2 as a terminal oxidant. The reaction is extremely desirable as only benign water is generated as a by-product, compared to the use of stoichiometric oxidants. However, there are severe flammability and explosive hazards associated with aerating organic solvents, which must be overcome before they can be implemented industrially. In recent years, progress has been made in the development of catalytic flow processes on laboratory-scale reactors (Table 1.3).

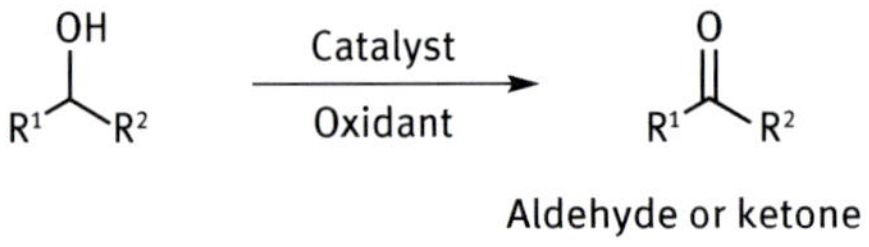

Fig. 1.19: Selective oxidation of alcohols to aldehydes and ketones.

Taking advantage of the small reactor volume of a catalyst cartridge offered by a X-Cube reactor, aerobic oxidation of primary and secondary alcohols to aldehydes and ketones can be achieved using Ru/Al_2O_3 as catalyst [68]. A variety of fourteen alcohols can be oxidized using either O_2 or air under recycle batch mode. The rate of the reaction can be accelerated by increased pressure, which increases the solubility of O_2 in the reaction solvent, and a high productivity value can be calculated for the oxidation of benzyl alcohol (Table 1.3, entry 1). More recently, 1% Au/TiO_2 was utilized using the same system for the oxidation of 1-phenyl ethanol to the corresponding ketone in a single-pass mode (Table 1.3, entry 2) [69]. No additives or extraneous reagents are required, thus offering a very clean product stream which can be directly utilized in a subsequent transformation.

Table 1.3: Oxidation of alcohols under flow.

Entry	Catalyst	Reaction conditions	Reactor	Productivity [a]	Ref
1	5 wt% Ru/Al$_2$O$_3$	Benzyl alcohol to benzaldehyde, O$_2$ (25 bar), 90 °C, toluene	X-Cube, recirculation flow	9324	[68]
2	1 wt% Au/TiO$_2$	1-Phenylethanol to acetophenone, O$_2$ (100 bar), 140 °C, acetone	X-Cube, single-pass	101	[69]
3	25	Benzyl alcohol to benzaldehyde, $p(CO_2/O_2{:}92/8)$ = 15 MPa, 60 °C, 1 mL/h, recirculation	Tubular stainless steel fixed-bed reactor	29.5	[70]

[a] mmol (product) mmol(cat)$^{-1}$ h^{-1}.

Utilization of CO$_2$ for the aerobic oxidation reactions is particularly beneficial as this eliminates the use of highly flammable organic solvents. Leitner and co-workers described the use of a np-Pd catalyst immobilized on mesoporous silica-supported functionalized with dipyridine ligands. The catalyst **25** was used in a packed-bed reactor, using sc-CO$_2$ to carry the substrate and O$_2$ [70]. A single-pass conversion of 41% was achieved, which can be maintained for 28 h on stream (Table 1.3, entry 3).

1.3 Conclusion

In this Chapter, the use of flow reactors for catalytic organic synthesis has been demonstrated with selected examples. Wherever possible, comparisons between different reactors were made to highlight pros and cons of each system.

Flow reactors will have an increasingly important role to play in the future development of catalytic methodologies. The high surface-to-volume ratio of the reactor enables multiphase reactions to be performed in a single-pass, while inhibitory effect of (by-)product(s) may be minimized. However, it is also important to acknowledge current limitations of performing catalytic reactions in flow, particularly when organic solvents are used. Leaching of metal catalyst into the product stream is a significant issue, which will demand more effective strategies for their immobilization/recovery.

Another bottleneck for the development of catalytic flow methodology is the formation of insoluble precipitates, typically inorganic salts (e.g., in cross-coupling reactions), which can lead to clogging of the reaction channels. This ultimately requires the development of not only robust, but also atom-economical reactions where (ideally) no by-product is formed (e.g., hydrogenation reactions); or, if this is not possible, low-molecular weight by-products that can be removed easily (e.g., ethylene in RCM reactions). The ability to attain clean conversions also allows the process to be telescoped with other reactions, to enable multiple transformations with one continuous flow. Additionally, the development of catalytic flow strategies also involves greater understanding of how reactions occur at the interface, and phase boundary behavior that may govern mass and heat transfer that are important for maximum efficiency. This invariably requires better 'mixing' between chemists and chemical engineers, in order to deliver innovative solutions for sustainable synthesis.

Study questions

1.1. Why are catalytic flow reactions of interest to a development chemist in the fine chemical/pharmaceutical industry from an environmental, efficiency and safety perspective?
1.2. What additional physical process design consideration need to be taken into account when using heterogeneous catalysts in continuous flow reactors?
1.3. Apart from catalyst properties, what is the key governing factor in a multiphase reaction?
1.4. What are the two most important C-C bond forming processes using solid catalysts?
1.5. Why are ring closing reactions of particular interest for continuous mode of operation?
1.6. What particular problem can solid Palladium catalysts suffer from and what can be done to prevent this?
1.7. Why are catalytic membrane reactors suitable for enzymatic reactions?
1.8. What is the safety advantage of a continuous hydrogenation reactor using hydrogen gas over its batch counterpart?
1.9. How can asymmetric hydrogenations best be facilitated in continuous flow reactors?
1.10. In an aerobic oxidation using oxygen, mass transfer has a similar influence on the rate of reaction as in the case of hydrogenations. How can the rate of reaction be accelerated?

Further readings

1. A good general textbook for the practice and application of catalysis:
   Bartholomew CH, Farrauto RJ. *Fundamentals of Industrial Catalytic Processes*, 2nd edn, Wiley, New York, 2006.
2. On catalysts, reactions and kinetics (particularly hydrogenations and oxidations):
   Augustine RL. *Heterogeneous Catalysis for the Synthetic Chemist*, Marcel Dekker, New York, 1996.
3. On catalytic flow (tubular or plug flow) reactor design:
   Levenspiel O. *Chemical Reaction Engineering*, Wiley, New York, 1999.

Bibliography

[1] Wirth T. Flow Chemistry: Enabling technology in drug discovery and process research. ChemSusChem 2012, 5, 215–216.
[2] Ager D, Poechlauer P. Sustainable, efficient and safe processes through the use of flow and catalytic reactions. Chim Oggi 2012, 30, 42–44.
[3] Newman SG, Jensen KF. The role of flow in green chemistry and engineering. Green Chem 2013, 15, 1456–1472.
[4] Yuryev R, Strompen S, Liese A. Coupled chemo(enzymatic) reactions in continuous flow. Beilstein J Org Chem 2011, 7, 1449–1467.
[5] Mak XY, Laurino P, Seeberger PH. Asymmetric reactions in continuous flow. Beilstein J Org Chem 2009, 5, 19.
[6] Zhao D, Ding K. Recent advances in asymmetric catalysis in flow. ACS Catal 2013, 3, 928–944.
[7] Jimeno C, Sayalero Sm, Pericas MA. Covalent Heterogenization of asymmetric catalysts on polymers and nanoparticles, in *Heterogenized Homogeneous Catalysis for Fine Chemical Production*, Barbaro P, Liguori F (eds), Springer Netherlands, 2010, 33, 123–170.
[8] Riisager A, Fehrmann R, Haumann M, Wasserscheid P. Supported ionic liquid phase (silp) catalysis: an innovative concept for homogeneous catalysis in continuous fixed-bed reactors. Eur J Inorg Chem 2006, 695–706.
[9] Hintermair U, Francio G, Leitner W. Continuous flow organometallic catalysis: new wind in old sails. Chem Commun 2011, 47, 3691–3701.
[10] Lozano P, Garcia-Verdugo E, Luis SV, Pucheault M, Vaultier M. (Bio) Catalytic continuous flow processes in $scCO_2$ and/or ILs: towards sustainable (bio) catalytic synthetic platforms. Curr Org Syn 2011, 8, 810–823.
[11] Lombardo M, Quintavalla A, Chiarucci M, Trombini C. Multiphase homogeneous catalysis: common procedures and recent applications. Synlett 2010, 1746–1765.
[12] http://www.nobelprize.org/nobel{_}prizes/chemistry/laureates/index.html (accessed October 2013).
[13] Lefebvre F. Ring Opening metathesis polymerisation and related chemistry, *NATO Science Series*, Vol. 56, Khosravi E, Szymanska-Buzar T (eds), pp 247–261, Springer Netherlands, 2002.
[14] Biermann U, Bornscheuer U, Meier MAR, Metzger JO, Schaefer HJ. Oils and fats as renewable raw materials in chemistry. Angew Chem Int Ed 2011, 50, 3854–3871.
[15] Marvey BB. Sunflower-based feedstocks in nonfood applications: perspectives from olefin metathesis. Int J Mol Sci 2008, 9, 1393–1406.
[16] Mukheijee Singh O. Metathesis catalysts: Historical perspective, recent developments and practical applications. J Sci Ind Res India 2006, 65, 957–965.
[17] Martin WHC, Blechert S. Ring closing metathesis in the synthesis of biologically interesting peptidomimetics, sugars and alkaloids. Curr Top Med Chem 2005, 5, 1521–1540.
[18] Yee N, Wei X, Senanayake C, Challenges and opportunity in scaling-up metathesis reaction: Synthesis of Ciluprevir (BILN 2061). In *Sustainable Catalysis: Challenges and Practices for the Pharmaceutical and Fine Chemical Industries*, Dunn PJ, Hii KK, Krische MJ, Williams MT (eds), John Wiley & Sons, 2013, pp 215–232.
[19] Monfette S, Fogg DE. Equilibrium ring-closing metathesis. Chem Rev 2009, 109, 3783–3816.
[20] Czaban J, Torborg C, Grela K. Olefin metathesis: From academic concepts to commercial catalysts. In *Sustainable Catalysis: Challenges and Practices for the Pharmaceutical and Fine Chemical Industries*, Dunn PJ, Hii KK, Krische MJ, Williams MT (eds), John Wiley & Sons, 2013, pp 163–214.
[21] Mayr M, Mayr B, Buchmeiser MR. Monolithic materials: new high-performance supports for permanently immobilized metathesis catalysts. Angew. Chem Int Ed 2001, 40, 3839–3842.

[22] Krause JO, Lubbad SH, Nuyken O, Buchmeiser MR. Heterogenization of a modified Grubbs-Hoveyda catalyst on a ROMP-derived monolithic support. Macromol Rapid Comm 2003, 24, 875–878.

[23] Lim J, Lee SS, Ying JY. Mesoporous silica-supported catalysts for metathesis: application to a circulating flow reactor. Chem Commun 2010, 46, 806–808.

[24] Michrowska A, Mennecke K, Kunz U, Kirschning A, Grela K. A new concept for the noncovalent binding of a ruthenium-based olefin metathesis catalyst to polymeric phases: Preparation of a catalyst on Raschig rings. J. Am. Chem. Soc. 2006, 128, 13 261–13 267.

[25] Borre E, Rouen M, Laurent I et al. A Fast-initiating ionically tagged ruthenium complex: a robust supported pre-catalyst for batch-process and continuous-flow olefin metathesis. Chem Eur J 2012, 18, 16 369–16 382.

[26] Keraani A, Rabiller-Baudry M, Fischmeister C, Bruneau C. Immobilisation of an ionically tagged Hoveyda catalyst on a supported ionic liquid membrane: An innovative approach for metathesis reactions in a catalytic membrane reactor. Catal Today 2010, 156, 268–275.

[27] Duque R, Oechsner E, Clavier H et al. Continuous flow homogeneous alkene metathesis with built-in catalyst separation. Green Chem 2011, 13, 1187–1195.

[28] Riva E, Rencurosi A, Gagliardi S, Passarella D, Martinelli M. Synthesis of (+)-Dumetorine and congeners by using flow chemistry technologies. Chem Eur J 2011, 17, 6221–6226.

[29] LeBlond CR, Andrews AT, Sun YK, Sowa JR. Activation of aryl chlorides for Suzuki cross-coupling by ligandless, heterogeneous palladium. Org Lett 2001, 3, 1555–1557.

[30] Zhang GL. Ligand-free Suzuki-Miyaura reaction catalysed by Pd/C at room temperature. J. Chem. Res 2004, 593–595.

[31] Liu C, Rao X, Zhang Y, Li X, Qiu J, Jin Z. An aerobic and very fast Pd/C-catalyzed ligand-free and aqueous suzuki reaction under mild conditions. Eur J Org Chem 2013, 4345–4350.

[32] Ramarao C, Ley SV, Smith SC, Shirley IM, DeAlmeida N. Encapsulation of palladium in polyurea microcapsules. Chem Commun 2002, 1132–1133.

[33] Leeke GA, Al-Duri B, Seville JPK et al. Continuous-flow Suzuki-Miyaura reaction in supercritical carbon dioxide. Org Process Res Dev 2006, 11, 144–148.

[34] Solodenko W, Wen HL, Leue S et al. Development of a continuous-flow system for catalysis with palladium(0) particles. Eur J Org Chem 2004, 3601–3610.

[35] Defined as kg of waste generated per kg of starting materials.

[36] Solodenko W, Mennecke K, Vogt C, Gruhl S, Kirschning A. Polyvinylpyridine, a versatile solid phase for coordinative immobilisation of palladium precatalysts – Applications in Suzuki-Miyaura reactions. Synthesis 2006, 1873–1881.

[37] Mennecke K, Kirschning A. Immobilization of nhc-bearing palladium catalysts on polyvinylpyridine; applications in Suzuki-Miyaura and Hartwig-Buchwald reactions under batch and continuous-flow conditions. Synthesis 2008, 3267–3272.

[38] de M. Muñoz J, Alcázar J, de la Hoz A, Díaz-Ortiz A. Cross-coupling in flow using supported catalysts: mild, clean, efficient and sustainable Suzuki–Miyaura coupling in a single pass. Adv. Synth. Catal. 2012, 354, 3456–3460.

[39] Ping H, Haswell SJ, Fletcher PDI. Microwave-assisted Suzuki reactions in a continuous flow capillary reactor. Appl Catal A-Gen 2004, 274, 111–114.

[40] Baxendale IR, Griffiths-Jones CM, Ley SV, Tranmer GK. Microwave-assisted Suzuki coupling reactions with an encapsulated palladium catalyst for batch and continuous-flow transformations. Chem Eur J 2006, 12, 4407–4416.

[41] Ceylan S, Coutable L, Wegner J, Kirschning A. Inductive heating with magnetic materials inside flow reactors. Chem Eur J 2011, 17, 1884–1893.

[42] Singh BK, Kaval N, Tomar S, Van der Eycken E, Parmar VS. Transition metal-catalyzed carbon-carbon bond formation Suzuki Heck, and Sonogashira reactions using microwave and microtechnology. Org Process Res Dev 2008, 12, 468–474.
[43] Tan LM, Sem ZY, Chong WY et al. Continuous Flow Sonogashira C-C Coupling Using a Heterogeneous Palladium-Copper Dual Reactor. Org Lett 2013, 15, 65–67.
[44] Tischer W, Wedekind F. Immobilised enzymes: methods and applications. Top Curr Chem 1999, 200, 96–124.
[45] Slotema WF, Sandoval G, Guieysse D, Straathof AJJ, Marty A. Economically pertinent continuous amide formation by direct lipase-catalyzed amidation with ammonia. Biotechnol Bioeng 2003, 82, 664–669.
[46] Ceynowa J, Rauchfleisz M. High enantioselective resolution of racemic 2-arylpropionic acids in an enzyme membrane reactor. J Mol Catal B-Enzym 2003, 23, 43–51.
[47] Long WS, Kow PC, Kamaruddin AH, Bhatia S. Comparison of kinetic resolution between two racemic ibuprofen esters in an enzymic membrane reactor. Process Biochem 2005, 40, 2417–2425.
[48] Babich L, Hartog AF, van der Horst MA, Wever R. Continuous-flow reactor-based enzymatic synthesis of phosphorylated compounds on a large scale. Chem Eur J 2012, 18, 6604–6609.
[49] Babich L, Hartog AF, van Hemert LJC, Rutjes FPJT, Wever R. Synthesis of carbohydrates in a continuous flow reactor by immobilized phosphatase and aldolase. ChemSusChem 2012, 5, 2348–2353.
[50] Demuynck ALW, Peng L, de Clippel F, Vanderleyden J, Jacobs PA, Sels BF. Solid acids as heterogeneous support for primary amino acid-derived diamines in direct asymmetric aldol reactions. Adv Synth Catal 2011, 353, 725–732.
[51] Ayats C, Henseler AH, Pericas MAA. Solid-supported organocatalyst for continuous-flow enantioselective aldol reactions. ChemSusChem 2012, 5, 320–325.
[52] Oetvoes SB, Mandity IM, Fueloep F. Highly efficient 1,4-addition of aldehydes to nitroolefins: organocatalysis in continuous flow by solid-supported peptidic catalysts. ChemSusChem 2012, 5, 266–269.
[53] Arakawa Y, Wennemers H. Enamine catalysis in flow with an immobilized peptidic catalyst. ChemSusChem 2013, 6, 242–245.
[54] Bakker JJW, Zieverink MMP, Reintjens RWEG, Kapteijn F, Moulijn JA, Kreutzer MT. Heterogeneously catalyzed continuous-flow hydrogenation using segmented flow in capillary columns. ChemCatChem 2011, 3, 1155–1157.
[55] Oetvoes SB, Mandity IM, Fueloep F. Highly selective deuteration of pharmaceutically relevant nitrogen-containing heterocycles: a flow chemistry approach. Mol Divers 2011, 15, 605–611.
[56] Sourelis SG. The hydrogenation process. J Am Oil Chem Soc 1956, 33, 488–494.
[57] Jones RV, Godorhazy L, Varga N, Szalay D, Urge L, Darvas F. Continuous-flow high pressure hydrogenation reactor for optimization and high-throughput synthesis. J Comb Chem 2006, 8, 110–116.
[58] Irfan M, Glasnov TN, Kappe CO. Heterogeneous catalytic hydrogenation reactions in continuous-flow reactors. ChemSusChem 2011, 4, 300–316.
[59] Stolle A, Gallert T, Schmoger C, Ondruschka B. Hydrogenation of citral: a wide-spread model reaction for selective reduction of α, β-unsaturated aldehydes. RSC Adv 2013, 3, 2112–2153.
[60] Broecker FJ, Kaibel G, Aquila W, Fuchs H, Wegner G, Stroezel M. Process for the selective liquid phase hydrogenation of α, β-unsaturated carbonyl compounds. European Patent EP0947493, 1999.
[61] Recchia S, Dossi C, Poli N, Fusi A, Sordelli L, Psaro R. Outstanding performances of magnesia-supported platinum–tin catalysts for citral selective hydrogenation. J Catal 1999, 184, 1–4.

[62] Madarasz J, Farkas G, Balogh S et al. A Continuous-flow system for asymmetric hydrogenation using supported chiral catalysts. J Flow Chem 2011, 1, 62–67.
[63] Stephenson P, Kondor B, Licence P, Scovell K, Ross SK, Poliakoff M. Continuous asymmetric hydrogenation in supercritical carbon dioxide using an immobilised homogeneous catalyst. Adv Synth Catal 2006, 348, 1605–1610.
[64] Hintermair U, Hoefener T, Pullmann T, Francio G, Leitner W. Continuous Enantioselective hydrogenation with a molecular catalyst in supported ionic liquid phase under supercritical CO_2 flow. ChemCatChem 2010, 2, 150–154.
[65] Shi L, Wang X, Sandoval CA et al. Development of a continuous-flow system for asymmetric hydrogenation using self-supported chiral catalysts. Chem Eur J 2009, 15, 9855–9867.
[66] Sandee AJ, Petra DGI, Reek JNH, Kamer PCJ, van Leeuwen P. Solid-phase synthesis of homogeneous ruthenium catalysts on silica for the continuous asymmetric transfer hydrogenation reaction. Chem Eur J 2001, 7, 1202–1208.
[67] Theuerkauf J, Francio G, Leitner W. Continuous-flow asymmetric hydrogenation of the beta-keto ester methyl propionylacetate in ionic liquid-supercritical carbon dioxide biphasic systems. Adv Synth Catal 2013, 355, 209–219.
[68] Zotova N, Hellgardt K, Kelsall GH, Jessiman AS, Hii KK. Catalysis in flow: the practical and selective aerobic oxidation of alcohols to aldehydes and ketones. Green Chem 2010, 12, 2157–2163.
[69] Sipos G, Gyollai V, Sipocz T et al. Important industrial procedures revisited in flow: very efficient oxidation and N-alkylation reactions with high atom-economy. J Flow Chem 2013, 3, 51–58.
[70] Hou Z, Theyssen N, Brinkmann A et al. Supported palladium nanoparticles on hybrid mesoporous silica: Structure/activity-relationship in the aerobic alcohol oxidation using supercritical carbon dioxide. J Catal 2008, 258, 315–323.

Claude de Bellefon
2 Catalytic engineering aspects of flow chemistry

The purpose of this chapter is to provide chemists with some basic elements from the field of chemical engineering to enable a better understanding and to benefit from using Flow reactors. To keep this chapter useful and not too lengthy, the proposed methods, numbers and developments are often shortcuts underlying many assumptions that could not all be discussed nor even presented, but that apply to most of the chemistries performed in Flow reactors. For demanding reactions (very fast, exothermic, …) specific strategies must be developed asking for more in depth reactor engineering which is out of the scope of this article. Other authors have tackled the issue of summarizing the many concepts used to describe chemical and catalytic reactors in useful and simplified approaches and it is worth reading some of these contributions to get another viewpoint [1–4]. Last, most of the chemical engineering concepts, methods and tools presented is textbook chemical engineering and the reader is strongly advised to enrich the basic knowledge this chapter is attempting to provide [5–8].

2.1 Introduction

The product distribution from a chemical reaction not only depends on the intrinsic properties of the molecules (and catalysts) involved but also on phenomena occurring in the reactor. This is the heart of the process, where reactants are transformed. Reactants are often present in different phases such as gas and liquid which leads to multiphase systems. In many cases, a catalyst is also required to accelerate the reaction and increase the selectivity of the desired product. The overall efficiency of the process is the result of the combination of numerous physical and chemical phenomena. To satisfy the production outcomes, understanding and coupling both kinds of phenomena to design the most suitable reactor or simply to understand results at bench scale is needed.

In flow chemistry, most chemistries are catalytic reactions either with a soluble homogeneous catalyst or with a solid catalyst performed in tube-like reactors, or in reactors where the liquid flow is laminar. Actually, the well-known continuous stirred-tank reactor (CSTR) that can be derived from a simple batch tank by adding inlet and outlet pipes is a bad choice for most reactions since it provides a lower reaction yield and selectivity with similar reaction volume. Thus, stirred-tank reactors are out of the scope of this paper and the reader is referred to other publications for further readings on this topic.

The use of homogeneous soluble catalysts is described for monophasic liquid processes include, for example, Pd catalyzed C-C couplings (Suzuki, Heck, Trost–Tsuji, …) [9], olefin metathesis, and so on. Other processes involving liquid soluble catalysts and a gas phase have been published, mostly for hydrogenations, hydroformylations, carbonylations and oxidations. Some liquid-liquid and gas-liquid-liquid catalytic processes are described and specific issues will be mentioned shortly.

Processes where a solid catalyst is used, generally as a fixed-bed reactor, are predominant with mostly biphasic liquid-solid processes such as solid acids catalyzed estherifications, etherifications, Mikael additions, and so on [10]. Last but not least, triphase gas-liquid-solid reactions are more complex but also represent a large class of processes (hydrogenations, oxidations, …) [11]. Also, solids are largely used as high area, high porosity supports for enzymatic microreactors [12].

Thus, catalytic reactors are complex systems where many phenomena are coupled (Figure 2.1). Chemistry is active at the catalytic sites (grey cubes), diffusion into the porous network of the catalyst (blue arrows), transport across the diffusion layers around the catalyst particles (grey arrows), convective transport of the fluid along the catalyst particle bed (black arrows) and, last but not least, heat transfer at the outer boundaries of the reactor, that is, the walls (not shown).

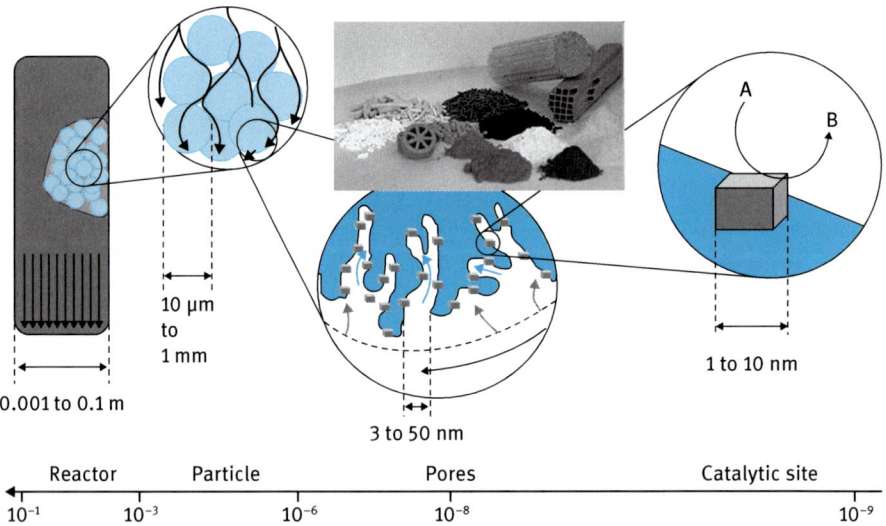

Fig. 2.1: Schematic representation (not at scale!) of the phenomena occurring in Flow reactors with a fixed bed of solid catalysts. From the left to the right, the typical dimensions from molecular sizes to reactor diameter (m). The photograph illustrates different types of powder or shaped solid catalysts.

Obviously, the hydrodynamic phenomenon – mixing, mass transfer between phases, and so on – will be rather different in the systems listed: liquid, gas-liquid, liquid-solid and gas-liquid-solid.

Last but not least, a key reagent is most often identified in Flow chemistry and is generally introduce as a liquid phase, either in pure form or, most often, as a solution in a solvent. This remark sounds quite obvious for synthetic chemists but tremendously helps to make lighter the Chemical Engineering toolbox for Flow reactors.

2.2 Basis of (catalytic) reactor engineering

2.2.1 Flow motion in reactors

The molecules dissolved in the liquid may experience different residence time in the reactor. In a Batch reactor, since the reactor is closed, all the molecules are experiencing the same time as well as the same neighboring providing that the reactor is perfectly mixed. That later assumption could be achieved using a stirrer as in a tank reactor (B) or by the external recycling of the liquid using a pump (RT) (Table 2.1). Both strategies are valuable and can be used to fit practical requirements. Fed-batch (FB) is largely used in the pharmaceutical industry as a way to control the reaction heat flux at the expense of processing time. The analogous, albeit not often described, set-up for tube-shaped reactors is the staggered feeding reactor (SFT stands for staggered fed tube). In a fed-batch, the molecules do not experience the same residence time since they are not introduced at once. The third category is the perfectly mixed continuous reactor. The tank version (CSTR stands for continuous stirred-tank reactor) looks attractive since adding a pipe for the inlet feed and a pipe for the outlet would

Table 2.1: Analogy between tank and tube reactors for a monophase liquid system.

Flow motion	Tank reactor	Tube reactor
Batch (closed, well mixed)	B	RT
Fed-batch	FB	SFT
Continuous mixed	CSTR	CRTR
Plug-flow	SeriesCSTR	PF

transform a Batch reactor into a Flow reactor. External recycling of the liquid using a pump will do the same in a Continuous recycled tube reactor (CRTR), a set-up that is often used in Flow chemistry. However, in the CSTR (and CRTR), the molecules cannot experience the same residence time since some will directly pass from the inlet to the outlet or, inversely, stay for an infinite time in the tank. Last, in a plug flow reactor, all the molecules ideally experience the same residence time. A good example is a chromatography column where it is fundamental that no backmixing occurs to get well separated fractions. In practice, the plug-flow behavior can be approached in a tube reactor equipped with internals or by a cascade of more than five CSTR in series.

Chemical engineering basis enlighten the differences between Batch and Continuous perfectly Stirred-Tank Reactors (CSTR) which, except for zero order reactions, will give lower conversions and lower selectivities than in a Batch or a Plug Flow reactor (PF) [6]. The Peclet number (Pe) is used to assess the plug-flow behavior of contactors. This is easy to understand. In a PF reactor, the fluid elements move forward as slices, ideally without cross-talking (as in chromatography). Thus at the entrance of the PF reactor, the concentration is high and a first order reaction will display a high rate. At the reactor outlet, the concentration is lower, let say 1/10 of the inlet concentration, that is, at 90% conversion, and the rate will be much lower. Thus overall, the global rate will be the sum of each of the rates in each of the slices of the fluid elements in the all PF reactor volume. In a CSTR, the concentration in the all reaction mixture is that at the outlet since it is perfectly mixed (no concentration gradient). Thus taking the 90% conversion, all the reaction volume is working at 1/10th of the inlet concentration. The first order rate of reaction over the all CSTR volume is low and the volume of the CSTR must be larger than that of the PF to compensate with a longer reaction time. In practice, while Pe > 1000 are generally required for chromatography, the CSTR could be seen as one reaction stage with Pe = 1, and Pe > 10 is generally enough to achieve high conversions (> 99%). Note that for specific applications, very high conversions could be required since the concentration of the reagent must drop 1000-fold requiring at least 99.9% conversion thus high Pe numbers. For example, in deep hydrodesulfuration of crude oil, the challenge could be as difficult as to lower the sulfur concentration from circa 5000 ppm to 5 ppm.

Beside monophase liquid reactions, multiphase systems are used involving a gas reagent or a second liquid phase. Gas-liquid reactions are mostly hydrogenations with hydrogen and oxidations with air, oxygen or ozone but hydroformylations, carbonylations, olefin methathesis, aminations, carbonatations may also involve permanent gases (CO, CO_2, NH_3, C_2H_4 ...). The case of liquid-liquid reactions is also quite popular. L/L catalysis presents the advantage of easy molecular catalysts separation and recycling. It is often composed of an organic/aqueous immiscible system such as those encountered in the transformation of acid catalyzed bio-based products, for nitrations where the organic phase transports the key reagents/products and the aqueous phase serves as the catalytic acidic medium (oleum) or for the H-transfer reduction with formate and a water soluble metal complex catalyst. Last, the catalyst may be composed of small solid particles or colloids that are transported with one of the liquid phase forming a pseudo homogeneous liquid, that is, that will not be deposited in the reactor within the time frame of the production.

For reactions involving a solid catalyst, the most popular design is the fixed-bed reactor where the solid catalyst can be fixed either as a compact bed of particles, as a coated layer on the reactor walls, as a monolith structure (honey-comb-like, foam-like) or in other more exotic shapes (fibers, etc.). All have advantages and drawbacks as it will be discussed later.

Whatever the solid catalyst, it should be placed homogeneously in the reactor. With this assumption, the global schematic set-up is very similar to those presented for soluble catalysts (Table 2.2). The specific issues to solve with solid catalysts are however challenging:
- High pressure drop if the particles are too small and/or if attrition leads to very small particles.
- Catalyst deactivation that would drive to catalyst change over thus lowering production.
- Shortcuts, by-pass or stagnation of the liquid if the bed is not homogeneous driving to partial utilization of the catalyst bed.
- Partial wetting of the catalyst in the case of reactions involving a gaseous reagent.
- Intraparticle diffusion limitation since diffusion in the porous network of a solid catalyst is not as easy as in the liquid phase
- Hot-spots formation at the catalyst sites in the case of exothermic reactions potentially leading to increased side-products formation.

Table 2.2: Most common reactor set-up for soluble or transported solid catalysts

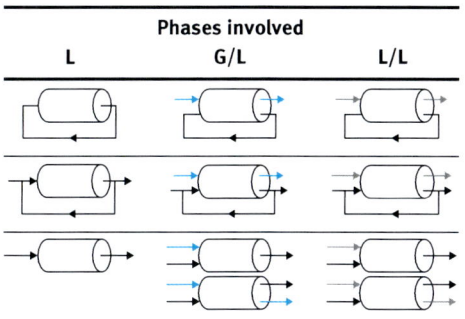

→ Black arrows stand for the liquid (catalytic) phase with the key reagent
→ Grey arrows stand for the second liquid phase
→ Blue arrows stand for the gas phase

2.2.2 Relevant physics

As presented above, the result, that is, the products distribution, depends on the way the concentration and temperature fields are controlled in the reactor. The concentration field is the result of the transport of molecules within the volume of the reactor to allow contact then reaction. One efficient way for mass and heat transfer is convection. Convective mass and heat transfer are governed by the forces and physical phenomena that apply to the fluid elements transporting heat and mass. Usually, viscous, inertial, gravitational and surface forces are considered in chemical reactors. Depending on the typical length scale imposed by the size of the reactor or by the size of internal structurations such as channels, monolith, foams and so on, the dominant forces can reverse (Figure 2.2).

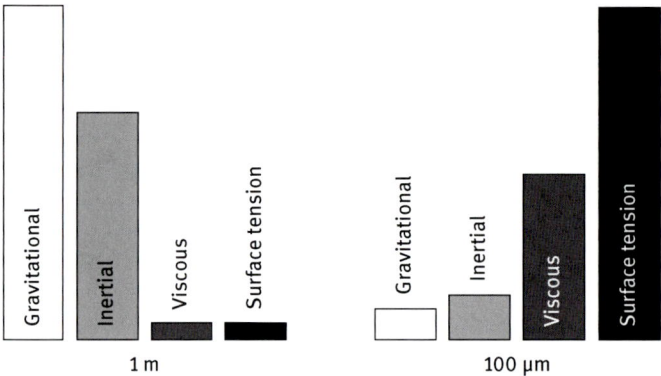

Fig. 2.2: Influence of length scale on the role of forces [13].

Flow reactors generally display rather small typical dimensions, either when simple tubes or capillaries are used, that is, in the range 2000–100 μm, or when monolith type materials are used such as hyper cross-linked polymers or silica rods in which the typical length to consider can be as short as 3 μm as it will be discussed later.

2.2.3 Characteristic times

Another yet less efficient way to transport mass (heat) is diffusion (conduction). Thus, in order to compare all the phenomena that help for mass and heat transfer along the reactor volume, characteristic times associated with each of these phenomena can be defined as well as the characteristic type for the chemical reaction (Table 2.3). Note that each of these characteristic times will be discussed and commented in the following.

Table 2.3: Characteristic times associated with most of the phenomenon occurring in flow reactors.

Relevance	Associated phenomenon or processes	time	Dependence with length***
Chemistry	Reaction (homogeneous & heterogeneous)*	C_{ref}/r	None
	Deactivation	$1/\alpha$	None
Mass transfer	Mixing homogeneous	L^2/D_m	L^2
	Liquid-solid mass transfer	L^2/D_m	L^2
	Gas-liquid mass transfer	L^2/D_m	L^2
	Intraparticle mass diffusion	L^2/D_{eff}	L^2
Heat transfer	Radial heat transfer to the wall	L^2/α_R	L^2
Forces	Gravity	$\sqrt{L/u}$	$L^{0.5}$
	Inertia	L/u	L
	Capillary	$\sqrt{\rho L^3/\sigma}$	$L^{1.5}$
	Viscosity	$\rho L^3/\mu$	L^3
Others	Residence time homogeneous	L/u	L
	Axial diffusion	L^2/D_{ax}	L^2

*In Flow chemistry, the key reagents are in a liquid phase thus the wording "homogeneous" refers to the liquid phase and "heterogeneous" refers to systems in which a solid catalyst is used. Chemistry is described by the intrinsic rate law, $r = f(C, T)$.

**Symbols: D_m is the mass diffusivity in the liquid (m^{2}s^{-1}), D_{eff} is the mass diffusivity in the liquid within the porous network of the catalyst (m^{2}s^{-1}), α_R is the radial conductive heat transfer (m^{2}s^{-1}), u is the liquid velocity (m · s^{-1}), ρ is the liquid density (kg · m^{-3}), σ is the surface tension (N · m^{-1}), D_{ax} is the axial dispersion (m^{2}s^{-1}).

***The length L used here is a generic wording. It may be very different in nature: the diameter of the capillary d_c, the diameter of the pores or the channels of a monolith, the diameter of the particles (d_p) and the reactor diameter (d_R) in a fixed bed, the length (h) of the reactor or channel and so on. This will be discussed later.

All these characteristic times depend on the properties of the fluids (viscosity μ, density ρ, surface tension σ, ...), molecules (diffusion D_m) and typical length (L), except for the chemical reaction time. Obviously, comparing the numerical value of these different times for a given chemical system in a given reactor allows identifying the limiting processes. Thus, ideally, the reaction time should be the largest so that the intrinsic chemical process would be limiting, placing the chemistry at the heart of the production process. In order to discuss the relative importance of the many phenomena and processes easily, dimensionless numbers have been created (Table 2.4).

Table 2.4: Some examples of dimensionless numbers relevant to Flow Reactors.

Number	Definition	Time ratio	Remark
Reynolds	$\text{Re} = \frac{\rho u L}{\mu}$	Inertial/viscous	Empty channel $L = d_c$ Packed bed $L = d_p$ Usually, laminar flow for $\text{Re} < 2000$
Weber	$\text{We} = \frac{\rho u^2 L}{\sigma}$	Inertial/surface tension	
Bond	$\text{Bo} = \frac{\rho g L^2}{\sigma}$	Gravity/surface tension	
Capillary	$\text{Ca} = \frac{u\mu}{\sigma}$	Viscous/surface tension	
Archimedes	$\text{Ar} = \frac{g\rho(\rho_p - \rho)L^3}{\mu^2}$	Gravity/viscous	L is the size of the solid particle (catalyst) in the fluid
Damköhler	$\text{Da} = \frac{r\tau}{C_{ref}}$	Residence time/reaction time	Other Damköhler numbers are used to compare the reaction time with mixing time, transfer times and so on
Peclet	$\text{Pe} = \frac{uL}{D_m}$	Convective/diffusive mass transport	Empty channel $L = d_c$ Packed bed $L = d_p$ No influence on chemistry for $\text{Pe} > 10$
Thiele	$\Phi = L\sqrt{\frac{k}{D_{eff}}}$	Intraparticle diffusion/reaction	Valid for first order kinetics only. L is the characteristic length of the catalytic layer.

2.2.4 Characteristic lengths

Many different lengths have been cited so far. Some lengths are very difficult to quantify since they depend on many phenomena. For example, the thickness of the diffusion layer around a solid particle in a fixed-bed reactor depends on the velocity of the liquid, the properties of the liquid (viscosity, density) and on the size of the particles. However, the typical lengths of solids are at least fixed and could be presented (Table 2.5). Examples of the diversity of length that can be found in catalytic reactors are illustrated in Figure 2.3.

Note that, albeit it is very convenient, it is always dangerous to represent 3D structures using a 2D scheme since much relevant information is lost. More detailed discussions about structured catalysts and catalytic reactors are available [16].

Solid materials with hierarchy ordered porosity thus featuring several typical length are also prepared to make the different phenomenon efficient at each of the scales which nicely illustrates the knowledge-based approach of material synthesis for catalytic applications (Figure 2.4) [17]. Other materials such as fibers, or microfibrous entrapped catalysts are also published [18, 19].

Table 2.5: Schematic presentation of solid catalytic objects and their associated characteristic lengths.

Shape	Comments
(d_p)	Particle of catalyst with homogeneous distribution of the catalytic sites over the all volume. Typical length is the particle diameter d_p. When the particle cannot be assimilated to a sphere, the typical dimension is the volume/surface ratio.
(e)	Particle of catalyst where the catalytic sites are deposited only at the outer boundaries of the support particle ("egg-shell" type). Beside the particle diameter which will impact the external liquid flow, the internal diffusion in the catalytic layer will depend on the catalyst layer thickness e.
(d_h)	Channel of a monolith reactor similar to those used in car exhaust gas catalytic converters. The dimension most often used is the hydraulic diameter d_h. For complex channel cross-section geometries, d_h is computed from the ratio of the cross section area/the wetted perimeter. Note that he channels are not "cross-talking".
(d_o, d_t)	Schematic of open foam cells or other materials with similar topology: hyper cross-linked polymers, inorganic materials. Beside the diameter of the openings d_o from one cell to another, a further length appears here that is the thickness or diameter of the struts d_t. The impact of d_o and d_t on the mass and heat transfers has not been yet fully assessed.

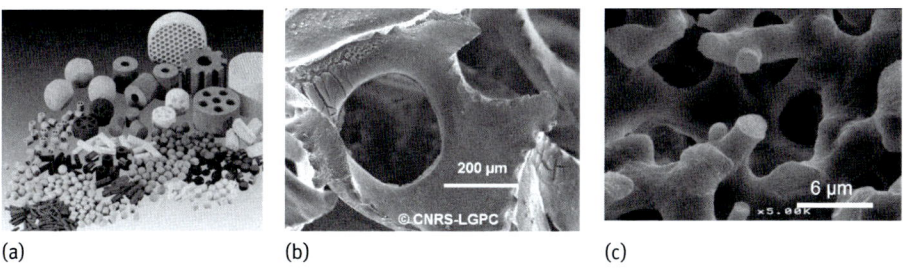

(a)　　　　　　　(b)　　　　　　　(c)

Fig. 2.3: Illustration of various types of solid catalysts and catalyst supports; (a) supports developed by Süd Chemie, (b) Silicon carbide foam supporting a TiO_2 catalyst [14], (c) inorganic monolith "MonoSil" from silica [15].

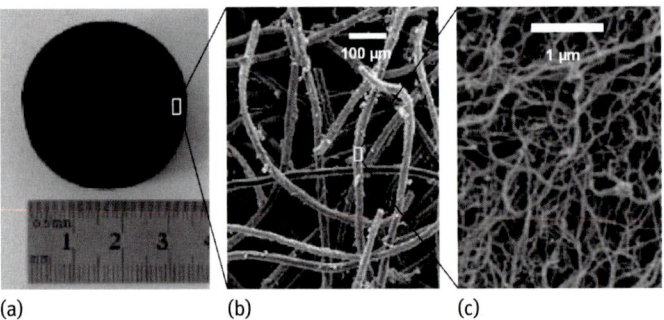

Fig. 2.4: Images of structured packing of carbon nanofiber composite. (a) Macroscopic shape and size of the composite. (b) SEM image of the graphite fibers covered with CNFs. (c) SEM image of the carbon nanofibers. Adapted from Cao *et al.* [17].

2.2.5 Surface area

Very much in the way characteristic lengths have been discussed, the different objects also display specific surface area. Surface areas A are very important for many phenomena. The heat flux transferred at the wall of the reactor will depend on the reactor outer exchange surface. The mass flux transferred from a gas bubble to the liquid will depend on the gas-liquid interfacial area. The rate of catalytic heterogeneous reaction depends on the number of active sites offered to the reagents thus to the surface of the catalyst crystallites. Thus, the higher the area, the higher the flux. Of course, the surface of a large object will be higher than a smaller one. Which matter is the specific area, that is, the unit of surface normalized by unit of volume.

As a general rule, the specific area a of any object is computed as the ratio between the surface of the object divided by its volume S/V (Table 2.6). For beads of 100 μm diameter, the solid-liquid interfacial specific area will be 60,000 $m^2 m_{Bead}^{-3}$. For a typical lab bench autoclave of ca. 10 cm diameter, the specific area for heat exchange at the outer walls is in the range 40–60 $m^2 m_{Reactor}^{-3}$ depending on the shape of the tank whereas for a tube reactor of half a centimeter in diameter, it will amount 800 $m^2 m_{Reactor}^{-3}$. Note that for catalysts, the internal surface area created by the porosity is most often given as the BET area expressed as square meters per gram of solid, most often in the range 1 up 1500 $m^2 g_{cat}^{-1}$. For a rapid first estimate and for many objects, the specific area could be taken as the reverse of the characteristic dimension related to the phenomenon concerned (see Section 2.2.4 on characteristic lengths).

Table 2.6: Typical specific area for mass transfer at the catalyst surface and heat transfer at reactor walls.

Type	Shape	S/V	Catalyst	Reactor
Sphere		$6/d$	d_p 2 mm a_s 3000 m²m$_s^{-3}$	Tank like d_R (tank diameter) 0.1 m a_{wall} 60 m²m$_R^{-3}$
Cylinder		$4/d$	d_p 1 mm a_s 4000 m²m$_s^{-3}$	Tube-like reactor d_R (tube diameter) 0.1 m a_{wall} 40 m²m$_R^{-3}$
Monolith Open cell foam			d_h, d_c or d_0 400 μm a_s 10,000 m²m$_s^{-3}$	
Microchannels				Heat exchanger like $a_{wall} \approx$ a_s 10,000 m²m$_R^{-3}$

2.2.6 Mixing

In the case of homogeneous catalysis, the efficient mixing of the liquid phase is key. This topic has been covered by many authors [2, 20, 21] and the decision to go for flask (Batch) or Flow reactors has been discussed in a very interesting way [3]. The key feature to check before going further with in depth hypothesis about discrepancies between Batch and Flow results in a monophase liquid system is the ratio between the intrinsic reaction time and the mass transfer time, which, in homogeneous mixture, reduces to the mixing time t_m. In turbulent flow, the mixing time is difficult to estimate. However, most reactors and conditions used in flow chemistry lead to laminar flow where the mixing time could be approached by Equation (2.1) in which d_c is the diameter of the channel or the capillary, D_m is the molecular diffusivity coefficient in the reaction mixture and Pe is the Peclet number [21]

$$t_m = \frac{d_c^2}{8D_m} \frac{\ln(1.52 \text{Pe})}{\text{Pe}} \ll \frac{d_c^2}{4D_m} \quad (2.1)$$

This requires the computation of the Peclet number (Pe) which, in turn, requires the knowledge of the liquid properties. As a first estimate, a maximum value for the mixing time can be obtained considering the radial diffusion timescale $d_c^2/4D_m$ which would correspond to the time required for mixing without laminar flow, that is, in a stagnant liquid. For a 500 μm tube and taking a typical diffusivity for liquids (10^{-9} m²s^{-1}), the time required to achieve a homogeneous concentration in laminar flow would be much below 60 s. In fact, with increasing liquid velocity, thus the flow rate, the mixing becomes much more efficient as the Reynolds (Re) number increases and mixing time in the range < 1 s can be easily reached even in rather large capillaries (d_c = 1 mm) at very low flow rate (0.1 cm^3 min^{-1}), that is, at Re ≈ 1.

Such mixing time will allow many reactions to be performed without mass transfer limitations since most chemical reactions are not as fast. However, some reactions may go to completion within a few seconds. Thus, a large effort has been made to design more efficient mixers based on multilamination. Analysis of such mixers where fluid elements experience splitting, rotation and recombining that approaches chaotic mixing is far behind the scope of this article and the reader is directed to more specific readings [22, 23]. A nice review paper embracing several type of mixers has been published [21].

Remark: Obviously, a good mixing is required for monophase liquid systems mainly at the inlet of the reactor, that is, to distribute the reagents and catalyst in an ideally homogeneous mixture, but must not be maintained all along the reactor length.

2.2.7 Heat issues

Reactions performed in Flow chemistry are very diverse. Some display almost no heat issues such as esterifications. Others could be quite exothermic and should deserve a specific reactor design (Table 2.7). Hydrogenation reactions are mild to highly exothermic. The highly exothermic nitroaromatic hydrogenations are generally performed with heterogeneous catalysts. Homogeneous hydrogenations concern almost exclusively the reduction of carbon-carbon double bonds, carbonyls and imines which are mildly exothermic. Furthermore, they are very selective so that in the case of large molecules featuring several chemical functions, only one is generally hydrogenated, with the corresponding exotherm.

The Benson group contribution method and more recent methodologies allow the computation of the heat of hydrogenation reactions, even for large molecules; note

Table 2.7: Heat of reaction for selected reactions.

Reaction	Example	ΔH kJ · mol^{-1}
Hydrogenation of alkenes	RCH=CHR → RCH$_2$–CH$_2$R	−117
Hydrogenation of nitriles	RC≡N → RCH$_2$–NH$_2$	−120
Hydrogenation of aromatics	Benzene → cyclohexane	−208
Hydrogenation of nitro-group	Ph–NO$_2$ → Ph–NH$_2$	−493
Alkylation with dialkylzinc	Et$_2$Zn + benzaldehyde	−190
Aldol reaction	Acetone + benzaldehyde	−50
Esterifications	ROH + RCOOH	~0

that the Benson method gives the reaction enthalpy assuming each species to be a perfect gas! Software and data base (e.g., NIST) are also available. With heat of reactions in the range 100–150 kJ · mol^{-1} (see Table 2.7), and considering that dilute (0.5 to 2 kmol · m^{-3}) solutions of the substrate are most often used, the maximum adiabatic temperature rise can be estimated (Equation (2.2)).

$$\Delta T_{adiab}^{max} = \frac{C_{A,L}^{in}(-\Delta H)}{\rho_L c_{pm}} \quad (2.2)$$

For most organic liquids, ρ_L and c_{pm} are in the range 800–1100 kg · m^{-3} and 2000–3000 J · kg^{-1} · K^{-1} respectively, which lead to maximum adiabatic temperature rise of 150 K. For $\Delta T_{adiab}^{max} < 20$, there is likely no particular reactor design and the actual temperature rise in Flow reactor should not be an issue, that is, often compensated by heat loss due to the large outer surface of the Flow system and insulating or heating cartridges may be requested. From $20 < \Delta T_{adiab}^{max} < 50$, the boiling temperature of the solvent must be checked for safety since a rise of the total pressure may be large depending on the solvent. A pressure-resistant liquid condenser may be used as a simple heat exchanger. From $50 < \Delta T_{adiab}^{max} < 150$, specifically design reactors displaying good to excellent heat transfer capacity must be used. For $\Delta T_{adiab}^{max} > 150$ K cautions must be taken and special reactor design must be applied.

2.3 Describing the chemistry

2.3.1 Kinetic rate laws

Reaction advancement: The advancement of a chemical reaction is the amount of chemical change away from the initial conditions. "Initial" may refer to the time at which the chemical system starts to change typically in a batch reactor or to the conditions (concentration, pressure, temperature) at the inlet of a flow reactor.

The advancement of a chemical reaction may be quantitatively described using a rate law. A rate law refers to the intrinsic chemistry or *intrinsic kinetics* i.e. that is free from any physical phenomenon such as mass or heat transfer. Different approaches can lead to different forms of rate laws but the common feature of rate laws is to describe how a reaction proceeds depending on the local species concentration C_i and the temperature T. Any rate laws can generally take the following form (Equation (2.3)).

$$r = f(C_i, T) \quad (2.3)$$

The constant k varies with the temperature only, except for some specific situations where it also depends on the electric current and/or voltage (electrochemistry) and on

Table 2.8: Examples of rate laws for simple reaction schemes.*

Reaction	Mass action	Empirical	Langmuir–Hinshelwood (LH)	
	Only for elementary steps	Easy to set. Does not reflect the mechanism. Hardly scalable.	Reflect the chemistry (mechanism) Quite scalable.	
A → B	$r = kC_A$	$r = kC_A^\alpha$, $0 \leq \alpha \leq 1$ $r = kC_A^\alpha C_B^\beta$ $-1 \leq \beta \leq 0$	$r = k\frac{K_A C_A}{1+K_A C_A}$ $r = k\frac{K_A C_A}{1+\sqrt{K_A C_A}}$	$r = k\frac{K_A C_A}{(1+K_A C_A)^2}$ $r = k\frac{K_A C_A K_B C_B}{1+K_A C_A + K_B C_B}$
A + B → C	$r = kC_A C_B$	$r = kC_A^\alpha C_B^\beta C_C^\lambda$ $0 \leq \alpha \leq 1$ $0 \leq \beta \leq 1$ $-1 \leq \gamma \leq 0$	$r = k\frac{K_A C_A K_B C_B}{1+K_A C_A + K_B C_B + K_C C_C}$ $r = k\frac{K_A C_A K_B C_B}{(1+K_A C_A + K_B C_B)(1+K_C C_C)}$	

*The rate unit is that of the rate constant k. It can be expressed in several ways depending on the target.

light (photochemistry). The variation of k with temperature is often described by the Arrhenius law but again other concepts could be used (Eyring …).

A very common misinterpretation is to consider a time dependence of the rate law. This mistake comes from the fact that the mass balance equation in a batch reactor at constant volume (Equation (2.4)) is often taken erroneously as the rate equation. Since the batch reactor is the most used and taught reactor in the synthetic chemistry community, this erroneous concept has sometimes spread to students, teachers and researchers.

$$\frac{dC_A}{dt} = r(C_i, T) \qquad (2.4)$$

As pointed above, different forms of rate laws may be encountered. For the sake of conciseness, examples for simple reactions are given here (Table 2.8) but more general situations could be easily derived.

Mass action rate laws stand only for elementary steps of a reaction mechanism. In such rate laws, the reaction order of species i is equal to the stoichiometric coefficient of that reagent in the reaction step. Power type rate laws are empirical in the sense that they do not reflect nor an elementary step nor a reaction mechanism. They are obtained by fitting the kinetic parameters (power exponent, kinetic constant) to experimental data. They cannot be extrapolated but are quite often used because the knowledge of the reaction mechanism is not required. On the contrary, Langmuir–Hinshelwood (LH) type kinetics are based on reaction mechanisms encountered in heterogeneous catalysis since they describe the chemisorptions of the reagents and products on solid surfaces. Note that the absorption constants K also depend on the temperature. Thus the variation of the rate with temperature in the case of LH rate laws could be more complex and often the Arrhenius plot is not linear [8].

2.3.2 Rate measurement and reaction time

The rate of a chemical reaction cannot be accessible directly but through the measurement of the rate of production (consumption) of the chemical species involved in the reaction medium under investigation. The relation between the rate of reaction and the rate of production/consumption of the products is given by the reaction scheme and the stoichiometric coefficients. For complex reaction schemes, that is, involving more than one reaction and leading to selectivity issues, the reader is referred to chemical engineering textbooks [1, 5]. Thus only simple systems will be discussed here but the method can be extended easily.

For a single reaction or for reacting systems that could be assimilated to a main reaction, the rate could be estimated from the slope of the concentration of the key reagent (A) versus time profile. Such experimental data can be measured in a batch reactor or in a Flow reactor. In the latter, a residence time is used instead of the chronological time in a batch reactor. The concept of residence time will be explained later but the method is basically the same. Often, the concentration versus time profile is not a straight line thus leading to many different slopes. It is advised to take the highest slope corresponding to the conditions for the highest rate of reaction. The highest rate will then lead the shorter reaction time thus providing a safe comparison with possible physical limiting phenomenon. Most often, the initial rate corresponding to the slope at initial time is used since most reactions are in positive reaction order with respect to the reagent concentrations which are at maximum values at initial time, that is, before they get consumed.

For catalytic reactions in homogeneous liquids (homogeneous catalysis), the amount of catalyst will change the rate of reaction. The reaction time must reflect this particular aspect. The different rate of reactions that could be estimated from the measured slope of the concentration versus time profiles can be easily related (Equation (2.5)).

$$\text{Slope} \cdot V_L = r_V V_L = \text{TOF} \cdot n_{cata} = \frac{1}{t_R} n_{cata} \qquad (2.5)$$

In which:
r_V is the rate per unit volume of liquid ($mole_A \cdot s^{-1} \cdot m^{-3}_L$)
V_L is the volume of the liquid phase
TOF is the turnover frequency ($mole_A \cdot mole^{-1}_{cata} \cdot s^{-1}$)
n_{cata} is the quantity of catalyst ($mole_{cata}$)
t_R is the reaction time which is the reverse of the TOF (s)

For heterogeneous catalytic reactions, a very similar approach can be used. However, it is more complicated since a heterogeneous catalyst is itself often composed of active sites distributed over a highly porous carrier. Good examples are hydrogenation catalysts composed of active noble metal crystallites (Pt, Pd, Ru, Rh ...) deposited on alumina, silica or charcoal. Thus, the rate will not only depend on the mass of catalyst used but also on the precious metal content and, ultimately, on the surface metal atoms. Thus different rates could be defined with respect to the specific process targeted (Equation (2.6)). Accordingly, several reaction times can be defined.

$$\text{Slope} V_L = r_V V_L = r_w m_{cata} = r_p V_p = r_M m_M = r_a a_S = \text{TOF} n_s = \frac{1}{t_R^{site}} n_s \quad (2.6)$$

In which:
r_w is the rate per mass unit of catalyst ($\text{mole}_A \cdot \text{s}^{-1} \cdot \text{kg}_{cata}^{-1}$)
m_{cata} is the mass of catalyst (kg_{cata})
r_p is the rate per volume unit of catalyst ($\text{mole}_A \cdot \text{s}^{-1} \cdot \text{m}_{cata}^{-3}$)
V_p is the solid of solid catalyst particles (m_{cata}^3)
r_M is the rate per mass unit of precious metal ($\text{mole}_A \cdot \text{s}^{-1} \cdot \text{g}_{Metal}^{-1}$)
m_M is the mass of precious metal (g_{Metal})
r_a is the rate per surface unit of precious metal crystallites ($\text{mole}_A \cdot \text{s}^{-1} \cdot \text{m}_{Metal}^{-2}$)
a_S is the surface of the precious metal crystallites (m_{Metal}^2)
n_s quantity (atoms) of active metal (sites) at the surface of crystallites (mole_s)
TOF is the turnover frequency ($\text{mole}_A \cdot \text{mole}_s^{-1} \cdot \text{s}^{-1}$)
t_R^{site} is the site reaction time which is the reverse of the TOF (s)

The determination of the "site" reaction time t_R^{site} is useful only when comparing an homogeneous catalyst and a supported catalyst but requires a detailed knowledge of the solid catalyst. When operating with the same solid catalyst, that is, same precious metal loading, same crystallites size, same dispersion, a reaction time for heterogeneous catalytic reactions t_R can be used safely. This time is computed from the ratio of the reference concentration of the key reagent C_A^0 and the volume rate of reaction r_p (Equation (2.7)).

$$t_R = \frac{C_A^0}{r_p} \quad (2.7)$$

The rate of reaction can also be taken from literature data. When a rate law is published it can be used in the same way as when it is measured. The reaction time can be computed in a similar way. Quite often, only a single data point is available, that is, a conversion number at a given time and under given reaction conditions (pressure, temperature, concentrations) in a batch reactor. An average rate of reaction could be estimated and the reaction time could be computed again in a similar way. Note that rate laws and reaction rates can be used in a more quantitative and efficient way for reactor design and modeling.

2.3.3 Catalyst deactivation

The activity and selectivity of a catalyst may vary with its turnover. The time (chronological) or age itself is not the source of deactivation but rather the fact that, very much like any living organisms, a catalyst can lose activity upon working. Several mechanisms that will be discussed below can be involved. The activity decay could also be characterized by a deactivation time. The time for deactivation can vary between a few seconds up to years depending on the phenomenon responsible for deactivation. Knowing, understanding and quantifying catalyst deactivation is mandatory to design a chemical process. Catalyst deactivation has been the topic of several textbooks [24–26].

Catalyst deactivation in Flow chemistry can have several effects. The lowering of activity can lead to lower conversion and, less often, to lower selectivity. In the case where no quantification of the deactivation is available, the time-on-stream during which the required conversion must be maintained cannot be predicted. Thus, the production must be followed regularly, a fresh catalyst charge must be available at any time and technician's or operators must also be available for catalyst change-over. A too fast catalyst deactivation will also hamper the possible use of structured catalytic reactors such as monoliths, microreactors, foam reactors where the catalyst is washcoated on the wall or the channels. Thus, it is generally mandatory to gain some knowledge about catalyst deactivation.

Deactivation must be first recognized and, second, quantified.

The causes for catalyst deactivation are numerous. They could be physical processes such as sintering at high temperatures of metal crystallites, mechanical attrition can lead to smaller catalyst particles flowing out of the reactor or leading to very high pressure drop. Chemical phenomena are also occurring such as fouling, cocking, oxidation and/or leaching of the active sites, and poisoning. Under typical operating conditions in flow chemistry, three major causes of catalyst deactivation may be considered: attrition oxidation and/or leaching of the active sites, and poisoning thus deactivation is case sensitive.

Mechanical attrition could result from a too high pressure drop through a fixed bed of solid particles. Spherical particles are much more resistant to attrition since the forces are spread over the all particles surface more equally. On the contrary, plate shaped particles will tend to break since all the pressure will exert on small areas such as corners or edges. This phenomenon has been extensively studied in high pressure chromatography.

Oxidation and/or leaching of the active sites is often encountered for molecular catalysts which must operate at a given oxidation state. For example, in the rhodium based hydroformylation process, oxidation of the phosphine ligands can lead to rhodium leaching. Also, it has been recognized that too low substrate concentrations can lead to "naked" metal centers, hence to metal clustering and loss of catalytic activity.

Poisoning is a very common cause for catalyst deactivation. It results from an almost irreversible adsorption of chemicals on the catalytic surface. Nickel-based catalysts are easily deactivated by sulfur containing chemicals. Heavy metals, phosphorous and arsenic containing chemicals, are poisons for reduced metal particles used in hydrogenations. In fine chemistry, most of poisoning comes from the side-products. In the hydrogenation of nitroaromatics, heavier side-products such as diazoaromatics form and could be strongly adsorbed on the Pd crystallites. In acid-based catalyzed reactions such as aldol condensations, polymerization may occur leading to irreversible active site coverage and deactivation. Thus the chemistry as well as potential side reactions must be known in order to anticipate for catalyst deactivation.

> **TurnOver Number:** The TON describes the work the catalyst can achieved before declining. An ideal catalyst would have an infinite turnover number in this sense, because it would never be consumed. The unit is $mole_{reagent} \cdot mole_{catalyst}^{-1}$.
>
> $$\text{TON} = \frac{\text{Number of molecule transformed}}{\text{Number of catalytic sites}} \quad (2.8)$$
>
> Pharmaceuticals, fine chemistry: $100 < \text{TON} < 10^4$
> Large chemical production: $10^4 < \text{TON} < 10^6$
> Oil refining, energy, pollution: $10^6 < \text{TON}$

For the proper quantification of deactivation, it is strongly advised to use the turnover number TON (Equation (2.8)). In the definition, the number of catalytic sites could be computed from the mass of the catalyst, the mass of the precious metal involved and, more thoroughly, from the exposed (active) catalytic centers. Thus, it is easy to determine, in the case of acid catalysis with phosphoric acid, toluenesulfonic acid, polystyrene sulfonate, heteropoly acids, zeolites and so on where the acid concentration ($meq \cdot g^{-1}$) is available. It is also easy for metal complex catalysis (homogeneous catalysis). For metallic crystallite catalysis, the metal dispersion must be known to enable the determination of the surface metal atoms. When the dispersion is not available, the total number of metals can be used. Other units and other names are often used in industry such as the mass of catalyst lost per ton of product which immediately translates into operating costs (OPEX). The very same information can easily be deduced from the larger definition of the TON. Sometimes, the number of catalytic sites cannot be determined and the ratio of mass of reagent transformed over the mass of catalyst is used.

The main difficulty is to provide a criteria that fixes the limit at which a catalyst is considered to be "deactivated". One "absolute" criteria is when the catalyst is dead, that is, when no activity is observed. This could be useful for fundamental studies but not for production nor kinetic studies. A useful criteria is to define a conversion number X under which the process is no longer realistic [28]. Obviously, high TON_X are seeking but it depends very much on the application field and on the catalyst cost. Ranges for TON are proposed, and as any rules, such ranges might not apply to some particular process. Maybe one of the most striking examples is that of the production of the herbicide (S)-Metolachlor using a sophisticated iridium based asymmetric catalyst displaying a TON of 10^6 in the industrial production unit [27]!

The procurement of the TON can be longsome and this is likely the reason why very few TON values are published. The first difficulty is the very large utilization of batch reactors for catalyst testing. In a batch reactor, transient phenomena are observed. The composition of the reaction mixture changes with time thus the rate of reaction will also change and any other transient phenomenon occurring will be superimposed. Decoupling those phenomena in a batch reactor is tedious. On the contrary, a Flow reactor is operating in a quasi steady-state mode in the sense that the composition of the reaction mixture at the outlet of the reactor is constant with time. Thus leading to decoupling chemical reaction and deactivation.

However, since batch reactors are widely used, some techniques could be applied albeit they are not free of pitfalls. As a first approach, consecutive additions of a given quantity of substrate after a first run can provide information on the deactivation (Figure 2.5 (a)).

Several pieces of information can be extracted from this simple experiment. First, a global TON value can be computed from the total quantity of the substrate that has been transformed divided by the quantity of catalyst in that experiment. Second, a *deactivation rate law* can be estimated considering the decay of the rate of reaction. The initial rate of reaction measured for the first run can indeed be taken as the rate *free of deactivation* (r^0). Then the following slopes (rate) provide lower values. The deactivation law ϕ is defined as a function with a numerical value between zero (complete deactivation) and unity (no deactivation) that multiplies the rate of a fresh catalyst (Equation (2.9)). The underlying assumption of this method using batch data is that only deactivation is responsible for the decrease of the measured rate. This might not be the case for rate laws that are negative order with respect to the product (see section above about rate law).

$$r = r^0 \phi \quad 0 < \phi < 1 \tag{2.9}$$

r is the current rate of reaction
r^0 is the rate free of deactivation (fresh catalyst)
ϕ is the deactivation law

Very much like a rate law, the deactivation law depends on the concentrations (of a poisonous species, of a side-product …) and temperature only and not on time. How-

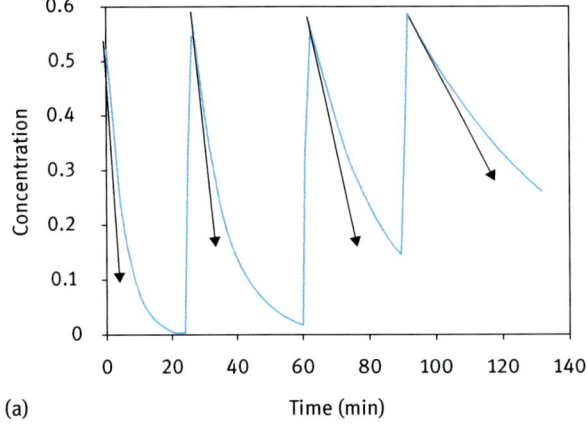

(a)

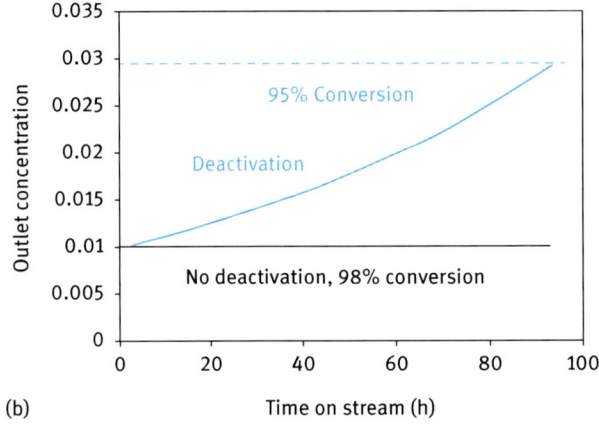

(b)

Fig. 2.5: Schematic presentation of how catalyst deactivation could be evidenced (a) in a batch reactor using the consecutive addition method. Red arrows represent the slope at initial conditions for each run. (b) In a Flow reactor.

ever, empirical rate laws based on time $\phi(t)$ are often proposed as a first shot (Equation (2.10)).

$$\phi(t) = 1 - \alpha t \qquad (2.10a)$$
$$\phi(t) = \exp(-\alpha t) \qquad (2.10b)$$
$$\phi(t) = 1/(1 + \alpha t) \qquad (2.10c)$$

In these deactivation laws, α is called a *deactivation parameter* that has no chemical meaning, that is, it bears no information on the source of the deactivation, it is not a constant from a chemical point of view and it may change its value upon changing the operating conditions (e.g., hydrogen pressure, substrate concentration) which is not the case with a true kinetic constant. Nevertheless, when the operating conditions of a process are fixed, it is a convenient method for deactivation quantification. A characteristic time may be attributed to deactivation as the reverse of the deactivation parameter, that is $t_{\text{deact}} = \alpha^{-1}$. The deactivation time will be useful to access the potential application of a process, anticipate for catalyst changeover and cost.

In the case where a Flow reactor is used (Figure 2.5 (b)), a simple parameter-fitting procedure provides the deactivation parameter.

In the published literature, mostly in the field of homogeneous catalysis, recycling experiments comparable to the method shown here, are performed. While the recycling experiment itself is of very good value, the way they are generally displayed may drive to misleading conclusions. For example, an experiment leading successfully to 12 runs without a significant drop in conversion cannot be conclusive about catalyst deactivation since the 12 runs may represent only a very small TON. Thus, the TON after the 12 runs must be computed and commented according to the purpose of the process as discussed above.

2.4 Methodology for Flow reactor dimensioning

2.4.1 Batch versus Flow reactor comparison

Most often, benchmarking the flow system versus a traditional batch type mode should be performed since it gives the proof that things have been performed thoroughly. Furthermore, such comparison may reveal discrepancies that could be very informative about the driving phenomena in both batch and flow reactors.

As a general remark, reactor comparison must be made on the basis of a similar composition of the reaction mixture, pressure and temperature since the intrinsic rate depends on these parameters. A popular practice, in particular in the field of homogeneous catalysis, is the use of the substrate/catalyst (S/C) ratio as a comparison criterion. This is meaningless since the rate law depends on concentrations and the same S/C ratio could be achieved with very different concentrations.

For a monophasic liquid chemical system and under the same experimental conditions of temperature and concentrations, the comparison between batch and flow systems can be made using the batch time and the residence time in the flow reactor (Equation (2.11)). Such comparison is built on the assumption that in the flow reactor, the volume occupied by the liquid reaction mixture is constant with time and the flow rate of the liquid is constant. Indeed, the residence time in the flow reactor, the wording "Flow time" will be used here, is given by Equation (2.11) which means that deviation in the reaction volume will drive to a deviation in the residence time.

> **Definition of time in Flow reactors:** In homogeneous reactors, that is, with one liquid phase, the contact time or flow time (symbol τ) is defined as the ratio between the volume V_L occupied by the liquid reaction mixture in the reactor and the volume flow rate of the feed Q_L.
>
> $$\tau = \frac{V_L}{Q_L} \quad (2.11)$$
>
> In practice, the cross-section Ω of the reactor is constant, which leads to a direct relation between the length of the tube L and the flow time.
>
> $$\tau = \Omega \frac{L}{Q_L} \quad (2.12)$$

The Batch time (symbol t) is defined as the time elapsed between the mixing of all reagents and catalysts required for the reaction to proceed and the measurement of the reaction mixture composition.

The Flow time (symbol τ) is related to the volume occupied by the liquid reaction mixture (V_L) in the reactor (Equation (2.11)). Caution must be taken to avoid dead volumes. Thus, depending on the reactor shape and geometry, it may reveal difficult to fulfill the assumption of full occupation of the reactor volume by the flowing reaction mixture. In a plug-flow reactor, the volume of the reaction mixture (V_L) is often taken as the geometrical volume of the reactor chamber.

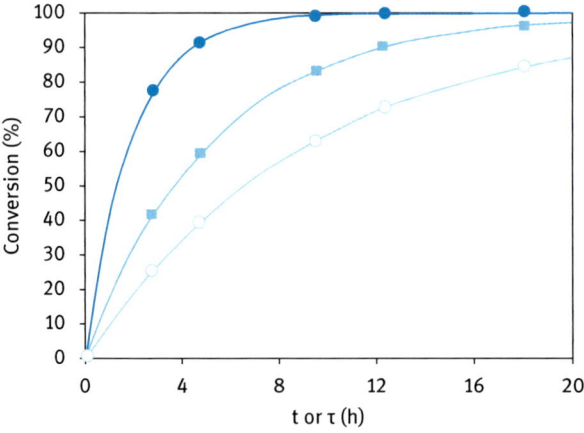

Fig. 2.6: Theoretical comparison between Batch (square symbols) and Flow (round symbols) reactors operating under the same conditions of temperatures and reagents and catalyst concentrations for a using time t or residence time τ for monophasic liquid systems or 1/WHSV for systems involving a solid fixed catalyst.

Thus, it can be immediately assessed if the Flow reactor performs better (•) or lower (○) than the Batch (■). The explanation why the two reactors perform differently

2.4 Methodology for Flow reactor dimensioning — 53

will be discussed later. Making comparisons at too high conversions are not informative since differences are hardly visible (see data point at ca. 18 h). In the example, the time features in the abscissa is the reaction time in the batch reactor or the residence time in the flow microfluidic chip reactor [3]. Note that the Flow time τ is the reverse of the Liquid Hourly Space Velocity (LHSV = Reactant Liquid Flow Rate/Reactor Volume), a wording frequently used in industry. Note that in the case of monophasic liquid flow reactor, the Flow time τ is the reverse of the Liquid Hourly Space Velocity (LHSV = Reactant Liquid Flow Rate/Reactor Volume).

Cautions must be taken with chemical reactions that will drive to a noticeable change in the density of the liquid. Albeit the mass flow will be constant, of course, the volume flow will change leading to a change in volume flow rate, thus in the Flow time hence to erroneous comparison with the Batch time and, ultimately, to false conclusions.

In the case where the homogeneous catalyst concentration is not the same in batch and Flow experiments, and since intrinsic rate laws are generally first order with the catalyst concentration in homogeneous catalysis, the time (t or τ) must be normalized by the quantity of catalyst introduced. This normalized time is easy to compute from the Flow time and the Batch time.

For Flow reactors fitted with a fixed solid catalyst, the reasoning to compute the Flow time is very much the same and follows Equation (2.11) or Equation (2.12) depending on the rector geometry. However, since the reactor volume is partially occupied by the solid catalyst and by the gas for triphase systems, the liquid volume is more tricky to assess. Indeed, in most catalytic flow reactors, the catalyst is inserted as a compact fixed-bed of particles which drive to a porosity in the range 40–50% (note that theory gives 58% of solid for a compact arrangement of spheres).

The major issue lies in the fact that the liquid/catalyst ratio is generally not as high in Batch reactors. Thus the Batch-Flow comparison on the basis of time only is not appropriate and, again, some kind of "normalized time" must be used instead. In the field of catalytic reactors with solid catalysts, the Weight Hourly Space Velocity (WHSV) is used. It is defined as the ratio of the mass flow rate of the key reactant (Q_m in e.g., $g \cdot h^{-1}$) divided by the mass of the catalyst in the reactor (m_{cata} in e.g., g) (Equation (2.13)). Note that the dimension of WHSV is the reverse of a time (h^{-1}).

Batch and Flow reactor comparisons for heterogeneous catalysis: The experiments must be performed under the same operating conditions (temperature, concentrations, pressure) since they can have a very strong impact on the intrinsic kinetics and would lead to meaningless comparisons.

The Flow time in a heterogeneous reactor is computed as the reverse of the Weight Hourly Space Velocity (WHSV). Thus one WHSV is associated with one conversion number.

$$\text{WHSV}_{\text{Flow}} = \frac{Q_m}{m_{\text{cata}}} (h^{-1}) \qquad (2.13)$$

> The corresponding normalized time in the Batch, that is $1/\text{WHSV}_{\text{Flow}}$ in the Flow reactor, is obtained by dividing the initial mass of the key reagent (m^0_{reagent}) by the mass of catalyst (m_{cata}) and by the time (t_X) needed to achieve conversion X in the Batch reactor.
>
> $$\text{WHSV}_{\text{Batch}} = \frac{m_{\text{reagent}}}{m_{\text{cata}} t_X} (h^{-1}) \qquad (2.14)$$
>
> Using WHSV will ensure that the quantity of key reagents experiencing a contact with the catalyst is the same in Batch and Flow modes.

Calculations are straightforward when the mass of catalyst inside the reactor and the incoming reactant mass flow rate are known. The later is easily computed from the volume flow rate, the reagent concentration and the density of the feed. Comparison with the batch mode could be achieved thoroughly (Figure 2.1). Thus, Batch and Flow reactors will be considered as equivalent from a chemistry point of view if they yield the same conversion at same 1/WHSV. Discrepancies between Batch and Flow side-reactions occurring in the liquid phase will be minimized in the flow reactor since it generally offers much more catalytic sites for the targeted main reaction whereas they will develop more in the Batch reactor where a smaller proportion of catalyst is generally used.

Other spatial velocities may be used. The mass of precious metals may be used in place of the total mass of catalyst. This could be useful to compare similar catalysts that have different metal loading. The molar quantity of catalyst (and key reagent) can be used advantageously in the case of supported homogeneous catalysts or with single site catalysts.

2.4.2 Checking for mass and heat transfer limitations

In a catalytic reaction, the reagent(s) and catalyst, molecular or solid, must get into contact at some point for the reaction to proceed. The molecules have to be transported from the bulk of the liquid to the catalytic sites. This mass transport can be ensured by convection and diffusion phenomena. Thus the catalytic reaction is the result of several steps in series or in parallel, each step having potentially large influence on the global process. The analysis made here is for three phase gas-liquid-solid systems but it can be easily derived for liquid-solid processes. The steps are as follows, assuming the gas reagent being the limiting reagent which is most often the case since the reacting gas species usually encountered in Flow chemistry are not very soluble (H_2, O_2, CO, ...) in most liquids.

1. Transport of the gas reagent from the gas phase to the liquid phase
2. Dissolution of the gas reagent into the liquid phase
3. Transport and diffusion in the liquid phase to reach the catalyst surface
4. Diffusion into the catalyst pore network to reach the active sites
5. Adsorption on the active site, reaction and desorption (intrinsic kinetics)

For a liquid-solid reaction, the reasoning is similar but limited to steps 3 to 5 and the limiting reagent has to be determined.

The apparent rate of the reaction will be determined by the rate of the slowest step. For slow intrinsic kinetics, the reaction is only dependent of the kinetic law and the amount of catalyst. Increasing the amount of catalyst leads to an increase in the reagent consumption rate until the characteristic reaction time becomes lower than the mass transport time. Under such conditions, the reaction rate starts to be solely dependent on resistance encountered by reactants to reach the active site. These resistances, corresponding to steps 1 to 4, could be represented as resistances in series (Figure 2.7). At steady state, that is, at equal transferred flow of the limiting reagent, the mass fluxes are equals.

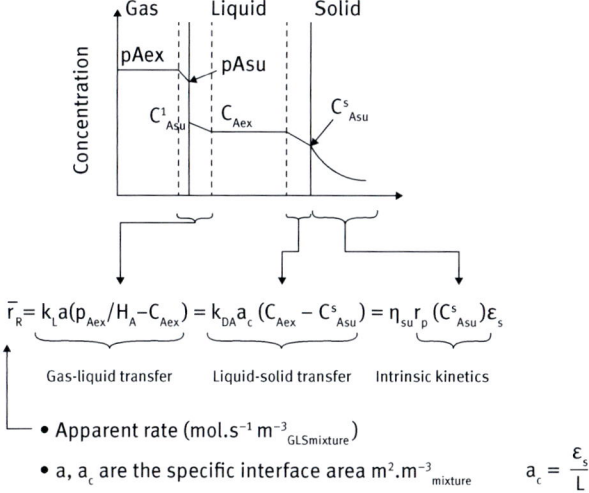

Fig. 2.7: Scheme for three phase mass transfer illustrating the three main mass transfer resistances and associated mass balance equation, gas A being the limiting reagent.

The mass balance equation of Figure 2.7 means that the mass flow of the limiting reagent, in general gas reagent A, is equal to the measured rate of reaction (apparent rate). This equation could easily rearrange to the equation (2.15) for a first order intrinsic kinetics. Note that the approach requires refinements for other kinetics but is quite a good approximation for any reaction that is positive order with respect to the limiting reagent and that displays no inhibition by none of the reaction products such as in Langmuir–Hinshelwood type kinetics.

$$\underbrace{\frac{P_{Aex}/H_A}{\bar{r}_R}}_{\substack{\text{Apparent} \\ \text{reaction time} \\ \bar{t}_R}} = \underbrace{\frac{1}{k_l a}}_{\substack{\text{Gas-liquid} \\ \text{transfer time} \\ t_{GL}}} + \underbrace{\frac{L}{k_{DA}\varepsilon_s}}_{\substack{\text{Liquid-solid} \\ \text{transfer time} \\ t_{LS}}} + \underbrace{\frac{1}{\eta_{su}}}_{\substack{\text{Catalyst} \\ \text{efficiency}}} \underbrace{\frac{1}{k_p \varepsilon_s}}_{\substack{\text{Intrinsic} \\ \text{reaction time} \\ t_R}} \tag{2.15}$$

This last equation (Equation (2.15)) displays several interesting features. First, it allows comparing directly the characteristic times of several potentially limiting phenomena. Computing these characteristic times separately can provide a way to asses the working regime of the reaction and reactor without any further experiments.

P_{Aex} is the pressure of the limiting gas reagent and H_A is the Henry constant that quantify the solubility of gas A in the liquid mixture.

$k_l a$ is the volumetric gas-liquid mass transfer coefficient (s^{-1}). Many correlations are published for different types of reactors including trickle beds, structured packings, monolith, foams, and some microstructured reactors. However, such correlations must be used with caution and are not easy to manipulate. Furthermore, it has been shown that they hardly apply to reactors displaying very small packing (random or structured) likely because surface forces are playing a large role at a small scale. For example, gas-liquid mass transfer coefficients in the range 2–15 s^{-1} have been reported in micropacked bed reactors [29, 30], and up to 6 s^{-1} in microtructured falling film reactors [31]. This is much more than values in the range 1–2 s^{-1} measured in the best lab scale tank pressure reactor equipped with a gas effect stirrer [32]. Thus, it could be said that gas-liquid mass transfer in small packed-bed reactors will be limiting only for very fast reactions, for example for hydrogenations with a reaction rate in the range 0.1 mol · s^{-1} · g_{metal}^{-1}.

L is the characteristic length of the system where the liquid to solid occurs. It is related to the particle diameter in a micropacked bed, that is $d_p/6$ for a spherical particle, but it is the channel diameter d_c or other hydraulic diameter for wall coated catalysts such as in monoliths.

k_{DA} is the liquid-solid mass transfer coefficient for the limiting reagent A (m · s^{-1}). Again, this coefficient cannot be estimated for micropacked bed reactors. As a matter of fact, $k_l a$ and k_{DA} are often lumped in a global gas-liquid-solid mass transfer coefficient $K_l a$ [27]. For liquid-solid reactors, k_{DA} could be computed safely using the well-known correlation of Kunii and Levenspiel [33].

The solid fraction volume ε_s is computed from the volume of catalyst introduced in the reactor, either as particles or coated on the wall. Note that the knowledge of the volume of the catalytic section, that is, the volume of the bed not that of the all tubing, is also required since $\varepsilon_s = V_s/V_R$.

2.4 Methodology for Flow reactor dimensioning — 57

Last, the catalyst efficiency η_s is determined from the Thiele modulus (Equation (2.16)).

$$\phi^2 = \frac{r(C_{As})L^2}{C_{As}D_{eff}} \tag{2.16}$$

$$\phi^2 = \frac{k_p L^2}{D_{eff}} \tag{2.17}$$

In the case of a first order intrinsic kinetics, the Thiele modulus simplifies and does not depend anymore on concentrations (Equation (2.17)). Here, L is the characteristic length of the catalyst layer (e) or particle ($d_p/6$) and k_p is the intrinsic kinetic constant in the proper units (mol·$m^{-3}_{catalyst}$·s^{-1}) and D_{eff} ($m^2 s^{-1}$) is the effective diffusivity in the porous catalyst solid. When $\phi > 2$ the catalyst efficiency is in turn computed easily (Equation (2.18)).

$$\eta_s = \frac{1}{\phi} \tag{2.18}$$

When $0.4 < \phi < 2$, more complex computations have to be performed [7].

When $\phi < 0.4$, the system can be said free of internal mass transfer limitations.

The estimation of the Thiele modulus can thus help assigning the working regime of the reactor. The mass balance equation shown in Figure 2.7 nicely illustrates the impact of the operating parameters. Thus, a more powerful agitation of the fluids in the reactor generally increases the mass transfer coefficients $k_l a$ and k_{DA}. It has no effect on H_A, η_s and k_p. The temperature has an effect on the gas solubility H_A, on the catalyst efficiency η_s and of course on k_p by virtue of the Arrhenius activation law for intrinsic kinetic constants. However, the temperature has a negligible impact on the external mass transfer coefficients $k_l a$ and k_{DA} since viscosities and densities changes little with small temperature variations ranging 10–20 K. The particle size impacts k_{AD} and the catalyst characteristic length (e or $d_p/6$) impacts η_s but has no effects on $k_l a$, H_A and k_p. Last, variation of the mass of catalyst will modify the catalyst hold-up ε_s and helps determine external gas-liquid versus external liquid-solid versus internal mass transfer limitations. Combining these tests as well as others leads to a global methodology to assess the working regime of the Flow reactor (Figure 2.8).

Some experiments are thus required [34]. Some are easy to perform, others are trickier. The easiest test is to vary the fluids flow rates keeping the Hourly Space Velocity (HSV) constant. That is achieved using a longer catalytic bed (Figure 2.8 (a)) while increasing the flow rate in proportion. The variation of the catalyst particle size (or thickness of washcoated catalytic layer) is more difficult to perform but safely provides a diagnostic on possible diffusion limitations inside the catalyst porous network (Figure 2.8 (c)). Another experiment can help assessing the effect of heat transfer (Figure 2.8 (b)). In the case of an exothermic reaction, the heat produced at the catalyst

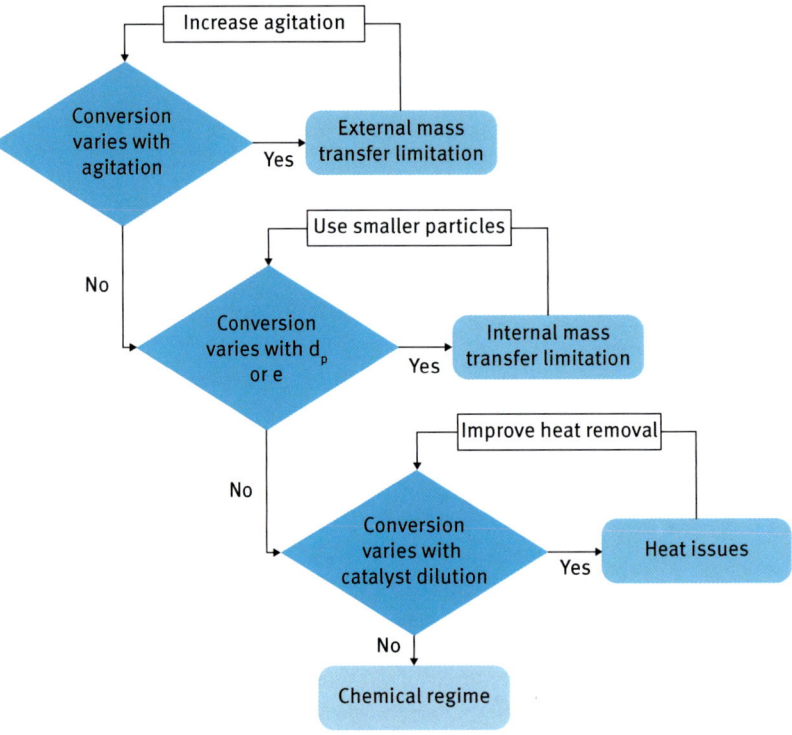

Fig. 2.8: Proposed methodology to help assessing the working regime in gas-liquid-solid & liquid-solid Flow reactors.

site must be removed at the outer wall of the reactor. Catalyst particles dilution with chemically inert particles of the same size will help to average the heat flux generated by the reaction over a larger section of the reactor. Note that measuring the inlet and outlet liquid temperature is often easier and can give some information about heat transfer. Catalyst dilution can be sometime cumbersome because segregation between inert and catalytic particles may occur. Last, further information can be obtained from pseudo Arrhenius plot (Figure 2.8 (d)). Since the external mass transfer process is almost not activated with temperature, the observation of a rather flat profile at high temperature likely indicates external mass transfer limitations. In the case where the activation energy estimated from the plot is ca. half that of the published values for the same reaction, it may be concluded that internal diffusion is limiting. That can be further checked with the particle size test.

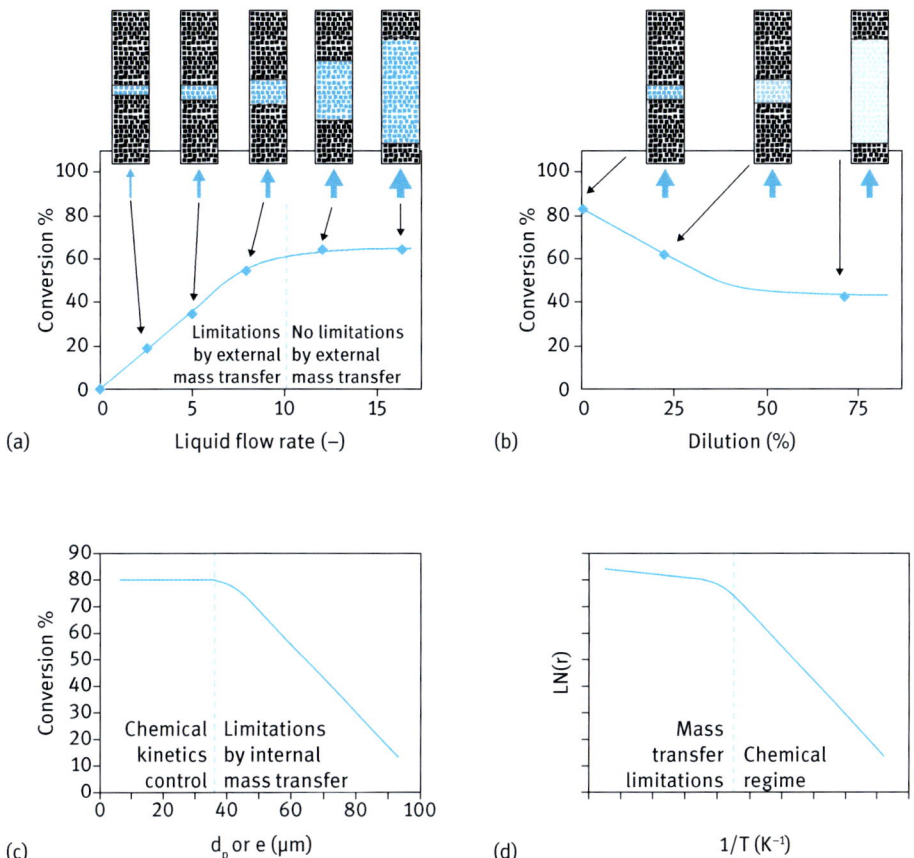

Fig. 2.9: Testing working regimes in catalytic flow reactors. (a) Impact of the variation of the mixing in the reactor by increasing the flow rate, keeping the space velocity constant (HSV or WHSV). (b) Effect of the dilution of the solid catalyst in the bed for exothermic reactions. (c) Impact of the variation of the catalyst particles diameter. (d) Impact of mass transfer limitations on Arrhenius type plots. The blue area sketched the catalyst bed.

2.4.3 Basis for reactor scale-up

When scaling up reactors in general and heterogeneous catalytic reactors in particular, a good knowledge of both laboratory and production reactors is required. Thus, the mass and heat transfer capabilities of both reactors must be known as well as other potential limitations such as pressure drop and plug flow behavior. Many correlations describing these properties can be found in textbooks for traditional reactor setup and in the literature for more recent reactor design [35, 36]. Procedure for reactor scale-up are also published [37].

Then, a rough idea of potential limitations can be drawn by using the characteristic time approach and the dimensionless numbers. For example, it is obvious that a reactor must accommodate a residence time long enough compared to the reaction time (Da number). Methods to assess the working regime of catalytic reactors will be presented in the following sections. Four main situations can be encountered.

1. The production is performed in the same reactor that was used for data acquisition (kinetics, selectivity, deactivation ...). Here, a longer time-on-stream is required to ensure the desired output of the product providing that the targeted time-on-stream is shorter than the deactivation time. The production could be computed from the following development (Equation (2.19)):

$$\text{Production} = \text{WHSV}_X \cdot m_{\text{cata}} \cdot X \cdot t_{\text{op}} (\text{kg}) \tag{2.19}$$

$\text{WHSV}_X (h^{-1})$ is the mass hourly space velocity at which the targeted conversion X was obtained during the laboratory trails, under the same operating conditions (same catalyst, concentration, temperature, pressure). m_{cata} (kg) is the weight of catalyst required. t_{op} (h) is the operation time (or time-on-stream) during which the reactor has to perform at the targeted conversion.

If the reactor is a fixed bed with a packing catalyst particles, the reactor volume could then be computed from the catalyst density ρ_{cata} hence the catalyst volume and the solid hold-up ε_s 0.6 for a compact fixed bed. For any reactor with a catalyst content ε_s, a similar procedure is used (Equation (2.20)).

$$V_R = \frac{V_{\text{cata}}}{\varepsilon_s} = \frac{m_{\text{cata}}}{\rho_{\text{cata}} \varepsilon_s} \tag{2.20}$$

2. The production is performed in a different reactor but both reactors are working in the chemical regime: the same procedure can be applied.
3. In the laboratory, reactor internal diffusion only is limiting. As a first approach, the same catalyst can be used with the same particle diameter and again a similar procedure may be used. If the size of the catalyst is changed, different results will be obtained and a more sophisticated procedure must be used with the expertise of the chemical engineering department.
4. When external limitations are playing in liquid-solid catalysis at laboratory scale, a popular procedure is to keep the same limitations at production scale. Thus it is advised to perform the production at the same Re number. Indeed, most of the liquid-solid mass transfer performance is related to the Re number. For a given chemical system, the Re number is proportional to the liquid superficial velocity v_{sL} (m · s^{-1}) (see Table 2.4). The liquid velocity is given by the liquid feed flow rate Q_L divided by the reactor cross-section (m^2). Thus, a degree of freedom is lost since one more design parameter (cross-section) must be fixed. While this procedure has proved to be very efficient in many cases, it can also result in unrealistic reactor aspect ratio, that is, a very long reactor.

When none of these methods could be applied or when other more complex phenomenon are playing and, eventually, coupled (heat transfer, pressure drop, residence time distribution ...), it is advised to work with chemical engineers.

2.5 Conclusion

The purpose of this chapter was to provide chemists with some basic elements from the field of chemical engineering to enable a better understanding and to benefit from using Flow reactors. After a short presentation of the flow motions in Flow reactors, the physical and chemical relevant phenomena, simple methods are proposed to help solving most issues of Flow reactors. Last, methods to help reactor scale-up are proposed.

Study questions
2.1. How can batch and flow reactors be compared?
2.2. Why is the substrate/catalyst ratio not used for reactor and catalyst comparisons?
2.3. What are the methods to check the chemical regime in Flow reactors?
2.4. Why are simple proportion rules for reactor scale-up not used?

Further readings
1. J. B. Butt. Reaction kinetics and reactor design. Marcel Dekker, New York, 2nd edn., 2000.
2. G. F. Froment, K. B. Bischoff. Chemical reactor analysis and design. Wiley, New York, 2nd edn., 1990.
3. O. Levenspiel. Chemical reaction engineering. Wiley, 1999.
4. G. Ertl, H. Knözinger, F. Schüth, J. Weitkamp. Handbook of Heterogeneous Catalysis, Wiley-VCH, 2nd edn., 8 volumes, 2008.
5. Ullmann's Encyclopedia of Industrial Chemistry – Heterogeneous Catalysis and Solid Catalysts O. Deutschmann, H. Knözinger, K. Kochloefl, T. Turek, Wiley-VCH, 2009.

Bibliography

[1]   I. R. Baxendale. *J. Chem. Technol. Biotechnol.* 88 (2013) 519–552.
[2]   R. L. Hartman, J. P. McMullen, K. F. Jensen. *Angew. Chem. Int. Edn.* 50 (2011) 7502–7519.
[3]   F. E. Valera, M. Quaranta, A. Moran, J. Blacker, A. Armstrong, J. T. Cabral, D. G. Blackmond. *Angew. Chem. Int. Ed.* 49 (2010) 2478–2485.
[4]   (a) V. Hessel, S. Hardt, H. Löwe. Chemical Micro Process Engineering. Wiley-VCH, Weinheim, p. 32, 2004; (b) V. Hessel, A. Renken, J. Schouten, J.-I. Yoshida. Micro Process Engineering. Wiley-VCH, Weinheim, 2009.
[5]   J. B. Butt. Reaction kinetics and reactor design. Marcel Dekker, New York, 2nd edn., 2000.
[6]   O. Levenspiel. Chemical reaction engineering. Wiley, 1999.

[7] P. A. Ramachandran, R. V. Chaudhari. Three-phase catalytic reactors. Gordon & Breach, Philadelphia, 1992.
[8] G. F. Froment, K. B. Bischoff. Chemical reactor analysis and design. Wiley, New York, 2nd edn., 1990.
[9] N. Nikbin, M. Ladlow, S. V. Ley. *Org. Res. Proc. Dev.* 11 (2007) 458–462.
[10] A. Kirschning, W. Solodenko, K. Mennecke. *Chem. Eur. J.* 12 (2006) 5972–5990.
[11] C. G. Frost, L. Mutton. *Green Chem.* 12 (2010) 1687–1703.
[12] P. He, G. Greenway, S. Haswell. *Microfluid Nanofluid* 8 (2010) 565–573.
[13] A. Lerclerc. PhD, University of Lyon, 2007.
[14] P. Rodriguez, F. Simescu-Lazar, V. Meille, T. Bah, S. Pallier, I. Fournel. *Appl. Catal. A: General* 427–428 (2012) 66–72.
[15] A. El Kadib, R. Chimenton, A. Sachse, F. Fajula, A. Galarneau, B. Coq. *Angew. Chem. Int. Ed.* 48 (2009) 4969–4972.
[16] A. Cybulski, J. A. Moulijn. Structured Catalysts and Reactors. CRC Press, 2nd edn., 2005.
[17] Y. Cao, P. Li, J. Zhou, Z. Sui, X. Zhou, W. Yuan. *Ind. Eng. Chem. Res.* 49 (2010) 3944–3951.
[18] R. R. Kalluri, D. R. Cahela, B. J. Tatarchuk. *Appl. Catal. B: Environ.* 90 (2009) 507–515.
[19] N. Semagina, M. Grasemann, N. Xanthopoulos, A. Renken, L. Kiwi-Minsker. *J. Catal.* (2007).
[20] L. Falk, J.-M. Commenge in "Micro Process Engineering" Vol. 1, V. Hessel, A. Renken, J. Schouten, J.-I. Yoshida eds., Wiley-VCH, 2009, chap 6.
[21] L. Falk, J.-M. Commenge. *Chem. Eng. Sci.* 65 (2010) 405–411.
[22] V. Hessel, S. Hardt, H. Löwe, F. Schönfeld. *AIChE J.* 49 (2003) 566–577.
[23] T. M. Floyd, M. A. Schmidt, K. F. Jensen. *Ind. Eng. Chem. Res.* 44 (2005) 2351–2358.
[24] J. B. Butt. E. E. Petersen. Activation, deactivation and poisoning of catalysts. Academic Press, 1988.
[25] R. Hughes. Deactivation of Catalysts, Academic Press, 1984.
[26] G. Ertl, H. Knözinger, F. Schüth, J. Weitkamp. Handbook of Heterogeneous Catalysis, Wiley-VCH, 2nd ed, 2008.
[27] H. U. Blaser. *Adv. Synth. Catal.* 344 (2002) 17.
[28] T. V. Le Doan, P. Stavárek, C. de Bellefon. *J. Flow Chem.* 2 (2012) 77–82.
[29] M. W. Losey, M. A. Schmidt, K. F. Jensen. *Ind. Eng. Chem. Res.* 40 (2001) 2555–2562.
[30] C. de Bellefon in "Micro Process Engineering", V. Hessel, A. Renken, J. Schouten, J.-I. Yoshida eds., Wiley-VCH, Weinheim, 2009, chap. 2.2.2.
[31] J.-N. Tourvieille, F. Bornette, R. Phillipe, C. de Bellefon. *Chem. Eng. J.* 227 (2013) 182–190.
[32] V. Meille, N. Pestre, P. Fongarland, C. de Bellefon. *Ind. Chem. Eng. Res.* 43 (2004) 924–927.
[33] D. Kunii, O. Levenspiel. Fluidization engineering. Wiley, 1969.
[34] C. Perego, S. Peratello. *Catal. Today* 52 (1999) 133–145.
[35] K. Pangarkar, T. J. Schildhauer, J. R. van Ommen, J. Nijenhuis, F. Kapteijn, J. A. Moulijn. *Ind. Eng. Chem. Res.* 47 (2008) 3720–3751.
[36] M. P. Duduković, F. Larachi, P. L. Mills. *Catal. Rev.* 44 (2002) 123–246.
[37] J. P. Euzen, P. Trambouze, J. P. Wauquier. Scale-up Methodology for Chemical Processes, Technip, Paris, 1993.
[38] L. Shi, X. Wang, C. A. Sandoval, Z. Wang, H. Li, J. Wu, L. Yu, K. Ding. *Chem. Eur. J.* 15 (2009) 9855–9867.

Thomas H. Rehm
3 Continuous-flow photochemistry in microstructured environment

3.1 Environmental impact in view of *Green Chemistry*

The paradigm change from batch to continuous-flow chemistry is one consequence of the call for sustainable and environmentally friendly technologies. The necessity to translate well-known chemistry from the flask to a continuously running plant opened the door for defining new possibilities of industrial chemical processing in form of microstructured reactors with far smaller size than common batch vessels. With the advent of such equipment, high-pressure and high-temperature regimes as well as supercritical states of gases or solvents became manageable and were utilized for chemical reactions. But despite those advantages of microreactors for synthetic chemical synthesis under thermal conditions, the important class of light-driven reactions should not be neglected. Photochemistry in general describes the physical and chemical processes of material conversion initiated by the absorption of photons, often performed at or close to room temperature and under normal pressure [1]. Under such sustainable and environmentally friendly conditions (*Biological Process Windows*) reagents and follow-up products can be obtained which are rarely available via thermal treatment. This alternative draft for doing chemistry with (solar) light power of appropriate wavelength as the only activation energy is another step forward to understand and implement *Green Chemistry* [2–4].

This chapter will give an overlook on the physical, technological and chemical parameters of photochemistry in microstructured environments under continuous-flow conditions. As described earlier in this text book, contacting between gaseous, liquid and solid components is very successful in microstructured equipment. In case of photochemical reactions, the photon arises as a further "reactant". Beside the individual selection of the wavelength of the photons, their flux can be dosed like a stream of gas or liquid. This is of particular importance since the energy input by the photon flux can then be adjusted to the reaction conditions and consequently to the stream of gas and liquid (and vice versa). But one specific feature distinguishes the photon from the other reaction components: photons themselves do not leave any by-products, they just initiate reactions. Hence, they can be considered as the perfect "chemical" reagent.

3.2 Physical considerations – reasons why microstructured equipment is preferred for flow photochemistry

3.2.1 Absorption of light by molecules in solution

Photochemistry and photocatalysis, respectively, base upon the interaction of electromagnetic radiation with molecules. Depending on their molecular structure, light of different wavelength (equivalent to energy) can be absorbed. Benzene being a prominent example of an aromatic molecule with 6 π-electrons absorbs light of the UVA range between 220 and 276 nm. Anthracene, a linear homologue of benzene with 14 π-electrons absorbs ultraviolet (UV) light at longer wavelength (280 to 390 nm) [1]. In general, the interplay between molecules and light of appropriate wavelength leads to the conversion of the molecules into their excited state. Depending on the reaction parameters, the excited state can tread to different reaction paths [5]. In case of a photochemical reaction, the light-absorbing molecule itself performs a structural (chemical) conversion with bond cleavage and/or bond formation except for any isomerization as is the case for trans to cis conversions of, for example, stilbene or azobenzene. In case of a photocatalytic reaction, the energy is absorbed by a molecular sensitizer (e.g., Rose Bengal) or bulk material (e.g., TiO_2) and subsequently transferred to the real reaction substrate, whereas the light-absorbing molecule or material returns to its structural and electronic ground state.

The Lambert–Beer law describes the process of light absorption by a solution of molecules with defined concentration along a defined path length [6]:

$$A = -\lg\left(\frac{I}{I_0}\right) = \varepsilon_\lambda \cdot c \cdot l \tag{3.1}$$

The equation expresses the absorbance A, thus the decrease of light intensity I_0 along the path length l (cm) through the solution, depending on the molar decadic extinction coefficient ε_λ (L mol^{-1} cm^{-1}) and the molar concentration c (mol/L). The molar extinction coefficient is a specific constant, which describes the ability of a molecule to absorb light of specific wavelength. The higher this value, the stronger is the absorption of light by the molecule. The extinction coefficient is an intrinsic value for every substance, but it can be influenced by the type of solvent, temperature, pressure or the pH value of the solution. Additionally, the aggregation of the light-absorbing molecules can influence both the extinction coefficient and the absorption range of the material. With regard to the technical realization of a photochemical application, both concentration of the light-absorbing material and light path length have a similar important impact. In Figure 3.1 three graphs are plotted with different progression illustrating the impact of concentration and extinction coefficient on the light transmission depending on the path length. The first, green-blue graph represents a solution of riboflavin tetraacetate (0.1 mM) with approx. $\varepsilon_\lambda = 12{,}500$ L mol^{-1}cm^{-1} (in acetonitrile/water, 50 : 50, v/v, measured at 446 nm) [7]. As one can see from the curve, the light must pass 8 mm through the solution in order to gain absorption of 90% (corresponding to a residual transmission of 10%). A slight increase in concentration to 0.5 mM results in a shortened path length of 1.6 mm (light-blue graph). These examples should illustrate that an effective transmission of a conventional glass flask by light is limited. Moreover, if light-absorbing materials are used which have a molar extinction coefficient beyond 300,000 L mol^{-1} cm^{-1} (e.g., tetraphenylporphyrine, measured at 420 nm), the light path length decreases even stronger to some ten micrometers at 90% absorption (dark-blue curve). In this case, the incident light will only be absorbed by the solution being closest to the light source, whereas deeper layers cannot be irradiated. A simple solution for this problem would be the application of a more powerful light source. But this way might lead to an increased occurrence of by-products or even decomposition of the product by over-irradition.

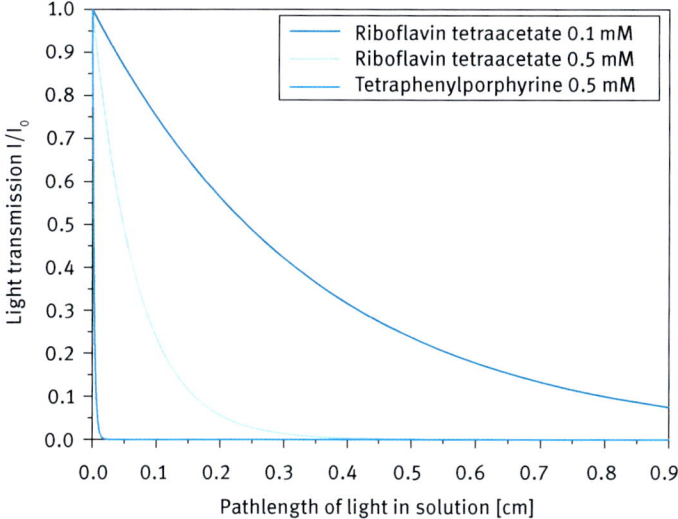

Fig. 3.1: Plotting the Lambert–Beer law for different concentrations of riboflavin tetraacetate (green-blue, light-blue) or with the extinction coefficient of tetraphenylporphyrine at 0.5 mM (dark-blue).

Beside the photochemical decomposition by repeated intense irradiation with light, the thermal impact has to be considered as well. The reaction flask itself is heated by the light source. Hence, the absorption of light by a molecule can also lead, in part, to a warming by the release of thermal energy. The electronic excitation of a molecule can lead to a more or less strong nonradiative deactivation at which vibrational energy is transferred via the collision with other molecules (usually solvent) to the surroundings. This process accompanies with heating of the solution [6].

Another description of the interaction between light and matter is the quantum yield at a given wavelength Φ_λ [8, 9]. The quantum yield describes the ratio between the amount of molecules formed and the number of absorbed photons:

$$\Phi_\lambda = N_{\text{product molecules}} / N_{\text{photons absorbed}}. \qquad (3.2)$$

The theoretical maximum of the quantum yield is 1. This is the case when each absorbed photon yields one molecule of the desired product. Usually photoreactions do not reach this optimal value and show quantum yields below 1 due to other photophysical processes as alternative reaction paths (e.g., deactivation, fluorescence). Some reactions can lead to quantum yields being larger than the theoretical maximum ($\Phi_\lambda \ll 1$). Here, the absorbed photons have initiated a chain reaction which propagates without a further ignition by light.

3.2.2 Role of solvent

Especially for photochemical reactions in solution, the specifications for the solvent are manifold. Beside its real use as transport medium for the dissolved reactants, the solvent must not absorb the incident light and quench any excited state of the substrate. But some reactions also need the characteristic of a solvent being able to transport intermediate reaction species, for example protons, in order to pass the desired reaction pathway. Therefore it is important to know all physical and chemical parameters of the solvent as main chemical substance being irradiated during a photochemical reaction. Table 3.1 shows common solvents for organic synthesis enumerated by their cut-off wavelength reflecting the lower limit of their spectral window for irradiation without light absorption [10]. Another important condition for the solvent is the strict absence of dissolved molecular oxygen. With the exception of photooxidation and photooxygenation reactions, the presence of molecular oxygen can lead to the quenching of excited states of light-absorbing molecules with follow-up reactions like hydrogen peroxide formation being able to initiate the formation of further by-products.

3.2.3 Micrometer-sized structures as key elements of reactor equipment for flow photochemistry

Several approaches based on microstructured equipment are available for solving the challenges regarding the irradiation depth and the heat removal as discussed above. The first and most important advantage of microstructured flow technology is the opportunity that the complete volume of a batch vessel can be transported continuously in form of a thin film or streamed phase inside a channel or a capillary. The micrometer-sized diameter and the structure of such external boundaries define the thickness of the volumetric flow of the reaction solution [8, 11–16]. In consideration of the resulting multiple decreased pathway, which the light has to pass through,

Table 3.1: Common solvents used for organic synthesis and their cut-off wavelength [10].

Solvent	Cut-off wavelength [nm]
water	185
acetonitrile	190
n-hexane	195
ethanol	204
methanol	205
cyclohexane	215
diethyl ether	215
1,4-dioxane	230
methylene chloride	230
chloroform	245
tetrahydrofuran	245
acetic acid	250
ethyl acetate	255
carbon tetrachloride	265
dimethylsulfoxide	277
benzene	280
toluene	285
pyridine	205
acetone	330

and the Lambert–Beer law, a complete and homogeneous irradiation of the reaction solution is feasible. In reverse, this also provides the possibility to use a very thin liquid film either containing the highly concentrated light-absorbing material or a light absorber with a very high extinction coefficient, for example Rose Bengal with $\varepsilon_\lambda > 100{,}000\ \text{L}\ \text{mol}^{-1}\text{cm}^{-1}$. The second advantage of microstructured equipment is the possibility to define exactly the volume which has to be irradiated. It is possible to control the period of irradiation time for each reaction solution volume by the rate of the liquid flow. This parameter essentially contributes to the reaction control and the prevention of by-product formation. The third advantage of microstructured equipment is the superior control of the heat management during the photochemical reaction. Due to the small volumes of the reaction solutions inside a microchannel, the ratio between the contact area of the channel walls and the appropriate solution volume is very large compared to a conventional flask. A portion of hot reaction solution located in the center of the microchannel can reach its thermal equilibrium with the cold channel wall much faster than is the case for a 10 mL flask and its glass boundary. This advantage is again of fundamental importance since the formation of by-products inside or near hot spots can be diminished very efficiently. Some types of microreactors also exhibit integrated heat exchangers. This design no longer necessitates the immersion of the reactor in a cooling medium with the possibility of providing the desired heat and exact temperature as close to the reaction space as possible.

Equations:

$$A = -\lg\left(\frac{I}{I_0}\right) = \varepsilon_\lambda \cdot c \cdot l \quad \text{Lambert–Beer law.}$$

Light absorbance along the path length l (cm) through a solution of molecules expressed in decrease of light intensity I_0, depending on the molar decadic extinction coefficient ε_λ (L mol^{-1} cm^{-1}) and the molar concentration c (mol/L) of the dissolved molecules.

$$\Phi_\lambda = N_{\text{product molecules}}/N_{\text{photons absorbed}} \quad \text{Quantum yield at a given wavelength.}$$

Ratio between number of product molecules formed and number of absorbed photons; ideal case $\Phi_\lambda = 1$; realistic case: $\Phi_\lambda < 1$; if $\Phi_\lambda > 1$, then photons initiated chain reaction which propagates without help of light.

Photochemical reaction: The light-absorbing molecule itself performs a structural (chemical) conversion with bond cleavage and/or bond formation. Cis/trans isomerization is not included.

Photocatalytic reaction: A photon is absorbed by a molecular sensitizer or bulk material and the energy is subsequently transferred to the real reaction substrate, whereas the light-absorbing molecule or material returns to its structural and electronic ground state.

- Molecules convert into their excited states upon light absorption of appropriate wavelength.
- Every molecule has its characteristic absorption profile due to its molecular and electronic structure.
- The extinction coefficient is a measure for the ability of a molecule to absorb light of a specific wavelength.
- The higher the extinction coefficient, the shorter is the light path length until full absorbance.
- Every solvent has its own spectral window limited by its cut-off wavelength. Inside this window, solvents do not absorb light and allow photochemical conversions.
- In most cases, the solvent should be degassed and oxygen-free in order to minimize undesired by-product formation or quenching processes.
- Microstructured reactor equipment allows the full illumination of an exactly defined reaction volume while minimizing multiple irradiations and resulting by-product formation.

3.3 Technological considerations for flow photochemistry

3.3.1 Light sources

The light source is the essential part of any reactor set-up for photochemical conversions. Beside solar irradiation, two types of artificial light sources have made their way to the academic and industrial labs for photochemical synthesis: a) lamps with a broad emission over a wide range of the electromagnetic spectrum; b) light sources with a narrow emission close to one defined wavelength. Depending on the type of light sources, high energy UV light, or the visible part of the electromagnetic spectrum, and even near-infrared radiation, can be made available for photochemical conversions. Of course, the use of solar irradiation is most advantageous considering the aspect of *Green Chemistry*. But with regard to continuous photochemical processes running over days or even weeks, a light source must be available that produces a steady photon flux of the desired wavelength and intensity.

3.3.1.1 Metal vapor and gas-discharge lamps

One important representative is the mercury vapor lamp. It is a nonthermal radiation emitter and produces polychromatic light as a mix of several emission wavelengths [17]. The physical principal of the light generation and emission is based on the excitation of valence electrons of neutral Hg atoms in the metal vapor. While the electrons return back to the ground state, the Hg atoms emit light, which corresponds to the energy difference between the excited state and the ground state. Uncoated versions of low-pressure Hg lamps have beside other emission bands one intense band in the UVC range with a maximum at approx. 254 nm making them ideal for the photochemistry of carbonyl- and small arene-based compounds [14]. Similar to household fluorescence lamps, phosphor-coated types of low-pressure Hg lamps emit at longer wavelengths in the UVB- and UVA-range. Such types work at low power with a maximum of 2000 W, whereas medium- and high-pressure versions of Hg lamps can have a power of up to 60,000 W with emission bands between 250 nm and 600 nm. They are commonly used as light sources for industrial photochemical reactors and applications that need a high power irradiation within a main spectral window of 290 nm to 400 nm. One drawback of these high power lamps is their emission in the IR region which necessitates, in consequence, an efficient heat management. Beside mercury vapor, sodium vapor lamps are also used as nonthermal radiation emitters for photochemical applications. Due to their emission bands in the visible part of the electromagnetic spectrum (450 nm–750 nm, high-pressure version) they are commonly used for dye-sensitized reactions. Another light source for photochemical synthesis uses highly pressurized Xenon gas, in some cases doped with mercury to provide high intensities in the UV range [17]. Such a Xenon arc lamp is a special type of gas-discharge lamp and can have a power of up to 30,000 W with a broad and panchromatic emission range between 200 nm to 1500 nm. Contrary to the metal vapor lamps, Xenon gas lamps are thermal emitters which have a continuous light emission without the spectral band fine structure.

3.3.1.2 Filter equipment

As described above, the variety of lamps regarding emission power and spectral range allows a wide application area for photochemical conversions. But one has to keep in mind that such lamps need both careful handling and a very efficient heat management for a long life time. And even more important, every experimenter must take care of the production of intense radiation and heat which can lead to severe damages of skin and eyes in the worst case. Beside these general rules, the emission of the lamps must be adapted to the photochemistry and the necessary wavelength. Neither lamp type emits at only one defined wavelength. Their emission is broad and the intensity of each spectral region can differ in a strong way. Therefore it is absolutely necessary to use filters to cut off any light with a wavelength that is either not compatible with the photochemical reaction or even destructive to the reactants. For example, quartz glass allows the transmission of UVC light down to a wavelength of 200 nm, whereas a filter tube made of Pyrex or Duran glass filters all high energy light with a wavelength smaller than 300 nm [14]. In consequence of this necessary constraint, the overall efficiency of such light sources is reduced, since the lamp power and emission, respectively, cannot be fully used for the desired photochemical conversions.

3.3.1.3 Light emitting diode (LED)

Compared to the already introduced lamps, light sources with a narrow emission band close to one defined wavelength bypass the necessary filtering. Beside laser sources, Light Emitting Diodes (LED) are the most prominent representatives of this type of nearly monochromatic emitters [18]. They are nonthermal emitters and offer a wide range of emission wavelengths depending on the semiconducting material used for the emitter. As described above for nonthermal emitters, the emission of light is based on the excitation of electrons from the valence band to the conduction band. In case of LEDs, the combination of elements from the 3rd and 5th main group of the periodic table allows the production of III/IV semiconductors [19, 20]. Also depending on the dopant of the semiconducting material, different emission bands can be obtained (Table 3.2) and selectively used for the irradiation during photochemical reactions.

Beside this main advantage, LEDs offer a very high lifetime of up to 100,000 hours and feature a very low heat production compared to metal vapor or pressurized gas lamps. Of course, the power of LEDs is not comparable with the output of a Xe gas

Table 3.2: Examples for III/IV semiconductors with dopants for LED production with their appropriate emission wavelengths [19, 20].

Semiconductor	Emission wavelength [nm]	Color
Aluminum gallium indium phosphide (AlGaInP)	$590 < \lambda < 610$	Orange
Aluminum gallium phosphide (AlGaP)	$500 < \lambda < 570$	Green
Zinc selenide (ZnSe)	$450 < \lambda < 500$	Blue
Indium gallium nitride (InGaN)	$400 < \lambda < 450$	Violet
Boron nitride (BN)	215	Ultraviolet

lamp, but the small size and the plane design enable the formation of LED arrays with tens of single light emitters whose power summarizes to a sufficient amount for the irradiation of the reaction solution. In addition, the versatility of a single LED point emitter offers the opportunity to design not only plane arrays, but also curved arrays. With this degree of freedom, LEDs can be perfectly adapted to many kinds of microreactors whereas the conventional lamps are radial emitters and force the reactor into a dedicated design.

For the sake of completeness, one last type of light source has to be mentioned as well. In recent years, the development of light emitting materials has made a huge step towards the application of carbon-based materials instead of conventional inorganic compounds. So-called organic LEDs (OLEDs) use the same physical principle for light generation as conventional LEDs [21]. Polyaromatic molecules like pentacene or metallated phthalocyanines are used as p-type organic semiconductors whereas their perfluorinated derivatives, or fullerenes and perylenes, act as n-type organic semiconductors. In this context, organic materials have two most important advantages. First, these compounds are synthesized, which allows the fine-tuning of the semiconductor by the toolkit of an organic chemist. With this possibility, the emission wavelength of the organic semiconductor can be rather easily controlled. The second advantage is their solubility in common organic solvents. This property allows the solution-processed fabrication of OLEDs by common printing techniques. Here, not only small surfaces can be printed, but already larger areas are possible enabling very homogeneous light emission intensities all over the panel. This advantage especially distinguishes the OLED technology from any spotlight. Further developments of the OLED technology show first results in form of flexible light emitting panels which might be perfectly adapted to any photochemical reactor equipment available so far.

3.3.1.4 Solar light

In the early days of photochemistry, sunlight was the only available light source with sufficient intensity. At this time, Giacomo Ciamician was the pioneer applying sunlight for the photochemical redox reaction between 1,4-benzochinone and ethanol as solvent leading to 1,4-hydrochinone and acetaldehyde [22]. The irradiation time was as long as five months. Since these first experiments, the development of appropriate equipment for solar light-driven reactions lead to the use of parabolic solar light concentrators which focus the incident light on a glass tube as flow reactor [23, 24]. Of course, one has to take into account that terrestrial sunlight contains just a minimum of UV light (Figure 3.2). Fortunately, the larger part of this spectral component is filtered and blocked by earth's ozone layer [25].

In consequence, any photochemistry based on sunlight must use the visible part of the solar spectrum ranging approx. from 400 nm to 700 nm [26]. Therefore, either

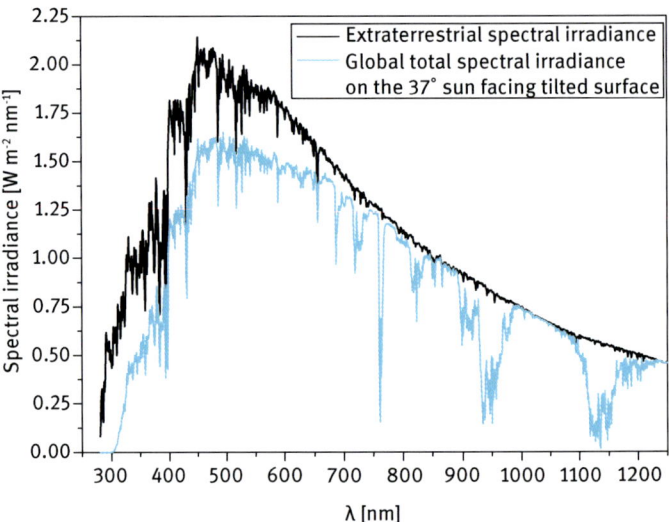

Fig. 3.2: Part of the solar irradiance as produced by the sun (black line) and filtered by earth's atmosphere (light-blue line). The data for preparation of the graph were taken from [25].

the reaction substrate itself is able to absorb visible light with subsequent formation of an excited state or a photocatalyst is applied as mediator for the transfer of solar energy to the substrate. In general, sunlight is free for everybody and can be used for photochemistry across the world, but its intensity and continuous availability depends on the weather, the time of the day and year. These constitutional constraints cannot be changed and are still seen as major drawbacks of this most ecologically friendly approach. But with the steadily increasing energy demand of industrial nations and the accompanying costs for the energy production itself, it might be necessary to consider the "outdoor" version of a light source as a real alternative to artificial lamps for doing photochemistry.

- Metal vapor and gas discharge lamps are high power light sources with polychromatic or panchromatic emission. Often a strong UV light emission is available.
- High power lamps need an elaborated energy, heat and safety management.
- Filter equipment is used to cut-off spectral regions of undesired wavelength.
- Light emitting diodes can produce nearly monochromatic light without high energy consumption and heat generation.
- Single LEDs can be assembled to arrays of different sizes and shapes according to the specification of the reactor concept.
- Solar light can also be used for photochemistry. With the help of sensitizers, the available visible light can be used for photocatalytic conversions.

3.3.2 Reactor concepts for flow photochemistry

The technological prerequisites of the reactor engineering are closely connected to the choice of the light source. Well-known batch reactors for photochemical synthesis are mostly based on the design of an immersion well reactor having an outer vessel for the reaction solution, in which an immersion well is placed in the center for cooling an appropriate light source. With the advent of flow photochemistry, a huge variety of easy-to-use reactor concepts appeared which are mainly custom-built. These designs and developments redound upon the flexibility to change, for example, the dimension of the microchannels or capillaries as easily as possible. Basically every material can be used that is transparent to the light used for photochemistry and able to form a micrometer-sized flow of reactants. Depending on the type of the photochemical reaction, a pressure-stable environment is necessary as well. Due to the broad variety of custom-built reactor concepts for photochemistry, only a selection of the most commonly used types will be discussed in the following chapters.

3.3.2.1 Chip and glass microreactors

Glass and polymeric materials were often used for designing microscale chip-like reactors. One example can be seen in Figure 3.3. Matsushita *et al.* designed a microreactor made of quartz glass to also allow the transmission of UV light [27]. The linear channel has an all-over length of 50 mm with a width of 500 µm and a possible depth of 10 µm to 500 µm enabling an easy control of the irradiated reaction solution volume.

The channel itself was inserted by micromilling. The resulting roughness on the walls was not removed by polishing in order to increase the mixing of the liquid flow by the formation of turbulences. Additionally, the bottom and the side walls of the microchannel were coated by a sol-gel process with a photocatalytic active TiO_2 thin film. Moreover, the rough surface of the channel walls increases the irradiated area for the heterogeneously catalyzed photoreaction.

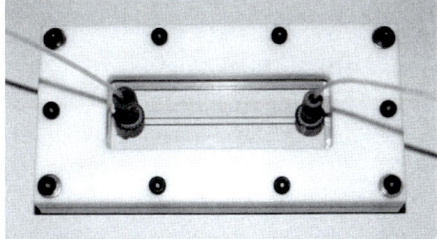

Fig. 3.3: Quartz glass microreactor with a single microchannel filled with methylene blue solution [27]. Reprinted from Chem. Engin. J., 135, Matsushita Y, Ohba N, Kumada S, Sakeda K, Suzuki T, Ichimura T, Photocatalytic reactions in microreactors, S303–S308, Copyright (2008), with permission from Elsevier.

Small-scaled chip reactors with a single channel offer only limited productivity and are mostly used as lab reactors for academic research. In order to increase the amount of product, the numbering-up of both channels and reactors is feasible. At best, one reactor with parallel aligned microchannels of the same design as the single channel can result in the multiple amount of product compared to the productivity of a single channel. An even greater increase can be achieved by the parallel use of such single reactor units of the same design. The scale-up approach, meaning the increase of reactor dimensions like channel length, is not easily possible since an increased channel length, for example, results in an increased irradiation time. In case of UV light especially, a prolonged irradiation time can lead to an increased formation of by-products or even degradation of the product.

Glass microreactors are similar to chip-based reactor types, but their size and design are significantly different. The microchannels are larger in diameter and can have a serpentine structure to increase the channel length while keeping the over-all size of the microreactor small. Even a mixing zone can be integrated prior to this main channel. There are also versions available which allow an efficient heat exchange for a secure reaction control. In analogy to the chip microreactors, the plane design of the reactor zone with the microchannels allows the placing of two light sources from the top and the bottom side. The dual irradiation from both sides of the channel structure can be very useful for reactions with high concentrations of light-absorbing materials or materials with a high extinction coefficient. In general, the concept of glass microreactors follows a multipurpose approach which also allows a large variety of gas/liquid and liquid/liquid reactions. Hence, commercialized versions of glass microreactors are already available and possess appropriate light sources well adapted for photochemical applications.

3.3.2.2 Falling film reactor

Griesbeck *et al.* designed a falling film reactor concept with a XeCl excimer lamp (60 cm in length, 3 kW) as central light source surrounded by a glass barrel [28]. The reaction solution is pumped from a cooled reservoir to the upper edge of this glass barrel from which it runs down by gravity along the surface. The reactor is designed for the irradiation of larger reactant quantities. Depending on the solvent used for the photochemical reaction this concept of a falling film reactor can provide a liquid film with thicknesses in the range of a millimeter. Photochemical reactions like photooxygenations or photooxidations rely on a sufficient amount of molecular oxygen dissolved in the reaction solution. Hence, the microreactor needs to be optimized with respect to the contacting between the gas phase and the liquid reaction solution. The concept of a microstructured falling film reactor was designed at the *Institut für Mikrotechnik Mainz GmbH, Germany* (now: Fraunhofer ICT-IMM) offering the possibility to generate very thin liquid films in the range of some ten micrometers [29, 30]. The design of the reaction plate as the main reactor component exhibits parallel channels with, for example, 600 μm in width and 200 μm is in depth. The plate stands upright inside the stainless steel reactor housing and allows both gravity and capillary forces to drive the reaction solution into the microchannels. Since the channels are not covered by a mask, the liquid thin films have excellent contact to the surrounding gas phase which can be pressurized and fed either with the liquid flow or in countercurrent flow leading to a strong diffusion of the gas inside the solution (Figure 3.4 (a)).

(a) (b)

Fig. 3.4: (a): Disassembled falling film microreactor displaying the back housing with heat exchanger unit, the reaction plate with the integrated microchannels, covered by the zone mask and the front housing with the inspection window; (b): the falling film microreactor with a magnetically fixed LED array on the front housing ©Fraunhofer ICT-IMM.

The design of the open channels leads to a very high surface-to-volume ratio with a specific phase interface of up to 20,000 m^2/m^3. Additionally, inside the back housing a heat exchanger unit is located with direct contact to the backside of the reaction plate. This integrated design allows a very efficient heat withdrawal or offers the opportunity to reduce the reactor temperature down to −60 °C. In the center of the front housing of the microreactor, a pressure-stable window made of glass quartz is located. It can be used as inspection window in order to control the liquid flow nearly all-over the open reaction plate. But it can be also used as entrance for irradiation from a plane light source closely located to the glass window (Figure 3.4 (b)). In contrary to batch vessels, only a small gas volume is available inside the reactor housing. In consequence, this reactor design minimizes the formation of explosive or highly reactive gas mixtures as it might be the case with oxidizing gases like oxygen, ozone and chlorine.

3.3.2.3 Capillary-based flow reactors

As remarked at the beginning of this chapter, the majority of flow reactors for photochemistry are custom-built using any available material from flow chemistry equipment. In literature, many examples can be found which describe a rather unorthodox reactor design. Light-transparent capillaries are somehow wrapped around glass de-

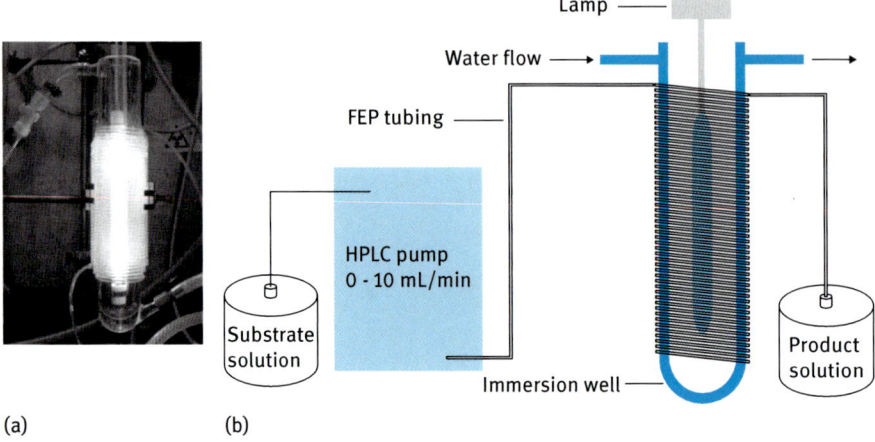

Fig. 3.5: (a): a FEP-based flow reactor with a mercury vapor lamp burning in the center of the immersion well [14]. Reprinted (adapted) with permission from Beilstein J. Org. Chem. 2012, 8, 2025–2052. Copyright 2012 Beilstein-Institut; (b): schematic illustration of continuous flow photochemical reactor based on a FEP capillary wrapped around an immersion well [31]. Reprinted (adapted) with permission from J. Org. Chem, 2005, 70, 7558–7564. Copyright 2005 American Chemical Society.

vices as holders and cooled with an air stream or a water bath. Often LEDs mounted on heat sinks are placed randomly close to the capillary. These constructions allow only basic heat management and rather unreproducible illumination of the reaction solution flowing through the capillary. Of course, there is no doubt about the functionality of such flow reactors. The photochemical reactions proceed in the expected way, but these designs are not adequate for a serious intention in both numbering-up and scaling-up.

A more suitable reactor for photochemical synthesis on a larger scale was designed by Booker-Milburn *et al.* based on an immersion well as used in the conventional batch photochemical reactor (Figure 3.5) [31]. A UV/Vis-transparent capillary made of fluorinated ethylenepropylene (FEP) polymer was tightly wrapped around the immersion well. The radial emission of the central Hg vapor lamp inside the immersion well allows a very homogeneous illumination of the reaction solution inside the FEP capillary. Other versions of this reactor concept replace the conventional Hg vapor lamp with high-power LEDs which minimize the costs for the electrical current of the lamp and the thermostat due to lower heat production. Depending on the inner diameter of the capillary and number of layers around the glass body, the reactor volume can be controlled. For example, a FEP capillary with an inner diameter of 2.7 mm and an outer diameter of 3.1 mm results in approx. 80 turnings per layer if 25 cm of the immersion well is covered. A flow reactor with three layers exhibits about 210 ml of total inner volume.

Hence, the wide bore of the capillary allows high flow rates without great pressure drops and minimizes the possibility of clogging due to precipitating reaction products. Efficient contacting of gaseous reactants with the liquid phase is also possible. The mixing of the two phases is possible by using, for example, a T-junction to dose and mix both reactants prior to the main capillary. With an appropriate flow rate the liquid and the gas phase alternately traverse the capillary in defined and stable droplets (*Taylor-Flow*). This flow regime allows efficient gas diffusion into the liquid phase due to internal turbulences inside the liquid droplet arising from the friction with the capillary wall.

The examples presented above give an impression on the variety of light sources and reactor concepts used for photochemical conversions. The majority of the reactors is still custom-built and only those reactor designs are commercially available which can be applied also to other liquid-liquid and gas-liquid reactions excluding photochemical ones. Except for the FEP flow reactor, all concepts provide only a low productivity compared to the industrial scale. However, some photochemical processes have entered the industrial field, although they are not carried out in microstructured equipment but often in conventional immersion well reactors or process-adapted versions. The *Toray* process, for example, describes the large scale production of ε-caprolactam (as precursor for *Nylon 6*) by irradiation of cyclohexane in presence of NOCl and HCl [32–34]. Just recently, pharmaceutical producer *Sanofi* initiated the large scale production of artemisinin being the essential part of the most efficient anti-malaria therapy. The key step in the synthesis of this active pharmaceutical ingredient needs the *in situ* generation of singlet oxygen, which can be easily done in a photocatalytic step [35–37]. Starting in 2014, *Sanofi* plans to produce up to 60 tons of artemisinin per year, which corresponds to 30% to 50% of the global demand. These processes show the necessity and applicability of large scale photochemical conversions both for bulk ware as well as for highly valuable pharmaceuticals. Interestingly, very few years ago academic research started to focus on the use of visible light for photochemical synthesis. Due to the increased knowledge in the preparation and use of efficient photocatalysts as new tool for highly specific organic synthesis, many publications also report on the exclusive advantages of microscale continuous-flow equipments for conducting and controlling photochemical reactions. Most importantly, new applications have been developed which allow the selective introduction of, for example, fluorinated groups under mild and controlled conditions [38]. Other labs report on efficient carbon-carbon [39, 40] and carbon-phosphor [41] bond formation established by photochemical synthesis with visible light in continuous-flow reactors. Furthermore, a two-stage continuous-flow synthesis of activated vitamin D_3 was achieved as well [42, 43]. The increased number of publications on this topic seems to reflect the necessity of intelligent and secure reaction control with microstructured equipment.

> - Continuous-flow reactor concepts exist in many variations:
>   - single channel chip reactors for process development
>   - multi channel reactor units for the production of small product quantities
>   - glass microreactors for multi-purpose applications
>   - capillary flow reactors with highly individual irradiation areas for the production of medium product quantities
>   - specialized reactor concepts like the falling film microreactor for gas/liquid contacting
> - Micrometer-sized structures allow the generation of thin liquid films and streams, which can be fully irradiated for maximum of light absorption.
> - Microchannels can be coated with photocatalyst material allowing continuous-flow heterogeneous photocatalysis.

3.4 Chemical considerations for flow photochemistry

3.4.1 Photochemical reactions without catalyst material

This chapter discusses various examples of photochemical reactions with reactants that interact with the incident light on their own. No catalyst material is used for the energy uptake and transfer to the substrate. Most reactions of this class need UV light generated by, for example, mercury vapor lamps in order to provide enough energy initiating the reactions. In the first example Jähnisch *et al.* show the photochlorination of toluene-2,4-diisocynate [44, 45]. The desired radical reaction path leads to the side-chain chlorinated product 1-chloromethyl-2,4-diisocyanatobenzene via the homolytic cleavage of molecular chlorine by UV light (Figure 3.6). The gas-liquid reaction is carried out in a microstructured falling-film reactor. In this case it was necessary to substitute the microstructured stainless steel plate by a nickel plate in order to avoid the formation of iron chloride on the plate surface. $FeCl_3$ is a Lewis base and catalyzes the electrophilic ring chlorination, which is an undesired reaction path leading to toluene-5-chloro-2,4-diisocyanate.

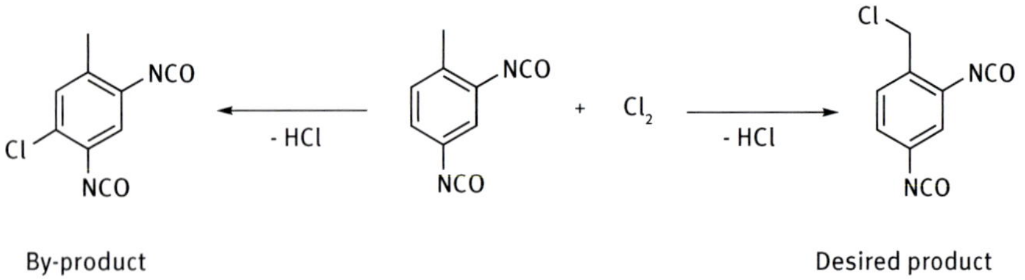

Fig. 3.6: Possible reaction pathways during the chlorination of toluene-2,4-diisocynate with molecular chlorine [44, 45].

In batch mode, a 30 mL reactor yielded, after 30 minutes, a conversion of 65% with a product selectivity of 45%. With a residence time of 14 seconds, the falling-film microreactor yielded a lower conversion of 54%, but a significantly increased selectivity of 79%. The vast difference in selectivity can be attributed to the improved irradiation of the reaction solution in the microchannels. Both the high penetration of light through the thin film as well as the defined irradiation time in the falling film microreactor have a positive impact on the selectivity. In batch mode, a large volume of the reaction solution is not irradiated leading at a reaction temperature of 130 °C to a higher amount of by-product via the thermal reaction pathway.

A second example describes the conversion of a nitrite functionalized steroidal compound to its oxime derivative (*Barton* reaction) [46]. The reaction is based on the photolysis of the nitrite group by UV light into an alkoxy and a persistent nitric oxide radical (Figure 3.7). After an intramolecular radical translocation from the oxygen to the neighboring methylgroup, the NO radical is incorporated subsequently at the methyl radical leading to the desired oxime.

Ryu *et al.* designed two stainless-steel reactors with different microchannel sizes and hold-up volumes for the continuous synthesis of the oxime: Type A with 0.2 mL (1 mm width, 107 µm depth, 2.2 m length) for the optimization of the reaction parameters and Type B with 4 mL (1 mm width, 500 µm depth, 0.5 m length, 16 lanes) for

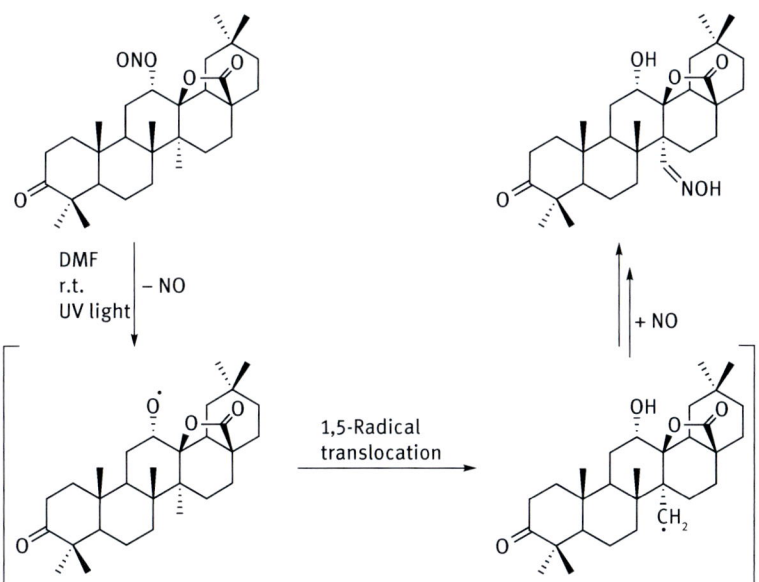

Fig. 3.7: Reaction mechanism for the nitrite photolysis [46]. Reprinted from Tetrahedron, 65, Sugimoto A, Fukuyama T, Sumino Y, Takagi M, Ryu I, Microflow photoradical reaction using a compact light source application to the Barton reaction leading to a key intermediate for myriceric acid A, 1593–1598, Copyright (2009), with permission from Elsevier.

the (multi) gram-scale synthesis. Quartz glass, Pyrex glass or soda lime glass were considered for the reactor window. Beside a high pressure mercury lamp (300 W), a black light (15 or 20 W) was tested as well as a UV-LED (1.7 W). After optimization of the process, the final lab plant consists of two serially connected reactors Type B (2 × 4 mL hold-up volume) with Pyrex glass windows (thickness: 10 mm). Eight black lights (each 20 W) were used as source of irradiation with an emission maximum of 352 nm, which fits perfectly to the necessary wavelength for initiating the nitrite photolysis. Due to lower power, the black lights do not produce so much heat like the mercury lamp. Therefore, the black light can be placed closer to the reactor while preventing heat-triggered decomposition of the starting material and the product. Neither quartz glass nor soda lime glass were chosen, since the first one does not filter any detrimental high energy UV light, while the latter one partially blocks the necessary light for the nitrite photolysis. Pyrex glass is transmissible for light with $\lambda >$ 300 nm, hence it cannot filter light with approx. 315 nm which causes the decomposition of both the starting material and the product. But the emission of the black light is very low in this spectral region, so that the formation of any by-products is already minimized by the right choice of light source and glassware. By the way, the UV-LED provided the same results as the black light with reactor Type A, and showed the best energy efficiency compared to the other light sources. But the UV-LED was not used for the scale-up, since the commercially available black lights might have been easier to adapt to the larger reactor Type B. With this equipment at hand it was possible to produce 3.1 g of product after 20 h of continuous processing (flow rate: 15 mL/h, residence time: 32 minutes) with subsequent purification by silica gel chromatography. Finally, an automated system with one reactor Type B and six black lights (each 15 W) was used to synthesize 5.4 g of product after 40 hours of continuous processing (flow rate: 12 mL/h, residence time: 20 minutes). This example gives insight into the route which has to be taken in order to select the right equipment for the reaction. Each component has a distinct impact on the process either on the productivity (reactor size) or the selectivity (light source and glassware).

3.4.2 Heterogeneous flow photocatalysis

Reactions can be catalyzed homogeneously with the catalyst material molecularly dissolved in the reaction solution. This type of catalysis often provides high conversion rates and selectivity, but involves a full catalyst separation from the reaction solution afterwards. In case of heterogeneous catalysis, the catalyst material is physically separated from the phase containing the reactant. Although heterogeneous catalysis is not able to provide as high a selectivity as homogeneous catalysts (especially in the case of enantioselectivity), they are easily separable from the reaction solution by filtration or centrifugation. Applying heterogeneous catalysis to flow chemistry offers the possibility to fix heterogeneous catalyst material on the wall of, for example, microchannels.

The reaction solution flows along the catalytically active channel surface allowing perfect contact between the liquid and solid phases by the optimized surface-to-volume ratio of microreactors. Additionally, the increased surface roughness of the channel walls causes turbulences and improved mixing inside the liquid phase.

The crystalline modification *anatas* of titanium dioxide (TiO_2) is a well-known and often used inorganic material for heterogeneous photocatalysis [47–52]. Its non-toxicity, chemical stability and high oxidizing power are ideal characteristics for a photocatalyst for wastewater treatment or decomposing atmospheric pollution [53, 54]. TiO_2 is an inorganic semiconductor which absorbs light with a wavelength ≤ 390 nm. This wavelength is equal to the energy of approx. 3.2 eV which is at least needed for the promotion of an electron from the valence band to the conduction band of TiO_2 (*anatas*-type). Every promoted electron (e^-) leaves a positively-charged hole (h^+) in the valence band which is the reason for the strong oxidizing power of this material (Figure 3.8).

Since the recombination of the separated charge carriers is quite low in TiO_2, other reactions can take place utilizing both the electron for the reduction or the hole for the oxidation of a substrate. In case of wastewater treatment, a water molecule can be oxidized to a hydroxyl radical ($\cdot OH$) and a proton (H^+), whereas dissolved oxygen is reduced to the superoxide ion ($\cdot O_2^-$). Both species are highly reactive enabling the decomposition of organic material in the wastewater. TiO_2 became very prominent with the advent of dye-sensitized solar cells, so-called *Grätzel* cells [55, 56]. By sensitization of TiO_2 with another appropriate light-absorbing material like polypyridyl complexes of ruthenium, the light absorption can be extended to the visible spectral region enabling the use of solar light for electricity production. Beside metal complexes noble

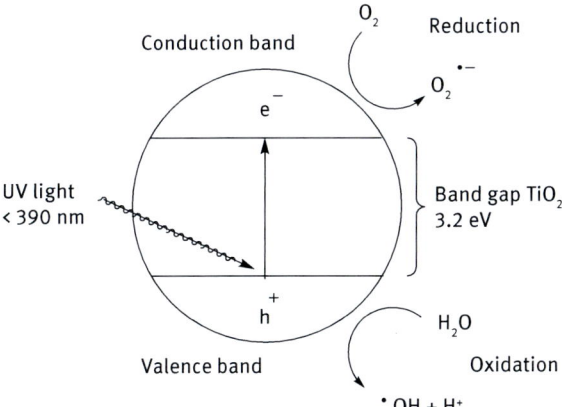

Fig. 3.8: Schematic illustration of charge carrier formation (e^-/h^+) by irradiating TiO_2 with UV light [47]. Reprinted (adapted) with permission from Langmuir, 2010, 26, 3031–3039. Copyright 2010 American Chemical Society.

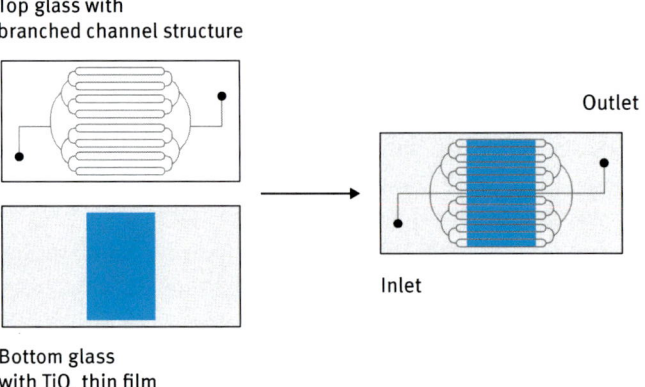

Fig. 3.9: Schematic illustration of a chip-based microreactor with branched microchannels covering a TiO$_2$ thin film [58]. Reprinted from Catal. Commun., 6, Takei G, Kitamori T, Kim H.-B, Photocatalytic redox-combined synthesis of L-pipecolinic acid with a titania-modified microchannel chip, 357–360, Copyright (2005), with permission from Elsevier.

metal nanoparticles like platinum can also be impregnated into the TiO$_2$ surface [57]. Both can be then used as part of a redox-combined photocatalytic process for the synthesis of fine chemicals as described in the following example. Kitamori, Kim et al. designed a chip-based microreactor assembled from two Pyrex glass substrates (Figure 3.9) [58, 59]. On the bottom glass, a TiO$_2$ thin film (thickness: approx. 300 nm, particle size: approx. 100 nm) was attached with a Sol-Gel technique. The top glass was modified by photolithography and wet etching in order to implement a series of branched microchannels (width: 770 μm, depth: 3.5 μm). Both glass slides were thermally bonded at 650 °C, resulting in a chip microreactor with branched microchannels covering a TiO$_2$ thin film as catalytically active surface. X-ray diffraction of the TiO$_2$ thin film proved the desired *anatas* structure and photodegradation experiments of methylene blue confirmed the photocatalytic activity of the thin film. Finally, platinum nanoparticles (approx. 0.2 wt%) were loaded on the thin film via photodeposition of H$_2$PtCl$_6$ dissolved in aqueous ethanol.

The chip microreactor was used for the continuous photocatalytic synthesis of L-pipecolinic acid from L-lysin. An oxygen-free, argon-saturated aqueous solution of L-lysin (2 mM) was pumped through the chip reactor with a syringe pump, while a high-pressure mercury lamp was used as light source including a UV transmitting filter. The illumination of Pt/TiO$_2$ with UV light generates the separated charge carriers (h$^+$ and e$^-$). The positively-charged holes are located at the TiO$_2$ surface and oxidize the amine groups to imines which are then converted with water into aldehydes and ketones, respectively. The formation of a cyclic *Schiff* base leads to the precursor of the desired product. The photoexcited electrons are located at the Pt nanoparticles and serve as reductant of the cyclic imine to an amine (Figure 3.10).

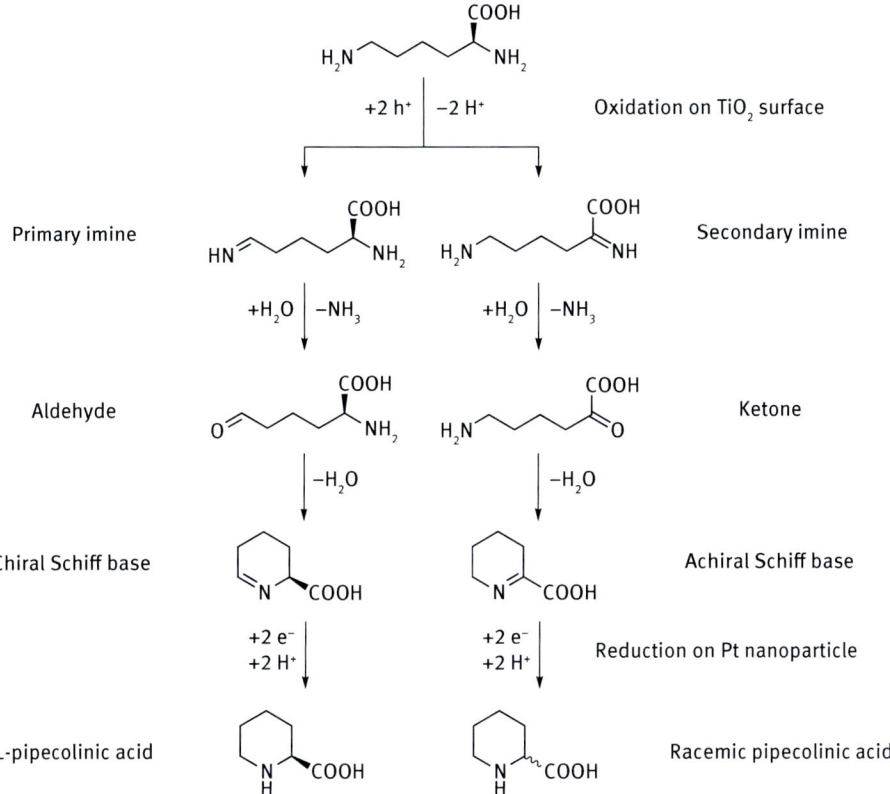

Fig. 3.10: Redox-combined reaction mechanism for the cyclization of L-lysin yielding pipecolinic acid [59]. Reprinted from J. Catal., 217, Pal B, Ikeda S, Kominami H, Kera Y, Ohtani B, Photocatalytic redox-combined synthesis of L-pipecolinic acid from L-lysine by suspended titania particles: effect of noble metal loading on the selectivity and optical purity of the product, 152–159, Copyright (2003), with permission from Elsevier.

At a flow rate of 1 µL/min and a resulting residence time of 52 seconds, the chip microreactor provided a conversion of 87% with a selectivity of 22% for all pipecolinic acid enantiomers. The enantiomeric excess (*ee*) for L-pipecolinic acid was 50%. The batch synthesis of L-pipecolinic acid (4 mL reaction solution, 20 mg 2 wt% Pt/TiO$_2$, reaction time: 60 min) showed similar conversion (86%) and *ee* (47%). The chip microreactor has a lower productivity of L-pipecolinic with 4.6×10^{-9} mmol/h compared to the batch synthesis with 3.1×10^{-7} mmol/h. This is of course the crux of such small-sized chip microreactors. But as discussed before, the fixation of catalyst material eliminates its separation from the reaction solution as it is the case in batch reactions. Additionally, the use of several continuously running chip microreactors (numbering-up approach) also allows the synthesis of the product on a larger scale.

3.4.3 Flow photocatalysis with organic dyes or noble metal complexes

The examples presented in the chapters above describe photochemical conversions of organic molecules without catalyst material or photocatalytic reactions with heterogeneous, inorganic material as catalysts. Both types need high-energy light in the UV region of the electromagnetic spectrum generated by powerful metal vapor or gas arc lamps with all their advantages and disadvantages. In recent years, several organic and organometallic dyes have proved their capability as photocatalytic active material (Figure 3.11). Metal-free organic dyes like Eosin Y [60, 61], Rose Bengal [62] or riboflavin tetraacetate [63, 64] and ruthenium or iridium based complexes like [Ru(bpy)$_3$Cl$_2$] or [Ir(ppy)$_3$] absorb at longer wavelengths in the visible region of the solar spectrum [65–71].

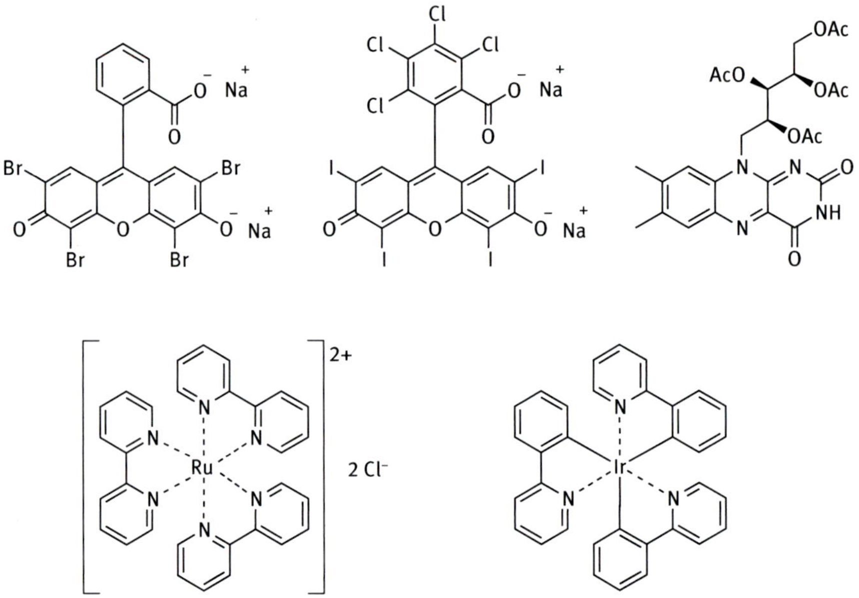

Fig. 3.11: Commonly used organic sensitizers (upper row: Eosin Y, Rose Bengal, riboflavin tetraacetate) and organometallic complexes for photoredox catalysis (bottom row: [Ru(bpy)$_3$Cl$_2$], [Ir(ppy)$_3$]) [60–71].

The absorption of appropriate light by the dyes generates long-living excited states which can act as either electron acceptors or electron donors [72]. Single electron transfer (SET) processes between the excited dyes and a substrate allows reaction pathways which are known from common radical chemistry under very mild conditions, but without the need of a chemical radical starter since the reaction is triggered externally by light. Concerning the necessary technical equipment, household common fluorescent lamps or LED stripes or arrays can be used as light sources. Both types are cheaper and easier to handle than, for example, mercury lamps. In respect to *Green Chemistry*, those new types of photocatalysts also allow the use of solar irradiation making them very valuable for sustainable chemical synthesis. Due to the use of less energetic light, the formation of by-products or even the decomposition of starting material and product is significantly minimized.

In the first example, Oelgemöller *et al.* describe the photooxygenation of 1,5-dihydroxynaphthalene with *in situ* generated singlet oxygen ($^1O_2$) resulting in 5-hydroxy-1,4-naphthoquinone (*Juglone*) [73]. Singlet oxygen as the reactive species is formed by interacting with Rose Bengal as photocatalyst (Figure 3.12). This triplet sensitizer absorbs visible light with a maximum at 548 nm (in 2-propanol) and initiates the conversion of triplet oxygen ($^3O_2$) to its reactive species. Subsequently, the singlet oxygen and 1,5-dihydroxynaphthalene react within a [4+2] cycloaddition to the 1,4-*endo*-peroxid, which converts to the hydroperoxide and finally to the desired product upon dehydration.

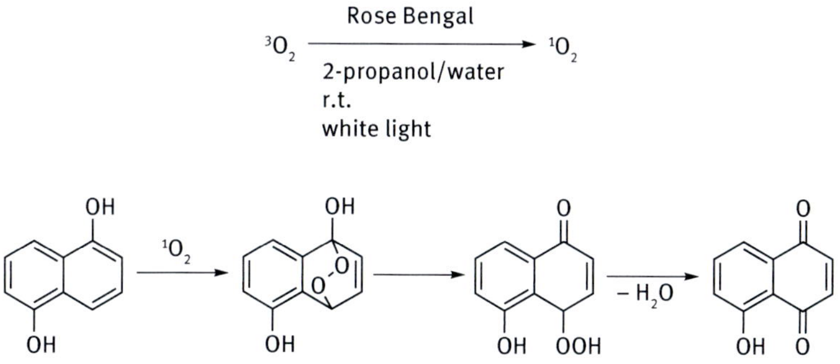

Fig. 3.12: *Juglone* synthesis with *in situ* generated singlet oxygen with Rose Bengal as triplet sensitizer [73]. Reprinted (adapted) with permission from J. Flow Chem, 2012, 2, 52–55. Copyright 2012 Akadémiai Kiadó Zrt.

The continuous synthesis of *Juglone* was performed in a falling film microreactor with an integrated reaction plate having microchannels of 1200 μm in width and 400 μm in depth. The starting material (10 mM) and Rose Bengal (0.49 mM) were dissolved in a mixture of 2-propanol with water (9 : 1). The liquid thin film was cooled on the reaction plate to room temperature via the integrated heat exchanger while air was blown over the liquid phase in countercurrent flow. A fluorescent lamp (18 W) or a LED array (3 W) was used as light source. With a flow rate of 0.08 mL/min and a resulting residence time of 32 seconds, the falling film microreactor achieved with the fluorescent lamp, a conversion of 31% after 6 cycles (160 seconds of irradiation). In case of the LED array, only 9% conversion was achieved. The batch synthesis was performed in a *Rayonet* chamber reactor equipped with 16 visible lamps of 8 W power (in sum 128 W). A Pyrex Schlenk flask (50 mL) was used as batch vessel with a central cold finger, whereas air was lead into the reaction solution via a polytetrafluoroethylene (PTFE) tubing equipped with a nozzle head. After 10 minutes reaction time, the batch set-up achieved a conversion of only 14%. Taking this data into relation with the power of the light source, the falling film microreactor combined with the LED array offers the best energy efficiency. With a value of 69% conversion per W and hour (% $W^{-1}h^{-1}$), the LED array is superior to the fluorescence lamp with 39% $W^{-1}h^{-1}$ and the batch set-up with 1.1% $W^{-1}h^{-1}$. This example clearly shows the impressive performance of microstructured equipment for photochemical conversions, of course in combination with an appropriate and energy saving light source. Beside this advantage, the use of aqueous 2-propanol as an environmentally friendly solvent and air as oxidizing reagent also contributes to the *Green Chemistry* approach.

The following example combines continuous-flow photocatalysis with chiral organocatalysis in order to step into the field of enantioselective synthesis [74]. Zeitler *et al.* performed a detailed comparison between batch vessel, microreactor (microflow) and a tube reactor (macroflow) for the asymmetric α-alkylation of octanal with diethyl bromomalonate (Figure 3.13) using a chiral imidazolidinone-based organocatalyst [75, 76]. The reaction traverses two combined cycles. The organocatalytic cycle starts with the reaction between the aldehyde and the cyclic organocatalyst under formation of an iminium species which converts into an electron-rich enamin. In parallel, an electron-deficient alkyl radical is formed via single electron transfer from the radical anion of the photosensitizer to the alkyl halide. The electron-rich olefin of the enamin reacts with the alkyl radical to an intermediate amino radical which is oxidized by the photoexcited sensitizer in the triplet state. The resulting iminium species is the precursor of the desired product which evolves upon the release of the organocatalyst.

For the batch reaction, a N,N-dimethyl formamide (DMF) solution containing all reagents calculated for a diethyl bromomalonate concentration of 0.5 M was transferred into a capped glass-vial (5 mL), cooled to −5 °C and illuminated from the transparent bottom side with a green light emitting diode (λ = 530 nm). After 18 hours reaction time, a conversion of 85% with an enantiomeric excess of 88% was determined. With the transfer of the reaction to a continuous-flow process using a glass microreactor with an internal volume of approx. 100 µL, the reaction time was shortened from 18 hours to 45 minutes (flow rate: 0.0016 mL/min) while keeping the conversion rate (86%) and *ee* (87%). This example proves once more the superior performance of microreactors for photochemical reactions. Beside this valuable proof of concept, a scale-up approach was also tested using a tube reactor similar to the capillary-based reactor concept. A FEP polymer capillary (ID = 0.8 mm) was wrapped around a beaker in two layers resulting in an allover length of 21 m and an internal volume of 10.5 mL. Under standard conditions as used for the microreactor and the batch reaction, the FEP capillary clogged upon the precipitation of 2,6-lutidine hydrobromide. With a high flow rate of 0.1 mL/min and a reduced bromoalkyl concentration

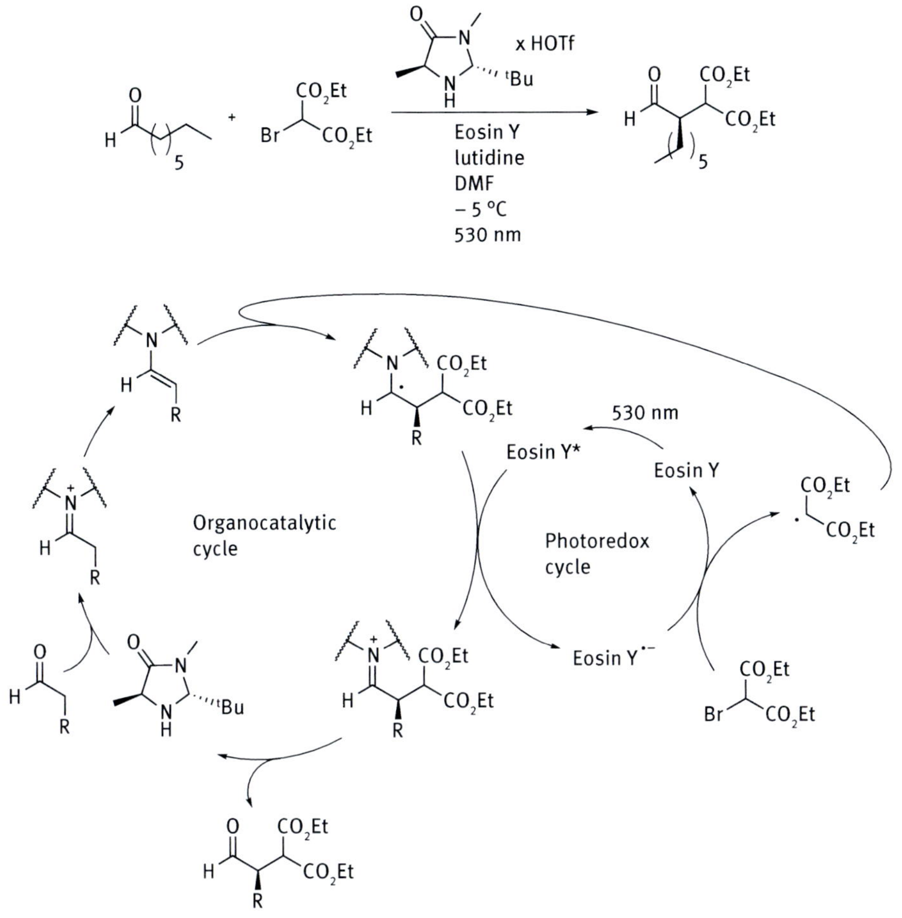

Fig. 3.13: Asymmetric α-alkylation of octanal with diethyl bromomalonate: the catalytic mechanism comprises two combined organocatalytic and photoredox cycles [76]. Reprinted with permission from Angew. Chem. Int. Ed., 2011, 50, 951–954. Copyright 2011 Wiley-VCH.

of 0.4 M, the macroflow reactor achieved a conversion of 92% at 82% ee. Based on the data from all three processes, the productivity of the tube reactor (1.92 mmol/h) was superior to the microreactor (0.037 mmol/h) and the batch reaction (0.018 mmol/h). This example shows the synergistic approach between flow photochemistry and asymmetric organocatalysis. Mild photochemical reaction conditions for asymmetric synthesis with chiral organocatalysts can be perfectly adapted to continuous-flow reactor equipment allowing the production of valuable compounds with high ee.

One example describes the photocatalytic synthesis of allylic trifluoromethanes from styrene derivatives (Figure 3.14). Carreira *et al.* use a cobalt-based photocatalyst being a more extraordinary metal for photochemical synthesis of fine chem-

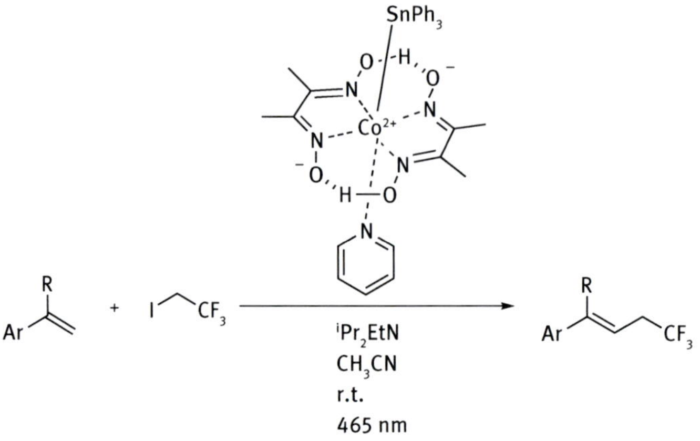

Fig. 3.14: Photocatalytic trifluoroethylation of styrene derivatives [38]. Reprinted (adapted) with permission from Org. Lett. 2013, 15, 1634–1637. Copyright 2013 American Chemical Society.

icals [38, 77]. Actually, cobalt is more often applied to the efficient production of hydrogen from water [78, 79]. Beside this novelty, the incorporation of fluorinated groups into small molecules is highly important for the pharmaceutical and agrochemical industry [80–82]. In recent years, drugs containing fluorinated groups were studied in great detail, since their metabolic stability is quite different compared to their non-fluorinated derivatives. Fluorinated alkyl groups often increase the lipophilicity of the compounds, whereas a single fluorine atom can change the electronic properties of an aromatic ring system. Both parameters can have a strong impact on the physiochemical behavior of the drug, for example on the binding to a receptor moiety.

The batch synthesis was performed in a *Schlenk* tube which was placed in the center of a cylindrical holder with LED stripes wrapped along the inside of the cylinder. The reaction solution was irradiated with blue light ($\lambda = 465$ nm) for 24 hours at room temperature. With this set-up at hand several styrene derivatives were screened with moderate (51%) to good conversion rates (83%), whereas only the *trans*-configured diastereomer was obtained as assigned by nuclear magnetic resonance (NMR) spectroscopy. Afterwards, the reaction was performed in a glass microreactor (inner volume 8 mL) with a flow rate of 0.267 mL/min and a LED array as light source. Except for one example (Table 3.3, entry 4) the conversions of the batch synthesis could not be reached, but the reaction time has been significantly decreased from 24 hours to 30 minutes. Beside this advantage, the mild reaction conditions and the easy handling of the photochemical process might allow the incorporation of trifluoroethyl groups at a late stage of the synthesis of a valuable pharmaceutical substrate.

The last example presents one important advantage of microreactor equipment. Reactive species which tend to generate large amounts of gas as one by-product of the reaction can be handled more safely in microreactors as their volume and, in conse-

Table 3.3: Scope of the trifluoroethylation as batch reaction compared to continuous-flow synthesis [38].

Entry	Product	Isolated yield batch	Isolated yield microreactor
1	tBu-C6H4-CH=CH-CH2-CF3	83%	48%
2	MeO-C6H4-CH=CH-CH2-CF3	76%	47%
3	AcO-C6H4-CH=CH-CH2-CF3	68%	38%
4	Ph2C=CH-CH2-CF3	51%	58%

quence, the amount of dangerous material used inside the reactor, is far smaller than in common batch vessel. One reaction class of unpredictable risk is the formation of diazonium salts (R–N_2^+). The decomposition of such compounds can lead to the sudden release of a large volume of nitrogen gas with an incalculable outcome. In case of the Ziegler–Stadler reaction for the formation of carbon-sulfur bonds, diazonium salts are necessary. In the traditional reaction protocol, diazonium salts must be isolated making this step and the whole synthesis route quite risky. Alternatively, metal-catalyzed versions of the Ziegler–Stadler reaction are available as well, but harsh conditions and poisoning of the catalyst by the sulfur reagent limit their use. In order to overcome these constraints Wang, Noëll et al. developed a continuous-flow photoredox reaction protocol which allows the generation of the diazonium salt as well as the C-S coupling in one stream [83]. A capillary microreactor was built with a perfluoroalkoxy (PFA) polymer tubing having an inner diameter of 0.5 mm and 464 µL volume. A blue light emitting LED strip is used as source of irradiation. The reactor is fed by two syringe pumps whereas the first syringe provides a mixture of aniline, p-methoxythiophenol, p-toluenesulfonic acid and the ruthenium photoredoxcatalyst in acetonitrile. tert-Butyl nitrite is dosed separately via the second syringe in order to ensure that the acid-catalysed diazotation of aniline with the alkyl nitrite takes place inside the reactor. The resulting phenyl diazonium salt reacts with the thiol under formation of the diazosulfide which is reduced by the photoexcited ruthenium complex (Figure 3.15). The subsequent decomposition of the diazosulfide anion releases molecular nitrogen, a highly reactive phenyl radical and the thiolate. The latter two recombine to the radical anion of the desired sulphide which is finally oxidized by the Ru(III) complex under recovery of the photoredoxcatalyst. Interestingly, another possible reaction path over a disulfide species including a direct reduction of the diazonium salt by the ruthenium complex is less pronounced under continuous-flow conditions.

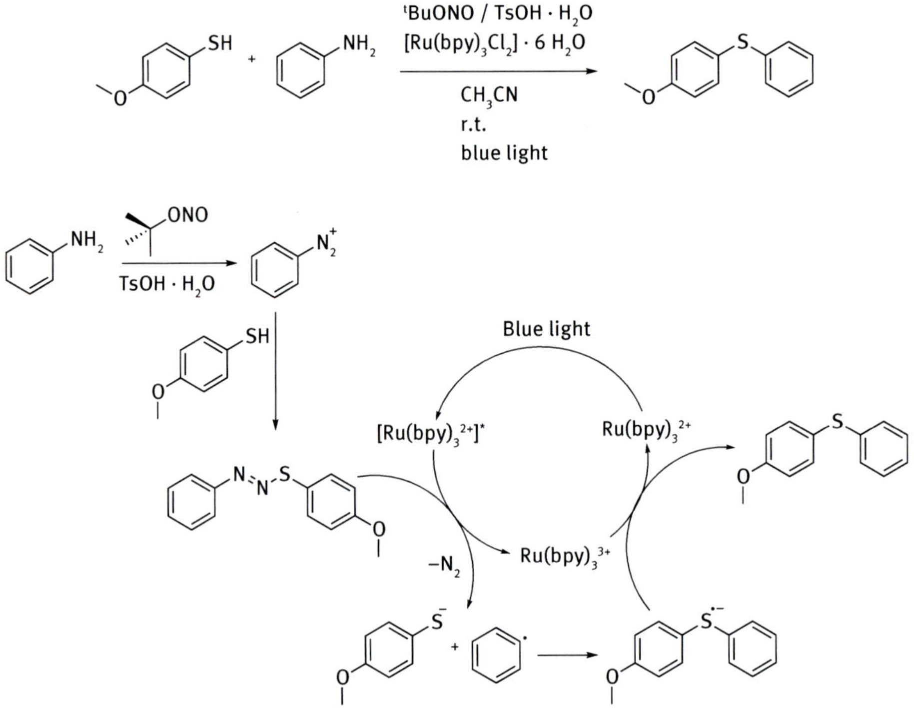

Fig. 3.15: One-stream Ziegler–Stadler reaction with *in situ* generation of phenyl diazonium salts and photoredox catalyzed formation of the C-S-bond [83]. Reprinted with permission from Angew. Chem. Int. Ed. 2013, 52, 7860–7864. Copyright 2013 Wiley-VCH.

With a residence time of 15 s, the microreactor achieved full conversion with 79% product yield. Remarkably, the reaction progress could be monitored by the formation of a segmented flow between the released nitrogen gas and the reaction solution. The batch photoredox reaction was performed under the same conditions yielding 85% product after five hours of irradiation. This value corresponds to a productivity of 0.17 mmol/h. With 13.2 mmol/h the microreactor trumps the batch synthesis by a factor of 78. The highly improved process reflects once again the obvious advantages of microreactor-based synthesis, not to forget the enhanced safety conditions while handling explosive intermediate species.

All photochemical reactions presented above give just a small insight into the great possibility of doing chemistry with visible light in optimized equipment. Efficient *in situ* generation of singlet oxygen is possible for photooxygenation reactions. Organocatalysis was combined with photoredox catalysis for asymmetric synthesis of small molecules. Fluorinated alkyl groups can be easily incorporated into a broad scope of molecules. Hazardous compounds were generated *in situ* and used directly for important C-S bond forming reactions. Despite this small selection, many other photochemical and photocatalytical reactions have been translated from batch to continuous-flow, and there is still a huge number of processes available which are worth to be tried in microstructured reactors. Most likely, the first great step has been made by changing from UV light to visible light. Here, the less-energetic irradiation allows the use of easily manageable light sources or even the sun, and minimizes in most cases the formation of by-products. In parallel, the focused research on new organic or organometallic visible light absorbing compounds escalated in the last years. This development can be attributed to the world-wide race for the most efficient organic solar cell or high performing organic devices for energy conversion in general [84–87]. Therefore it is absolutely worth to survey this field of research as well, since the physical principle is the same for the photochemical processes used in the synthesis of chemicals.

- Photochemistry performed without catalyst material in microreactors is often superior to the batch synthesis. The microenvironment exactly defines the irradiation volume and allows its full illumination. Although using high energy UV light, decomposition and by-product formation can be uttermost minimized.
- Immobilized TiO_2 on channel walls can be used as photocatalytic active substrate for other catalyst materials which in combination can broaden their application in synthetic organic chemistry.
- Highly reactive intermediates like photocatalytically generated singlet oxygen can be efficiently produced inside a safe microenvironment.
- Due to mild reaction conditions, photochemistry and photocatalysis allow the combination with sensitive reaction classes like organocatalyis or easily decomposing reactants.
- Carbon-based photocatalysts opened the door for the use of visible light for photochemistry. Sunlight, LEDs or household fluorescent lamps can be used as easy to handle light sources. Carbon-based photocatalysts can individually be synthesized allowing the adaptation to the desired reaction parameters like, for example, absorption range.

3.5 Summary and outlook

Although photocatalysis and photochemistry in general provide elegant routes for the synthesis of complex compounds, only a small number of reactions crossed the Rubicon and became established processes in industry. And even a smaller number is performed in continuous-flow mode. Flow chemistry with microstructured reactors and the necessity and possibility to use light from the complete solar spectrum are important steps on technological and chemical side that have been done to create a toolbox for sustainable photochemistry. Continuous-flow photochemistry is still a young field of research and many of today's reactor concepts are custom-built without a clear tendency on standardized components for industry. The focus is still on the development of the chemical reaction, than on the complete chemical process. But the complex linkages between the chemical, physical and technological parameters should not be considered as a problem or difficulty, but as great opportunity to adapt a reactor concept to a photochemical reaction and vice versa. In the future, the use of visible light will become even more important [88]. With the possibility to fine-tune an organic sensitizer molecule for solar irradiation, an even broader range of photocatalysts will become available. The immobilization of such tailor-made organic sensitizers on appropriate supports like, for example, polymers or metal oxides, will allow their use as heterogenized photocatalysts in continuous-flow processes without time consuming catalyst separation [89]. And chiral inductors incorporated at the sensitizer itself [90, 91] or combined with inorganic semiconductors [92] will allow a more detailed investigation of asymmetric photocatalysis in flow. First steps have been done in both fields and gave promising results allowing a look into the fascinating future of continuous-flow photochemistry.

Study questions
Photochemistry and photocatalysis base upon the interaction of electromagnetic radiation with molecules.
3.1. Which physical process takes place?
3.2. How does the molecular structure influence the absorption of light?
3.3. The solvent plays an important role for photochemical conversions. Why? Give both physical and chemical reasons!
3.4. What is the quantum yield of a photochemical reaction?
Light sources are an integral part of a photoreactor set-up.
3.5. Often filter equipment is necessary. Why? What are the consequences?
3.6. What are the main advantages of light emitting diodes?
3.7. Why is a photocatalyst necessary, in the most cases, for photochemical conversion with solar light?
Microreactors are perfectly suited for photochemistry and photocatalysis.
3.8. Which are the main advantages?
3.9. Describe the numbering-up approach! In view of photochemistry, what is the difference to the scale-up approach?
3.10. Photooxygenations are gas/liquid reactions. Tetraphenylprophyrine or Rose Bengal are often used as photocatalysts. Suggest a reactor concept and justify your decision!
3.11. Titanium dioxide is a commonly-used photocatalytic active material. How can it be used in microstructured reactor environment? Describe the advantages of this approach!
3.12. TiO_2 absorbs mainly light from the UV region of the electromagnetic spectrum. Is it possible to extend the absorption range of this semiconductor?
3.13. In recent years, visible light-absorbing molecules were examined for photocatalytic conversions. Which are the main advantages of this approach regarding their applicability in general and reaction selectivity?

Further readings
1. Klán P, Wirz J. Photochemistry of organic compounds – from concepts to practice. Chichester, West Sussex, UK, John Wiley & Sons Ldt, 2009.
2. König B (ed). Chemical photocatalysis. Berlin, Germany, De Gruyter, 2013.
3. Zollinger H. Color chemistry: syntheses, properties, and applications of organic dyes and pigments. 3rd edn, Zürich, Switzerland, Helvetica Chimica Acta, 2003.
4. Prier C, Rankic D, MacMillan D. Visible light photoredox catalysis with transition metal complexes: applications in organic synthesis. Chem. Rev. 2013, 113, 5322–5363.
5. Ravelli D, Dondi D, Fagnoni M, Albini A. Photocatalysis. A multifaceted concept for green chemistry. Chem. Soc. Rev. 2009, 38, 1999–2011.
6. Fagnoni M, Dondi D, Ravelli D, Albini A. Photocatalysis for the formation of the C-C bond. Chem. Rev. 2007, 107, 2725–2756.
7. Raveli D, Fagnoni M, Albini A. Photoorganocatalysis. What for? Chem. Soc. Rev. 2013, 42, 97–113.
8. Nocera D. The artificial leaf. Acc. Chem. Res. 2012, 45, 767–776.
9. Barber J. Photosynthetic energy conversion: natural and artificial. Chem. Soc. Rev. 2009, 38, 185–196.

Bibliography

[1] Klán P, Wirz J. Photochemistry of organic compounds from concepts to practice. Chichester, West Sussex, UK, John Wiley & Sons Ldt, 2009.
[2] Mason B, Price K, Steinbacher J, Bogdan A, McQuade T. Greener approaches to organic synthesis using microreactor technology. Chem. Rev. 2007, 107, 2300–2318.
[3] Sheldon R. Fundamentals of green chemistry: efficiency in reaction design. Chem. Soc. Rev. 2012, 41, 1437–1451.
[4] Ravelli D, Dondi D, Fagnoni M, Albini A. Photocatalysis. A multifaceted concept for green chemistry. Chem. Soc. Rev. 2009, 38, 1999–2011.
[5] Fagnoni M, Dondi D, Ravelli D, Albini A. Photocatalysis for the formation of the C-C bond. Chem. Rev. 2007, 107, 2725–2756.
[6] Atkins P, de Paula J. Atkins' Physical Chemistry. 9th ed. Oxford, UK, Oxford University Press, 2009.
[7] Megerle U, Wenninger M, Kutta R.-J, Lechner R, König B, Dick B, Riedle E. Unraveling the flavin-catalyzed photooxidation of benzylic alcohol with transient absorption spectroscopy from sub-pico to microseconds. Phys. Chem. Chem. Phys. 2011, 13, 8869–8880.
[8] Oelgemöller M, Shvydkiv O. Molecules 2011, 16, 7522–7550.
[9] Riedle E, Wenninger E. Time resolved spectroscopy in photocatalysis. In: König B. ed. Chemical photocatalysis. Berlin, Germany, De Gruyter, 2013, 319–378.
[10] Reichardt C. Solvents and solvent effects in organic chemistry. Weinheim, Germany, VCH, 1988.
[11] Van Gerven T, Mul G, Mouljin J, Stankiewicz A. A review of intensification of photocatalytic processes. Chem. Eng. Process. 2007, 46, 781–789.
[12] Coyle E, Oelgemöller M. Micro-photochemistry: photochemistry in microstructured reactors. The new photochemistry of the future? Photochem. Photobiol. Sci. 2008, 7, 1313–1322.
[13] Oelgemöller, M. Highlights of photochemical reactions in microflow reactors. Chem. Eng. Technol. 2012, 35, 1144–1152.
[14] Knowles J, Eliott L, Booker-Milburn K. Flow photochemistry: old light through new windows. Beilstein J. Org. Chem. 2012, 8, 2025–2052.
[15] Noël T, Wang X, Hessel V. Accelerating photoredox catalysis in continuous microflow. Chim. Oggi 2013, 31, 10–14.
[16] Shvydkiv O, Oelgemöller M. Microphotochemistry. In: Griesbeck A, Oelgemöller M, Ghetti F., eds. CRC Handbook of organic photochemistry and photobiology. 3rd ed. Vol. 1, Boca Raton, FL, USA, CRC Press, Taylor & Francsi Group, 49–72.
[17] Heraeus Noblelight. (Accessed January 13, 2014 at: http://www.heraeus-noblelight.com/en/products_1/uvprozesstechnik_1/products_uvp.aspx).
[18] Kreisel G, Meyer S, Tietze D, Fidler T, Gorges G, Kirsch A, Schäfer B, Rau S. Leuchtdioden in der Chemie – Eine Hochzeit verschiedener Technologien. Chem.-Ing-Tech, 2007, 79, 153–159.
[19] Leuchtdioden Rechercheportal. (Accessed January 13, 2014 at: http://www.led-info.de/).
[20] Light-emitting diode. (Accessed January 13, 2014 at: http://en.wikipedia.org/wiki/Light-emitting_diode).
[21] Hild O. OLED – Licht der Zukunft. Kunststoffe, 2006, 6, 40–43.
[22] Ciamician G. Gazz. Chim. Ital. 1886, 16, 111–112.
[23] Oelgemöller M, Healy N, de Oliveria L. Jung C, Mattay J. Green photochemistry: solar-chemical synthesis of juglone with medium concentrated sunlight. Green Chem. 2006, 8, 831–834.
[24] Oelgemöller M, Jung C, Mattay J. Green photochemistry: production of fine chemicals with sunlight. Pure Appl. Chem. 2007, 79, 1939–1947.

[25] Reference Solar Spectral Irradiance: Air Mass 1.5. American Society for Testing and Materials (ASTM) Terrestrial Reference Spectra for Photovoltaic Performance Evaluation. (Accessed January 13, 2014, at http://rredc.nrel.gov/solar/spectra/am1.5/).
[26] Protti S, Fagnoni M. The sunny side of chemistry: green synthesis by solar light. Photochem. Photobiol. Sci. 2009, 8, 1499–1516.
[27] Matsushita Y, Ohba N, Kumada S, Sakeda K, Suzuki T, Ichimura T. Photocatalytic reactions in microreactors. Chem. Eng. J. 2008, 135S, S303–S308.
[28] Griesbeck A, Matue N, Bondock S, Oelgemöller M. The excimer radiation system: a powerful tool for preparative organic photochemistry. A technical note. Photochem. Photobiol. Sci. 2003, 2, 450–451.
[29] Wille Ch. Entwicklung und Charakterisierung eines Mikrofallfilm-Reaktors für stofftransportlimitierte hochexotherme Gas/Flüssig-Reaktionen. PhD thesis, TU Clausthal, 2000.
[30] Hessel V, Ehrfeld W, Golbig K, Haverkamp V, Löwe H, Storz M, Wille C, Guber A, Jähnisch K, Baerns M. Gas/liquid microreactors for direct fluorination of aromatic compounds using elemental fluorine. In: Ehrfeld W. ed. Proceedings of the Third International Conference on Microreaction Technology (IMRET3), Berlin, Springer-Verlag, 2000, 526–540.
[31] Hook B, Dohle W, Hirst P, Pickworth M, Berry M, Booker-Milburn K. A practical flow reactor for continuous organic photochemistry. J. Org. Chem. 2005, 70, 7558–7564.
[32] Metzger R, Fries D, Heuschkel U, Witte K, Waidelich E, Schmid G. Photo-Nitrosierung und -Oximierung gesättigter Kohlenwasserstoffe. Angew. Chem. 1959, 71, 229–236.
[33] Fischer M. Photochemische Synthesen in technischem Maßstab. Angew. Chem. 1978, 17, 16–26.
[34] Pfoertner K. Photochemistry in industrial synthesis. J. Photochem. Photobiol. A 1990, 51, 81–86.
[35] Lévesque F, Seeberger P. Continuous-flow synthesis of the anti-malaria drug artemisinin. Angew. Chem. Int. Ed. 2012, 51, 1706–1709.
[36] Kopetzki D, Lévesque F, Seeberger P. A continuous-flow process for the synthesis of artemisinin. Chem. Eur. J. 2013, 19, 5450–5456.
[37] Booker-Milbrun K. A light touch to a deadly problem. Nature Chem. 2012, 4, 433–435.
[38] Kreis L, Krautwald S, Pfeiffer N, Martin R, Carreira E. Photocatalytic synthesis of allylic trifluoromethyl substituted styrene derivatives in batch and flow. Org. Lett. 2013, 15, 1634–1637.
[39] Tucker J, Zhang Y, Jamison T, Stephenson C. Visible-light photoredox catalysis in flow. Angew. Chem. 2012, 124, 4220–4223.
[40] Andrews S, Becker J, Gagné M. A photoflow reactor for the continuous photoredox-mediated synthesis of C-glycoamino acids and C-glycolipids. Angew. Chem. Int. Ed. 2012, 51, 4140–4143.
[41] Rueping M, Vila C, Bootwicha T. Continuous flow organocatalytic C-H functionalization and cross-dehydrogenative coupling reactions: visible light organophotocatalysis for multicomponent reactions and C-C, C-P bond formations. ACS Catal. 2013, 3, 1676–1680.
[42] Fuse S, Mifune Y, Tanabe N, Takahashi T. Continuous-flow synthesis of activated vitamin D_3 and its analogues. Org. Biomol. Chem. 2012, 10, 5205–5211.
[43] Fuse S, Tanabe N, Yoshida M, Yoshida H, Doi T, Takahashi T. Continuous-flow synthesis of vitamin D_3. Chem. Commun. 2010, 46, 8722–8724.
[44] Ehrich H, Linke D, Morgenschweis K, Baerns M, Jähnisch K. Application of Microstructured Reactor Technology for the Photochemical Chlorination of Alkylaromatics. Chimia, 2002, 56, 647–653.
[45] Hessel V, Löb P, Löwe H. Gas-Liquid Reactions. In: Wirth T. ed. Microreactors in Organic Chemistry and Catalysis. 1st ed. Weinheim, Germany, Wiley-VCH, 2008, 139–183.

[46] Sugimoto A, Fukuyama T, Sumino Y, Takagi M, Ryu I. Microflow photoradical reaction using a compact light source: application to the Barton reaction leading to a key intermediate for myriceric acid A. Tetrahedron, 2009, 65, 1593–1598.

[47] Hu X, Li G, Yu J. Design, fabrication, and modification of nanostructured semiconductor materials for environmental and energy Applications. Langmuir, 2010, 26, 3031–3039.

[48] Ichimura T, Matsushita Y, Sakeda K, Suzuki T. Photoreactions. In: T. Dietrich, ed. Microchemical Engineering in Practice. Hoboken, New Jersey, USA, John Wiley & Sons, Inc., 2009, 385–402.

[49] Kubacka A, Fernández-García M, Colón G. Advanced nanoarchitectures for solar photocatalytic applications. Chem. Rev. 2012, 112, 1555–1614.

[50] Li N, Lang X, Ma W, Ji H, Chen C, Zhao J. Selective aerobic oxidation of amines to imines by TiO_2 as photocatalysts in water. Chem. Commun. 2013, 49, 5034–5036.

[51] Molinari A, Montoncello M, Rezala H, Maldotti A. Partial oxidation of allylic and primary alcohols with O_2 by photoexcited TiO_2. Photochem. Photobiol. Sci. 2009, 8, 613–619.

[52] Rueping M, Zoller J, Fabry D, Poscharny K, Koenigs R, Weirich T, Mayer J. Light-mediated heterogeneous cross dehydrogenative coupling reactions: metal oxides as efficient, recyclable, photoredox catalysts in C-C bond forming reactions. Chem. Eur. J. 2012, 18, 3478–3481.

[53] Wöhrle D, Kaneko M, Nagai K, Suvorova O, Gerdes R. Environmental cleaning by molecular catalysts. In: Okada T, Kaneko M., eds. Molecular catalysts for energy conversion. Springer series in material sciences. Berlin, Germany, Springer, 2009, 263–297.

[54] Murphy S, Saurel C, Morrissey A, Tobin J, Oelgemöller M, Nolan K. Photocatalytical activity of porphyrine/TiO_2 composite in the degradation of pharmaceuticals. Appl. Catal. B-Environ. 2012, 119–120, 156–165.

[55] Kaneko M. Molecular catalysts for electrochemical solar cells and artificial photosynthesis. In: Okada T, Kaneko M., eds. Molecular catalysts for energy conversion. Springer series in material sciences. Berlin, Germany, Springer, 2009, 200–215.

[56] Hara K. Molecular design of sensitizers for dye-sensitized solar cells. In: Okada T, Kaneko M., eds. Molecular catalysts for energy conversion. Springer series in material sciences. Berlin, Germany, Springer, 2009, 218–250.

[57] Füldner S, Mild R, Siegmund H, Schroeder J, Gruber M, König B. Green-light photocatalytic reduction using dye-sensitized TiO_2 and transition metal nanoparticles. Green Chem. 2010, 12, 400–406.

[58] Takei G, Kitamori T, Kim H.-B. Photocatalytic redox-combined synthesis of L-pipecolinic acid with a titania-modified microchannel chip. Catal. Commun. 2005, 6, 357–360.

[59] Pal B, Ikeda S, Kominami H, Kera Y, Ohtani B. Photocatalytic redox-combined synthesis of L-pipecolinic acid from L-lysine by suspended titania particles: effect of noble metal loading on the selectivity and optical purity of the product. J. Catal. 2003, 217, 152–159.

[60] Hari D, Schroll P, König B. Metal-free, visible-light-mediated direct C-H arylation of heteroarenes with aryl diazonium salts. J. Am. Chem. Soc. 2012, 134, 2958–2961.

[61] Liu Q, Li Y.-N, Zhang H.-H, Chen B, Tung C.-H, Wu L.-Z. Reactivity and mechanistic insight into visible-light-induced aerobic cross-dehydrogenationative coupling reaction by organophotocatalysts. Chem. Eur. J. 2012, 18, 620–627.

[62] Pan Y, Kee C, Chen L, Tan C.-H. Dehydrogenative coupling reactions catalysed by rose bengal using visible light irradiation. Green Chem. 2011, 13, 2682–2685.

[63] Schmaderer H, Hilgers P, Lechner R, König B. Photooxidation of benzyl alcohols with immobilized flavins. Adv. Synth. Catal. 2009, 351, 163–174.

[64] Lechner R, Kümmel S, König B. Visible light flavin photo-oxidation of methylbenzenes, styrenes and phenylacetic acid. Photochem. Photobiol. Sci. 2010, 9, 1367–1377.

[65] Prier C, Rankic D, MacMillan D. Visible light photoredox catalysis with transition metal complexes: applications in organic synthesis. Chem. Rev. 2013, 113, 5322–5363.

[66] Maity S, Zheng N. A visible-light-mediated oxidative C-N bond formation/aromatization cascade: photocatalytic preparation of N-arylindoles. Angew. Chem. Int. Ed. 2012, 38, 9562–9566.
[67] Condie A, González-Gómez J, Stephenson C. Visible-light photoredox catalysis: aza-Henry reactions via C-H functionalization. J. Am. Chem. Soc. 2010, 132, 1464–1465.
[68] Tyson E, Ament M, Yoon T. Transition metal photoredox catalysis of radical thiol-ene reactions. J. Org. Chem. 2013, 78, 2046–2050.
[69] Lin S, Ischay M, Fry C, Yoon T. Radical cation Diels-Alder cycloaddition by visible light photocatalysis. J. Am. Chem Soc. 2011, 133, 19 350–353.
[70] Bou-Hamadn F, Seeberger P. Visible-light-mediated photochemistry: accelerating Ru(bpy)$_3^{2+}$-catalyzed reactions in continuous flow. Chem. Sci. 2012, 3, 1612–1616.
[71] Xuan J, Xiao W.-J. Visible-light photoredox catalysis. Angew. Chem. Int. Ed. 2012, 51, 6828–6838.
[72] Tamke S, Paradies J. Photoredoxkatalyse. Nachr. Chem., 2013, 61, 1122–1127.
[73] Shvydkiv O, Limburg C, Nolan K, Oelgemöller M. Synthesis of Juglone (5-hydroxy-1,4-naphthoquinone) in a falling film microreactor. J. Flow Chemistry, 2012, 2, 52–55.
[74] Raveli D, Fagnoni M, Albini A. Photoorganocatalysis. What for? Chem. Soc. Rev. 2013, 42, 97–113.
[75] Neumann M, Zeitler K. Application of microflow conditions to visible light photoredox catalysis. Org. Lett. 2012, 14, 2658–2661.
[76] Neumann M, Füldner S, König B, Zeitler K. Metal-free, cooperative asymmetric organophotoredox catalysis with Visible Light. Angew. Chem. Int. Ed., 2011, 50, 951–954.
[77] Weiss M, Kreis L, Lauber A, Carreira E. Cobalt-Catalyzed Coupling of Alkyl Iodides with Alkenes: Deprotonation of Hydridocobalt Enables Turnover. Angew. Chem. Int. Ed. 2011, 50, 11 125–128.
[78] Artero V, Chavarot-Kerlidou M, Fontecave M. Splitting water with cobalt. Angew. Chem. Int. Ed. 2011, 50, 7238–7266.
[79] Kudo A, Miseki Y. Heterogeneous photocatalyst materials for water splitting. Chem. Soc. Rev. 2009, 38, 253–278.
[80] Purser S, Moore P, Swallow S, Gouverneur V. Fluorine in medicinal chemistry. Chem. Soc. Rev. 2008, 37, 320–330.
[81] Mizuta S, Verhoog S, Engle K, Khotavivattana T, O'Duill M, Wheelhouse K, Rassias G, Médébielle M, Gouverneur V. Catalytic hydrotrifluoromethylation of unactivated alkenes. J. Am. Chem. Soc. 2013, 135, 2505–2508.
[82] Yasu Y, Koike T, Akita M. Intermolecular aminotrifluoromethylation of alkenes by visible-light-driven photoredox catalysis. Org. Lett. 2013, 15, 2136–2139.
[83] Wang X., Cuny G, Noël T. A mild, one-pot Stadler-Ziegler synthesis of arylsulfides facilitated by photoredox catalysis in batch and continuous-flow. Angew. Chem. Int. Ed. 2013, 52, 7860–7864.
[84] Nocera D. The artificial leaf. Acc. Chem. Res. 2012, 45, 767–776.
[85] Frischmann P, Mahata K, Würthner F. Powering the future of molecular artificial photosynthesis with light-harvesting metallosupramolecular dye assemblies. Chem. Soc. Rev. 2013, 42, 1847–1870.
[86] Barber J. Photosynthetic energy conversion: natural and artificial. Chem. Soc. Rev. 2009, 38, 185–196.
[87] Zhan X, Facchetti A, Barlow S, Marks T, Ratner M, Wasielewski M, Marder S. Rylene and related Diimides for organic electronics. Adv. Mater. 2011, 23, 268–284.
[88] Yoon T, Ischay M, Juana Du. Visible light photocatalysis as a greener approach to photochemical synthesis. Nat. Chem. 2010, 2, 527–532.

[89] Ribeiro S, Serra A, Gonsalves A. Covalently immobilized porphyrines as photooxidation catalysts. Tetrahedron 2007, 63, 7887–7891.
[90] Müller C, Bauer A, Maturi M, Cuquerella M, Miranda M, Bach T. Enantioselective intramolecular [2+2]-photocycloaddition reactions of 4-substituted quinolones catalyzed by a chiral sensitizer with a hydrogen-bonding motif. J. Am. Chem. Soc. 2013, 135, 14 948–951.
[91] Wiegand C. Herstweck E, Bach T. Enatioselectivity in visible light-induced, singlet oxygen [2+4] cycloaddition reactions (type II photooxygenations) of 2-pyridones. Chem. Commun. 2012, 48, 10 195–197.
[92] Cherevatskaya M, Neumann M, Füldner S, Harlander C, Kümmel S, Dankesreiter S, Pfitzner A, Zeitler K, König B. Visible-light-promoted stereoselective alkylation by combining heterogeneous photocatalysis with organocatalysis. Angew. Chem. Int. Ed. 2012, 51, 4062–4066.

Julian Schuelein and Holger Loewe
4 Electrochemistry in flow

4.1 Introduction

For about 150 years, electrochemistry has been applied in industry to produce chlorine and sodium hydroxide from aqueous sodium chloride [1]. Other common industrial applications are potassium, aluminum and magnesium manufacturing via fused salt electrolysis [2]. Aqueous systems are also used to produce copper and zinc as basic chemicals for metallic materials. Newer techniques rely on the principle of electrochemistry, for instance, fuel cells which are used to produce electric energy directly from chemical reactions, and lithium-ion batteries for efficient storage of electrical energy [3, 4].

Electrochemistry is one of the most diverse techniques available in chemistry. Therefore, the manufacturing of chemicals is only a small area in the field of electrochemistry. Physical chemistry helped to develop models and methods to understand the underlying concept and to predict reactions at the electrodes. Electrochemistry is also utilized in analytics, where it can be used to detect low-dosed chemicals based on their electrochemical properties [5].

Basic applications of electrochemistry are shown in Figure 4.1. The left side outlines all industrial and synthetic aspects; the right side, however, shows the use of galvanic elements for energy storage purposes.

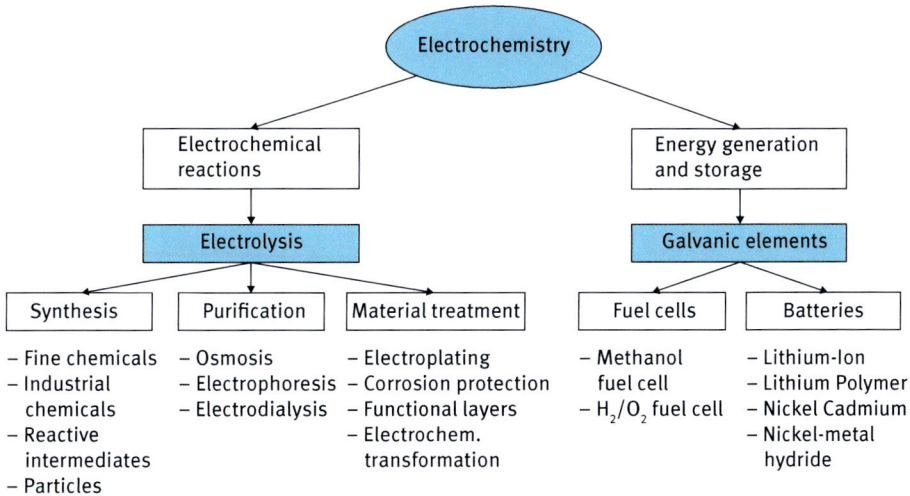

Fig. 4.1: Overview of possibilities with electrochemistry [77].

This chapter focuses on organic synthetic electrochemistry for fine chemicals, which is shown on the very left side of Figure 4.1. In general, there are two principles where electrochemistry can be used in synthesis: The first is to synthesize the desired product directly via oxidation or reduction (commonly used in industrial chemical production). The other more promising possibility, is to activate one or more educts via electrochemistry (for instance generating carbo-cations) and use those activated reagents in a subsequent reaction step.

Both principles can be applied using electrochemistry in batch reactors, but flow chemistry, especially in setups with microscaled internals, has some outstanding advantages. Due to the small proximity of electrodes in such a microscaled system, the uniformity of the electric current becomes better balanced, as well as the ohmic resistance of the electrolyte can be minimized between the electrode gap. Furthermore, the concentration of the conducting salt might be reduced and the high surface-to-volume ratio of a micro electrolysis cell allows one to easily control the temperature caused by increased amperage. However, disadvantages arise through hydrogen production at the counter electrode by overvoltage.

4.2 Electrochemistry in flow

There are multiple ways to conduct electrochemical experiments. Figure 4.2 shows a general process chart for direct synthesis of products valid for batch as well as flow

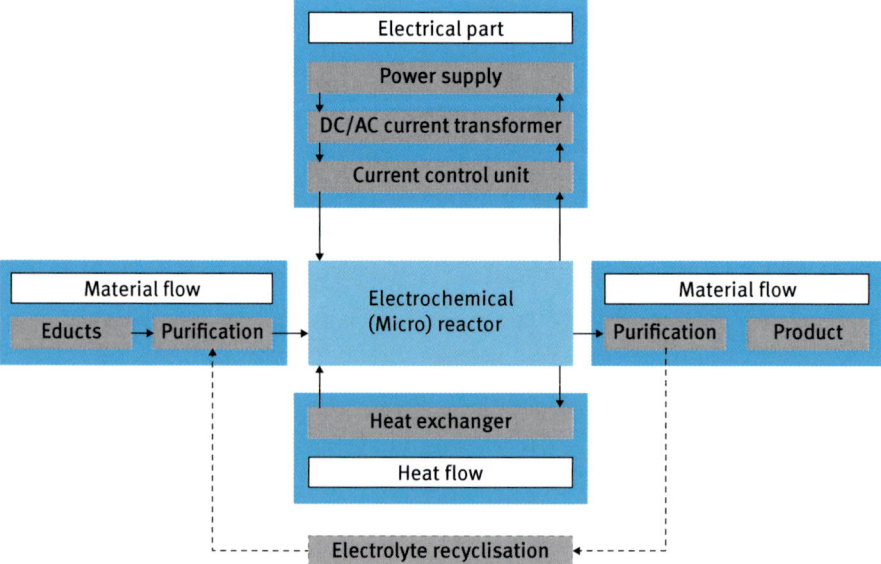

Fig. 4.2: Standard process chart of an electrochemical reactor for industrial-scale synthesis [77].

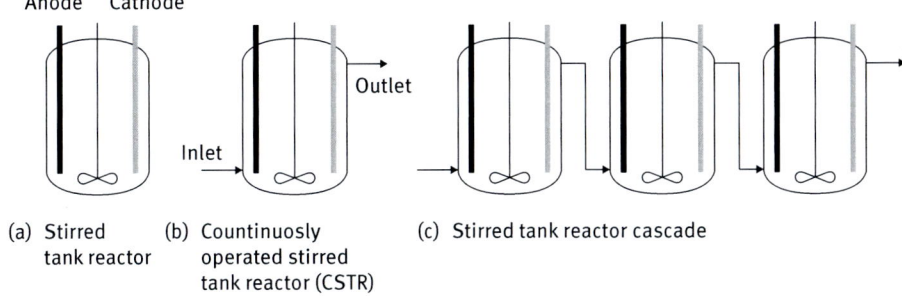

Fig. 4.3: Different batch reactor types for electrochemical reactions: (a) conventional stirred tank reactor with electrodes, (b) continuously operated stirred tank reactor (CSTR) with in- and outlet of reagent, (c) continuously operated stirred tanks in a cascade for demanding batch electrosynthesis [77].

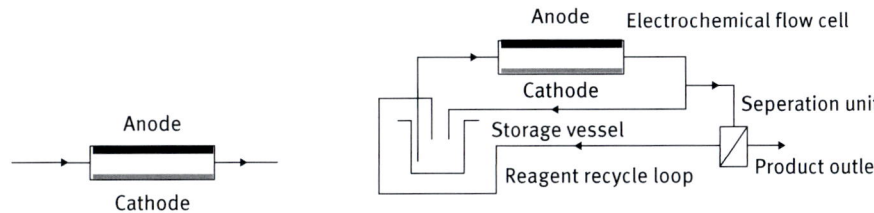

Fig. 4.4: Flow type reactor types for electrosynthesis. (a) Single pass with high conversion, (b) multi-pass with product separation and reagent recycle [78].

reactors. The three indicated streams show the material flow, the electrical setup and the heat flow. Optional electrolyte cyclization is also shown with a dashed line.

Most of the industrial applications use continuously operated stirred tank reactors (CSTR) or semi-batch processes because of their already applied role in industrial scale chemistry. However, there are a variety of applied reactor types (see Figure 4.3) [6–8]:
1. Batch reactor
2. Continuously operated stirred tank reactor (CSTR)
3. Stirred tank reactor cascade
4. Flow reactor

Batch reactors are seldom used for electrosynthesis of fine chemicals, whereas it is the reactor of choice for electroplating, surface modification and commodity production, because of the very highly concentrated salt solutions needed for surface modification and large scale synthesis [7].

The continuously stirred tank reactors are frequently used for metal production. Such reactors are basically batch reactors with inlet and outlet for a continuous phase.

This continuous phase can be used to extract soluble parts of metal ores and feed them into the reactor for reduction to solid metal. The reaction solution can then be used again to extract metal ores. A noteworthy characteristic of this reactor is that usually the inlet is at the bottom, on one side, and the outlet is on the top of the other side. This can be substantiated with the density of the incoming solution which is higher than the out tipping solution because of the dissolved metal ions. Metal ions are reduced during reaction and usually solidify at the cathode [6].

The stirred tank reactor cascade is a series of CSTRs where the inlet is at the very bottom of all tanks and the outlet at the top. The cascade reactor is suitable for demanding or multistep synthesis. The stirring speed and applied amperage as well as the applied voltage can be adjusted for each tank.

For fine chemical synthesis with high value educts and high demand for purity of the product, batch reactors can usually be substituted by flow reactors. Flow reactors consist of a pipe or rectangular-shaped plates which can be pressed together. Parts of

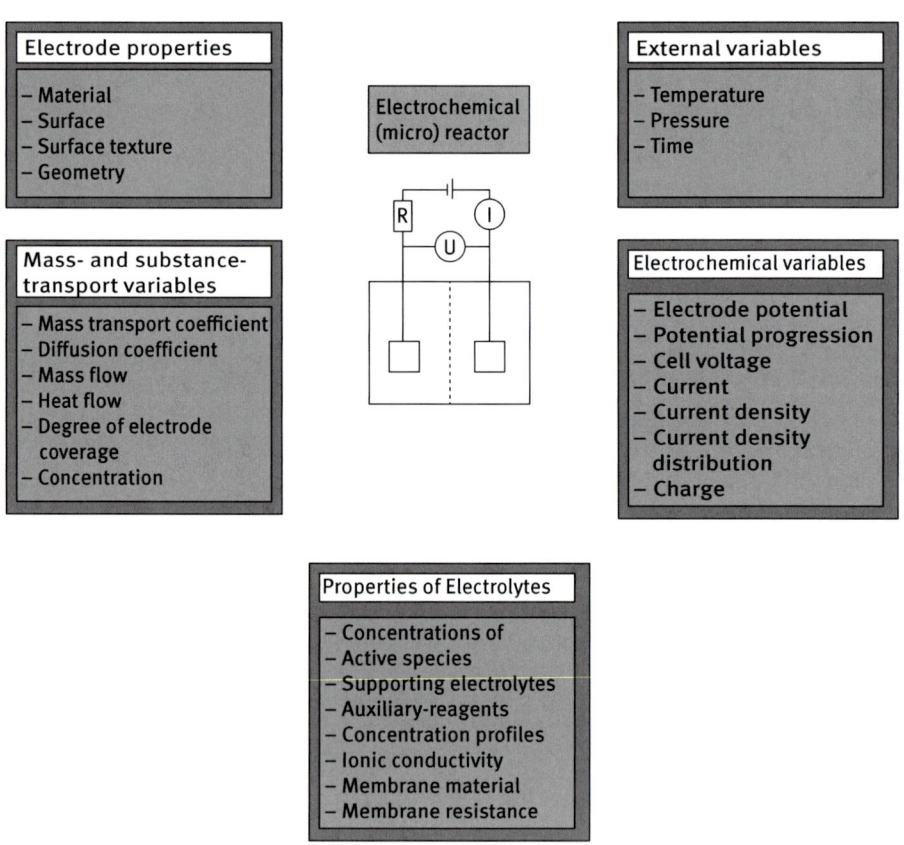

Fig. 4.5: Process parameters for an electrochemical reaction [77].

the pipes or the plates then represent the electrodes and depending on the reaction needs, a membrane can be integrated to separate the anodic and cathodic chamber [9].

Figure 4.5 indicates process parameters which should be taken into account when creating an electrochemical reactor. Properties of the electrode must be chosen wisely depending on the reaction conditions and potentials needed. Variables of the mass flow cannot be influenced directly, but can be adjusted in a margin by external parameters (e.g., temperature for the diffusion coefficient). Properties of the electrolyte are also only less easily influenced due to the general reaction conditions (use of the product prohibiting toxic ingredients, annual tons needed, security precautions, etc.) [10].

4.3 Microreactor design

- Microreactors have a very high surface-to-volume ratio and can provide superior mixing capabilities which could not only increase the yield but also enable highly exothermic or endothermic reactions.
- The excellent surface-to-volume ratio is of great advantage in electrochemistry, due to the maximization of the electrode area.
- Small sample volumes also enable to carry out the reaction in a potentiostatic controlled manner. Industrial synthesis usually have to be carried out in galvanostatic manner with high currents and a high probability to create side-products.

Electrochemistry in flow is not new; industrial applications are countless and reactor designs are diverse [10–20]. Nevertheless, electrochemistry suffers from high electrolyte concentrations, inhomogeneous electric fields and, therefore, high costs for process intensification.

During the last 20 years, microfabrication methods have become affordable for research and industrial uses. Microfabrication and microreactors are not very different from commercially available chemical engineering components, but they all share the small dimensions of the active area of interest. Usually those areas (mixing zones, extractors, evaporators, etc.) are in the micrometer range commonly below 500 µm, but the whole reactor can still be in centimeter-to-decimeter ranges of size depending on the annual targeted throughput.

In addition to the already known advantages of microstructured flow-through reactors, electrochemical reactors provide some important advantages. Conventional reactors for electrosyntheses are driven in a galvanostatic mode; a high electrical current is used, but the potential on the electrodes, which refers to the activation energy, cannot be controlled. Overvoltage and undefined current density distribution along the electrode surfaces, will result in formation of side-products as well as decomposition of the electrolyte itself.

By using a microscaled flow-through device, the electrochemical reaction might be really controlled in a potentiostatic mode, due to the small local current densities on the electrodes. For this kind of reaction control, a third electrode is needed as a reference electrode. Commonly, second type electrodes such as silver/silver chloride or calomel electrode are used. Due to the limited space in microreactors, those electrodes also have to be miniaturized, which was done by Küpper and Schmidt-Traub in 2003 [21]. A standard silver/silver chloride electrode with PVC housing with diameter of 1 mm was resized to a diameter of 50 µm which can easily fit into the reaction chamber of the electrochemical microreactor (ELMI). Those electrodes are known as "miniaturized solid-state-reference electrodes" (µSSRE) [21].

4.3.1 Thin gap cells

The basic setup of a microreactor for electrochemical synthesis is the so-called thin gap cell (see Figure 4.6). It consists of two conducting plates (working- and auxiliary electrode) and spacers to fix the proximity between the electrodes. Furthermore, at least one in- and outlet has to be present, usually done with polytetrafluoroethylene (PTFE) or silicone tubing. In the early stages of thin gap cells, adhesive tape was be-

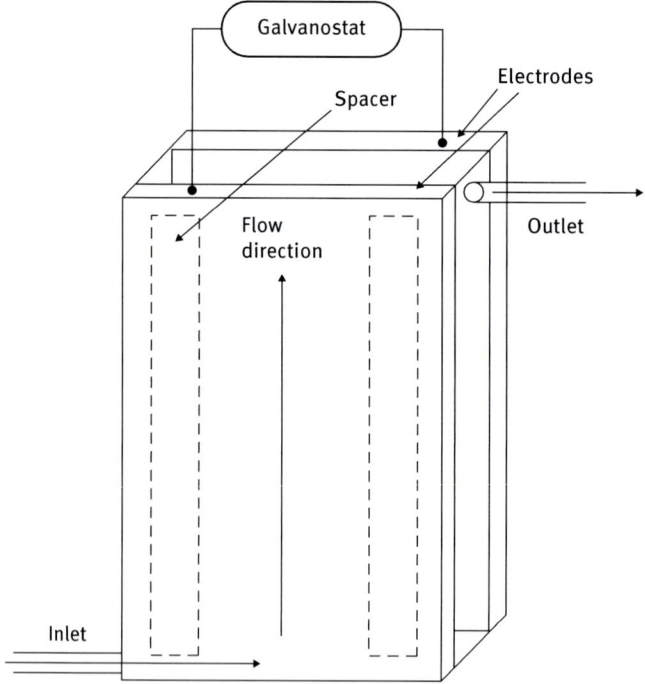

Fig. 4.6: Thin gap flow cell for electrochemical synthesis [79].

ing used as spacer (80–320 µm) and the sealing was done with epoxy resin [22–26]. The challenge in designing thin gap cells is the processing of the electrode plates itself. For a space of about 80 µm, the electrode plates have to be as smooth as possible. In general, this is vital to avoid overvoltage and local high current densities on edges or peaks of possibly rough surfaces. Depending on the electrode material, sophisticated methods have to be used for processing, for instance laser-cutting for smooth surfaces or lithography galvanic molding (LIGA) for high aspect ratios in microstructured regimes.

The drawback of electrochemistry is the broad variety of parameters which can be influenced. The main process parameters are concentration of educts, flow rates, electrode materials, solvents, heat exchange and layout of the cells, current and voltage. Prognosis of the optimum parameters is therefore not easily achieved without some assumptions and simplifications. To understand the behavior of thin gap flow cells Rode et al. created a model for the methoxylation of 4-methoxytoluene from the generally accepted reaction mechanism [27–29]:

$$A \rightarrow B + n_A e^- \qquad (4.1)$$

$$B \rightarrow C + n_B e^- \qquad (4.2)$$

$$C \rightarrow D + n_C e^-. \qquad (4.3)$$

Where A is 4-methoxytoluene and C is the desired intermediate product which can react further to D if the reaction is not stopped at the right point. The cathode reaction is the generation of hydrogen via reduction of methanol which also acts as a solvent and a methoxylation agent. Assuming a thin gap cell with an inter-electrode distance (d) and a total length (L) operated with the current (I) and an overall cell voltage (V).

For generalization of this model, all model parameters were reduced to their dimensionless counterparts:

$$x^* = \frac{x}{L} \qquad (4.4)$$

$$C_j^* = \frac{C_j(x^*)}{C_{A_0}} \qquad (4.5)$$

$$i_j^*(x^*) = \frac{i_j(x^*)}{n_A F k_m C_{A_0}} \qquad (4.6)$$

$$V^* = b_A V - \ln\left(\frac{k_m}{k_A}\right) - b_A \left[E_A^{\text{ref}} - E_C^{\text{ref}}\right] \qquad (4.7)$$

$$E_a^*(x^*) = b_A E_a(x^*) - \ln\left(\frac{k_m}{k_A}\right). \qquad (4.8)$$

> - Electrochemistry is a very complex topic due to the variety of process parameters which should be optimized (see Figure 4.5).
> - Estimation of the reaction concerning the electrochemical parameters can be done via the *Wagner* number (W_a) and the number of transfer units (NTU).
> - The *Wagner* number describes the distribution of the current distribution giving information about the controlled state of the reaction. High *Wagner* numbers are usually preferred.
> - The number of transfer units describes the ability of heat transport in the reactor. High NTUs lead to undercooling, moderate NTUs for "normal" reactions are usually preferred. For highly exothermic reactions the NTU has to be as high as possible.

The spatial coordinate is normalized to the reactor length (L) (Equation (4.4)), the spatially dependent concentrations $C_j(x^*)$ are standardized to the initial concentration of A (C_{A_0}, (Equation (4.5))) and the local current densities $i_j^*(x^*)$ are normalized to the limiting current density for the first oxidation step of reagent A (Equation (4.6)). Equations (4.7) and (4.8) deal with the cell voltage and local anodic potential difference normalization. The cell voltage is multiplied with the first exponential Tafel-coefficient for the first oxidation step (b_a) and also shifted with the ratio of mass-transfer coefficient (k_m) and pre-exponential Tafel-coefficient (k_a) as well as the reference potentials at both electrodes. Equation (4.8) is normalized in a similar way [30].

Based on *Tafel* type laws [31–33] for interfacial oxidation, Newton-type laws for mass transfer and the local dimensionless current densities for all three reactions, a set of independent dimensionless equations can be formed, which are used to characterize this reaction [32]. In total, there are ten equations of which six groups only depend on stoichiometry and kinetics of the system. The more important three groups ($\prod_1 - \prod_3$ in [30]) are dependent on parameters influenced by the reactor geometry and operation, whereas $\prod_3$ is a constant described by the mass transfer coefficient over the rate constant for *Tafel* kinetics for the oxidation of 4-methoxyltoluene.

$$\prod_1 = \frac{\kappa_{eq}}{n_A F k_m C_{A_0} b_A d} \tag{4.9}$$

$$\prod_2 = \frac{k_m L w}{Q_L}. \tag{4.10}$$

For electrochemical processes involving two electrons, ($n_A = 2$) in thin gap cells with a fully developed boundary layer and a concentration of A equaling unity. Hence, parameter $\prod_1$ is proportional to the conductivity between the electrodes (κ_{eq}), which also is described by the *Wagner* number. Small *Wagner* numbers stand for excellent uniformity of the current distribution along the electrode surface.

Parameter $\prod_2$ with Q_L describing the volumetric flow rate over the electrode distance (L) and (w) being the electrode width equals the Number of Transfer Units (NTU). This is because the mass-transfer flux between the solid electrode and the liquid electrolyte ($k_m L w$) is compared to an overall flux of the electrolyte (Q_L). NTUs are used to describe heat transfer quality of interfacial borders [34].

After those simplifications, the reaction system can be simulated by varying the *Wagner* number, the NTU and the dimensionless current I^*. Further limiting of those parameters is appropriate, because not all values are industrially accessible. The dimensionless current should be between 0.6 and 1.1, where 1.0 equals the electron flux necessary to yield 100% product. Higher currents are of course accessible, but the probability to gain more product from a dimensionless current over 1.5 is negligible. This is due to side-reactions which start to occur, for instance, in this reaction scheme, the conversion of desired product C to side-product D (Equation (4.3)). The *Wagner* number can be between 0.002, for an almost completely uniform current distribution, and infinity for a highly inconsistent current distribution (for instance due to outgassing hydrogen). The number of transfer units can accept values between 6 and 12. Hereby 6 denote the lower limit of heat flux where just about no mass transfer limitations occur. Higher NTUs suggest a nonoptimal use of the reactor.

For plotting purposes, the current density distribution ($i^*_{Gl}(x^*) = i^*_A(x^*) + i^*_B(x^*) + i^*_C(x^*)$) is calculated, which equals the sum of the three dimensionless current densities for each reaction step. Plotting those values enables one to estimate the reaction outcome according to mass-transfer limitations. Figure 4.7 (a) shows the current density distribution versus the spatial normalized coordinate with $I^* = 1$. For an infinite *Wagner* number $i^*_{Gl}(x^*)$ is greater than 1, implying that the second oxidation step to product C already appears. Therefore $n_A + n_B$ electrons are consumed, but the current density limit (see [30] Equation A-5) is calculated using the first oxidation step only, resulting in a number greater than unity. However, a decrease in NTU from 12 to 6 independent of the *Wagner* number results in an increase of the current density distribution, pointing out that the reaction may reach the mass-transfer limitations. Figure 4.7 (b) shows the factor (r), which stands for the local current density distribution over the local limiting current density. The graph shows that for high *Wagner* numbers, the current is evenly distributed over the complete electrode length. The fact that (r) is always below 1 suggests that no or very little undesired side-products are formed. For small *Wagner* numbers, the distribution is higher than 1 at the end of the reactor, stating that undesired reactions (for instance severe formation of D) occur at the reactor outlet. In this case, a segmented thin gap flow cell is essential for use, because current densities with this cell can be adjusted over the spatial coordinate (see Section 4.3.3).

Figure 4.8 (a) shows the concentration of all species along the reactor coordinate with infinite and small *Wagner* numbers. The empty symbols present high *Wagner* numbers showing a steady increase in product C to about 93% while side-product D is in a very low percentage range ($< 3\%$) at the outlet. This can be compared to small *Wagner* numbers (filled symbols), which show a maximum yield of 80% at 90% of the reactor length. In this case, product C reacts further to undesired side-product D rapidly. This matches the data shown in Figure 4.7 (b). Figure 4.8 (a) points out the

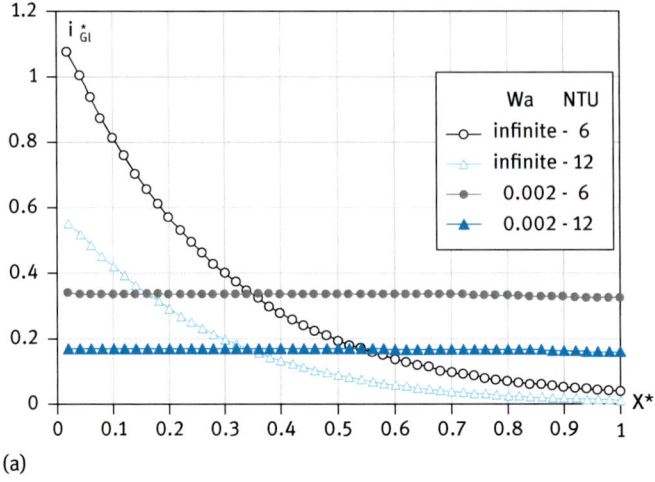

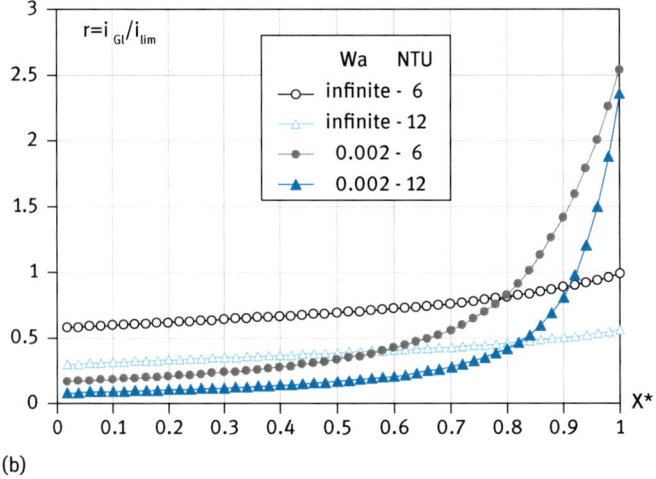

Fig. 4.7: (a) Dimensionless current density ($i_{Gl}^*(x^*)$) versus spatial normalized coordinate x^* using $I^* = 1$; (b) mass transfer limitation factor r versus normalized coordinate for different NTU and *Wagner* numbers using $I^* = 1$ [80] (Reprinted with permission from Journal of The Electrochemical Society 2008;155:E193, Copyright 2008 Electrochemical Society).

best reaction conditions possible: Current should be close to unity, NTU should be high and *Wagner* number should be infinite. An infinite *Wagner* number in this case describes a uniform potential distribution over the whole electrode length. This set of variables should yield about 96% product C with very little share of undesired product D (taken from Figure 4.8 (a)). Validation of the simulated data was done in Part 2 of the publication series [8] with a segmented flow reactor (see Section 4.3.3) where all segments were connected to one current generator.

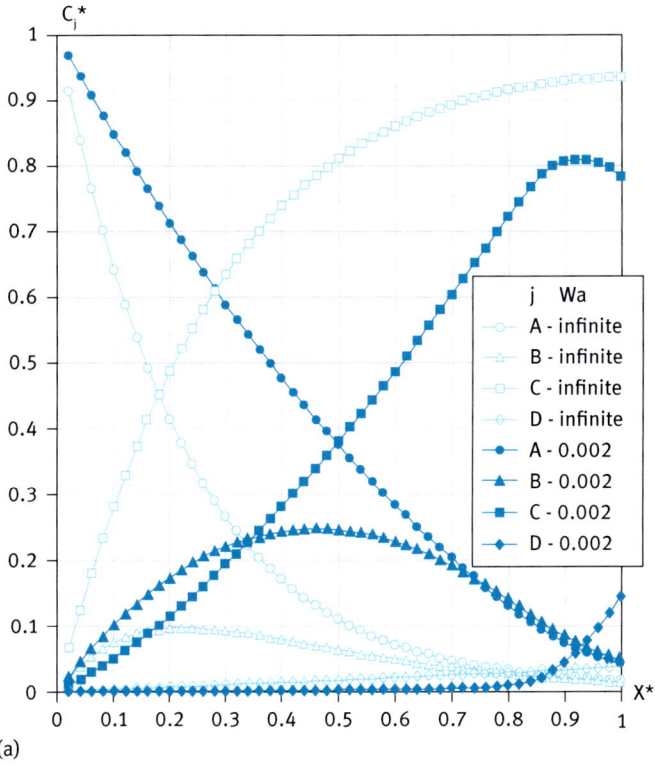

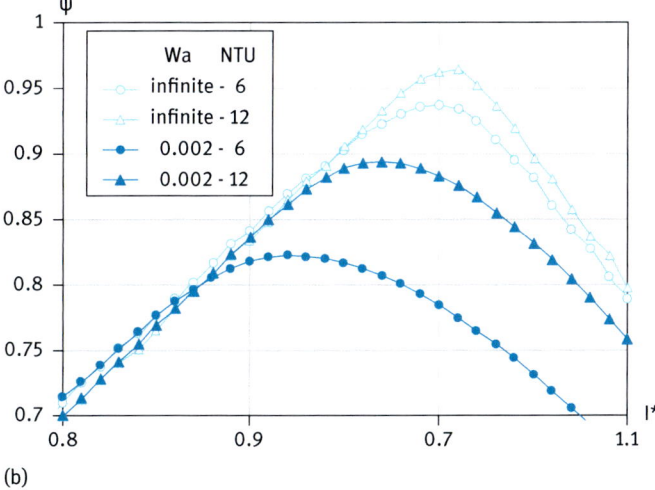

Fig. 4.8: (a) Concentration of species j C_j^* versus spatial reactor coordinate x^* with $NTU = 6$ and $I^* = 1$ for different *Wagner* numbers, (b) Yield Ψ of product C versus current I^* at $x^* = 1$ (reactor outlet) [80]. (Reprinted with permission from Journal of The Electrochemical Society 2008;155:E193, Copyright 2008 Electrochemical Society).

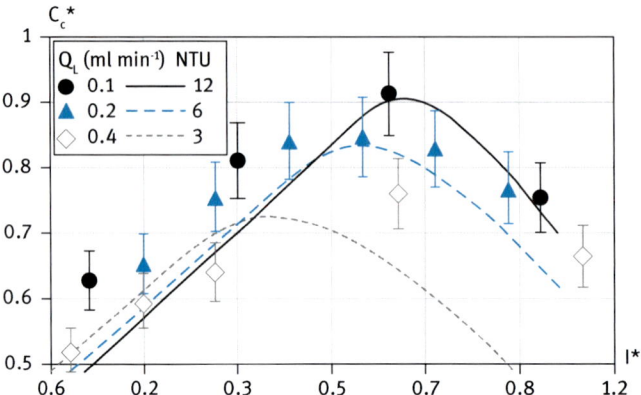

Fig. 4.9: Concentration of product C versus dimensionless current I^* at $x^* = 1$, lines represent simulated values according to Rode et. al. with Wa = 0.01, points denote measured values [81].(Reprinted with permission from Journal of The Electrochemical Society 2008;155:E201, Copyright 2008 Electrochemical Society).

The results in Figure 4.9 (points) show a good conformity to the simulated values (lines); especially for $NTU = 6$ the model can be validated. Higher NTUs tend to inaccuracies with small currents, whereas small NTUs tend to inaccuracies at high dimensionless current values. This can be explained with the "forgotten" aspect of hydrogen formation in the next paragraph. The concentration profiles of components A and D (not shown) are also consistent with the simulated values, whereas the concentration profile of B could not be predicted with the model. One possible reason for this can be found in the uncertainty of the reaction kinetics, which were not investigated but rather postulated.

Another aspect which was not taken into account until now is the cathodic reaction. The emerging hydrogen forms bubbles and blocking certain electrode areas (see Figure 4.10), the higher the spatial coordinate x^* the higher the volume fraction of hydrogen gets through the commencing reaction (see Figure 4.8). In fact, the hydrogen does not have any chemical effect but alters the conductivity and decreases the equivalent conductivity; therefore, emerging hydrogen significantly influences the efficiency of the flow cell. There are now three options to avoid the hydrogen influence: The safest option is using ultra low concentrations, approximately 0.01 M, so that even for complete conversion ($I^* = 1$) the solubility limit of hydrogen in methanol is not reached. Unfortunately, this is not acceptable for industrial-scale synthesis, due to the bad space-time yield. Hence, the second option can be used: The volume fraction of hydrogen over the liquid phase can be as high as 60 for a 1 M solution of 4-methoxytoluene. Calculations conclude that a pressure rise by a factor of 60 enables handling of solutions with concentrations up to 0.75 M without emerging hydrogen. An operating pressure of 60 bar is unusual for conventional industrial synthesis, but due to the small

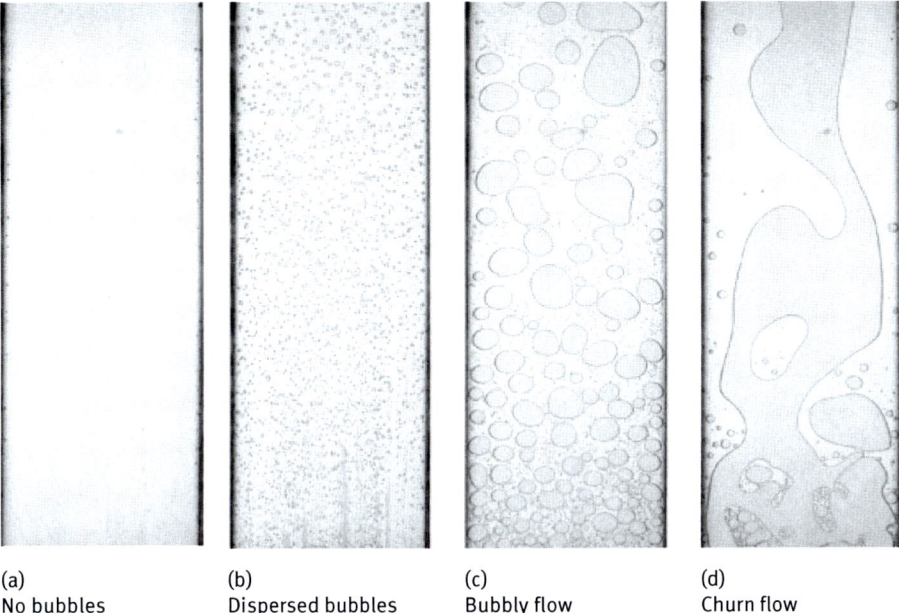

(a) No bubbles (b) Dispersed bubbles (c) Bubbly flow (d) Churn flow

Fig. 4.10: Flow patterns observed for methoxylation of p-methoxytoluene in methanol showing hydrogen formation at atmospheric pressure with an inter-electrode distance of 1 mm and flow velocity of 0.17 m^{-1} (A+B) respectively 0.017 m^{-1} (C+D) and current densities of 400 A m^{-2} (A), 1500 A m^{-2} (B+C) and 5500 A m^{-2} (D) [82]. (Reprinted with permission from Electrochimica Acta 2010;55:8172–8181, Copyright 2010 Elsevier).

volume and big surfaces, high pressures are, generally speaking, not a problem for thin gap flow cells. The early version of those flow cells cannot handle higher pressure because of the epoxy sealing, but the Institute fuer Mikrotechnik Mainz (IMM) invented an electrochemical microreactor (ELMI) with steel housing and pressure stability up to 60 bar [22].

4.3.2 ELMI – microstructured high pressure single pass thin gap flow cell

Figure 4.11 shows the electrochemical microreactor (ELMI V2) designed and built by IMM, which can hold higher pressures and has a built-in cooling system at the cathode (ELMI V2 only). Figure 4.11 (a) shows it assembled and ready for use while Figure 4.11 (b) shows the internal microstructure of the cathode as well as the copper foil contacts for the anode plate. This reactor was built with two different electrode gaps: ELMI V1 (Version 1, not shown) with 25 μm and ELMI V2 (Version 2) with 100 μm and integrated heat exchanger. The explosion view of the ELMI V2 is given in Figure 4.12. Depending on the type of reaction needed, the anode (Label number 7) can be made of platinum, lead/lead oxide or glassy carbon.

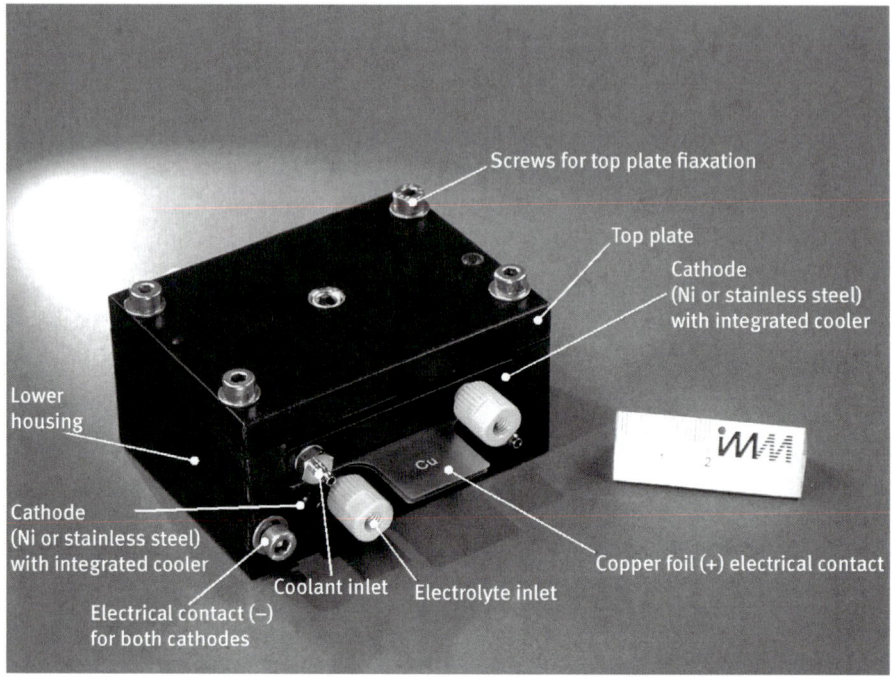

(a)

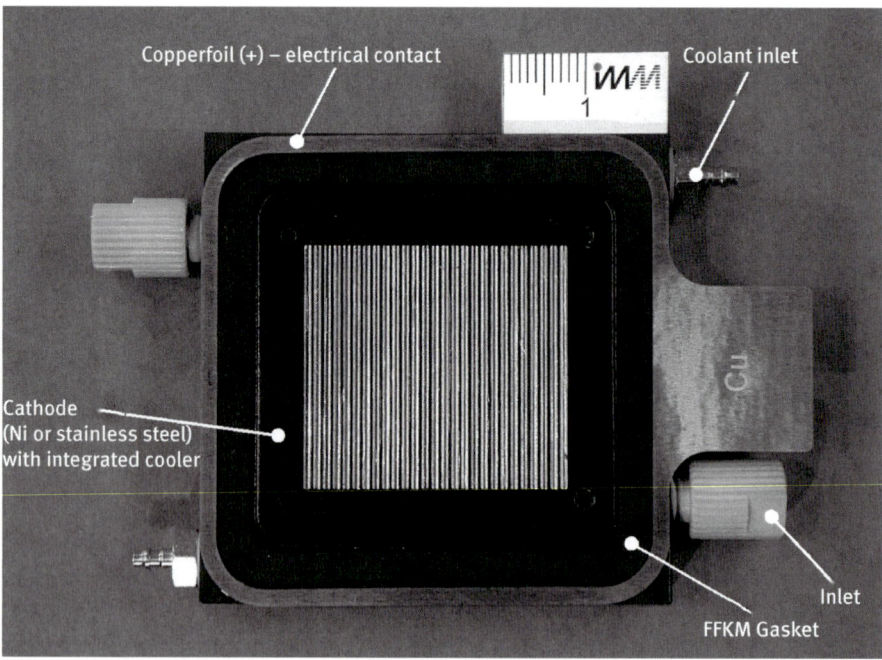

(b)

Fig. 4.11: (a) Photo of the assembled electrochemical microreactor (ELMI) designed and built by Institut fuer Mikrotechnik Mainz (IMM), (b) photo of the microstructured electrode [3, 7].

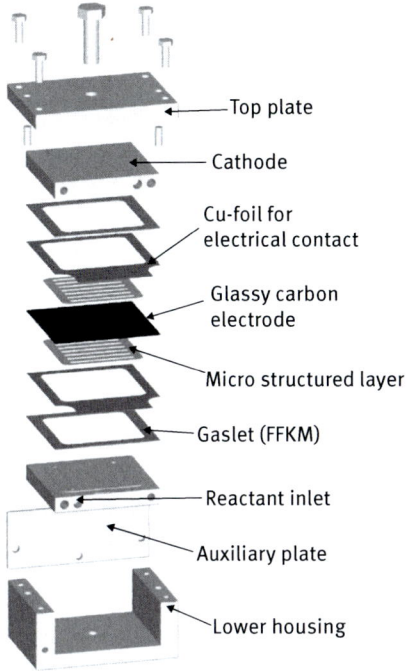

Fig. 4.12: Explosion drawing of ELMI V2 with most important parts labeled (source: IMM).

One of the main problems of organic electrochemistry, the heat generation, is addressed with coolant in- and outlets for temperature control of the cathodes (and the reaction). The generated heat in electrochemical reactors is rather complex and depends on many factors. One of those is the resistance drop (resistance penalty), which depends not only on the distance between the electrodes but also on the specific ionic conductivity:

$$\Delta U = IR_{\text{Drop}} = i\frac{d}{\kappa}. \tag{4.11}$$

ΔU is the potential difference of both electrodes, I is the current applied and R_{Drop} the resistance penalty. Transforming the equation displays the dependency of the resistance penalty towards the current density distribution I, the electrode distance d and the specific ionic conductivity κ. In organic electrochemistry, currents are usually high and conductivity is low, because auxiliary electrolytes are only added when absolutely necessary in the lowest amount possible. Combining this with microengineering and a small distance d still gives a reasonably small ohmic drop which mainly contributes to heat generation [22, 35]. A major engineering concern therefore is the design of a reactor with least possible ohmic resistance, which was pursued with ELMI setup [8].

Table 4.1: Comparison of the industrial BASF process [10] and the ELMI V2 reactor [79].

Process parameter	BASF process	Process with ELMI V2
Flow scheme	Continuous with recycle	Single-pass high-conversion
Inter-electrode gap	0.5×10^{-3}–1×10^{-3} m	0.1×10^{-3} m
educt inlet concentration	10–25 mass%	7.5 mass%
Supporting electrolyte concentration	0.3–3 mass% KF & $NaSO_3R$	0.07 mass% KF
Operating pressure	Atmospheric	5 bar
Current density	300–500 A m^{-2}	510 A m^{-2}
Overall conversion	90–99 %	90 %
Selectivity	85%	92 %
Material yield	> 80%	82 %

To compare ELMI V2 [36–38] with the respective industrial BASF process [15] for synthesis 4-methoxy-benzaldehyde, all important values are compared in Table 4.1. It can be shown that the performance of ELMI is equivalent to the industrial process while decreasing the concentration of supporting electrolyte by a factor of 10. Even completely without supporting electrolyte and educt concentrations as high as 2 M, a yield of 24% with a selectivity of over 90% was achieved. Compared to the single pass, high conversion thin gap cell the yields are slightly lowered, which can be explained with the lack of optimization work. Neither concentrations nor pressure has been optimized and the educt concentrations are higher than 0.01 M, so hydrogen generation occurs and the pressure of 5 bars is not sufficient to suppress gas bubbles evolving.

4.3.3 Segmented thin gap flow cells

- High-pressure stable thin gap flow cells can prevent hydrogen evolution at the cathode. Hence, educt concentrations can be increased to industrially needed levels.
- Microstructuring of the electrodes enables control of the flow pattern and further increase of the surface.
- Segmented thin gap cells allow complex current densities along the channel. For instance, a very high current at the beginning (high educt concentrations) which decreases over the reactor length to prevent side-reactions (the full squares in Figure 4.8 (a) show a possibility to use this setup. Highest yields are generated at 90% of the reactor length).

The remaining problem of electrochemical synthesis in thin gap flow cells with usually small *Wagner* numbers still remains unsolved. At the reactor outlet, where product concentration is high and educt concentration is low, uncontrolled high current density induces undesired side reactions. The calculations in Figure 4.8 (a) and the corresponding experimental data in Figure 4.9 show, that for small *Wagner* numbers the product yield starts to decrease at approximately 90% of the reactor length. This problem can be solved if the current densities are adjusted separately at certain reactor positions. For that reason, innovative electrode configurations were invented, using alternating anode cathode setups or segmented cathodes all referring to one anode [39].

Figure 4.14 (a) shows a reactor equipped with an interdigital-type electrode setup, where cathode and anode are alternately arranged, promoting the mass-transfer coefficient and limiting the ohmic resistance [4–6]. A disadvantage of this reactor is the still unsolved problem with high current density at high conversion. Figure 4.14 (b) shows a multisectioned flow-through reactor, where all anodes are connected to one central cathode in the vicinity of the reactor outlet. Each slice can be connected to a separated current generator for optimizing the electrical current flow. A laboratory scale synthesis of D-arabinose via oxidation of gluconate investigating different anode slice numbers (1, 2, 5 and 10) was done by Vallières (see Section 4.1) [40].

Figure 4.13 and Figure 4.14 (c) present the segmented thin gap flow cell. This cell combines the advantages of ELMI V2 with the ability to control the current density

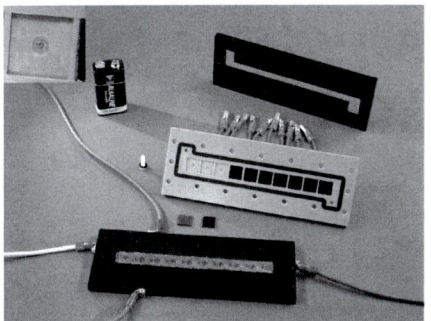

Fig. 4.13: Electrochemical microreactor with 10 individually controllable electrodes in PEEK/stainless steel housing, the upper left corner shows the gold connection of the electrode (source: IMM).

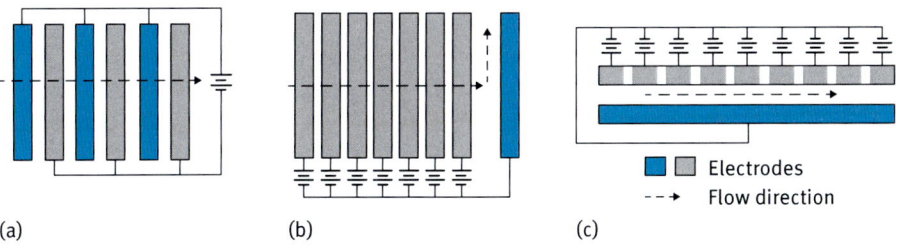

Fig. 4.14: Alternative electrode designs for configurable spatial resolved current density alteration: (a) alternating anode cathode setting; (b) cumulated spatial electrodes with one counter-electrode at the end, all electrodes can be set with different voltages and currents; (c) similar setup to (b) with smaller and equally distand gap between each electrode counter-electrode pair [78].

at almost any place in the reactor. Complex current density schemes are feasible with this reactor type [41]. If this segmented cell is used as high conversion single pass electrochemical reactor, the main advantage is an increased current density at the reactor inlet to speed up the reaction and allow higher educt concentrations as well as a decrease in current density near the reactor outlet to avoid side-products and therefor maximize the product yield. A disadvantage of this reactor, however, is the high number of current generators needed to enable the current density profiles.

A reactor model for the segmented flow cell was also developed, showing an upper limit in the concentration if the reactor is driven adiabatic [41]. However, applying isothermal conditions gives the opportunity to raise the number of products. The theory concludes that for higher conversions, the segmented cell reactor can supply up to eight times higher current densities [41].

Figure 4.13 shows a ten-fold segmented flow cell designed by IMM. The top housing of the flow cell is made of PEEK with milled holes for the electrodes with an insulating gap between. There are two different anode settings available: stainless steel or glassy carbon. The top left side of Figure 4.13 shows a magnified image of the cathode connection with golden spring connectors. Due to the very high current densities, gold springs are used to gently press against the electrodes and maximize the conductivity.

4.4 Electrochemistry in microreactors

4.4.1 Direct product synthesis

Usually, electrochemistry in organic synthesis is used to directly synthesize products using reduction or oxidation reactions. In some cases, hydrolysis or similar workup reactions may be applied to finalize the product. The most common products of everyday use are adipodinitrile for nylon synthesis with 300,000 tons per year as well as bleached montan wax from brown coal in the range of 11,000 tons per year. The difference in tons per year from the largest to the second largest process shows that there is still room for improvement in direct large-scale synthesis.

However, fine chemicals are produced in a variety of processes, mostly by BASF, Hoechst, 3M and Otsuka with a production volume between 10 and 100 tons per year [42, 43]. From this point of view, microreaction technology should be successfully applied for process intensification and optimization of the process' ecological balance. A few examples from literature are given below.

The anodic oxidation of 4-methylanisol to 4-methoxybenzaldehyde, an industrial process from BASF with 3500 tons per year [14, 44], was investigated using the ELMI V2. The process parameters could be optimized to 98% conversion with 82% yield and a selectivity of 95%. The electrolyte concentration could be decreased by a factor of 5, while the mass concentration also had to be reduced by a factor of 3 to avoid hydrogen evolution. Using numbering-up instead of scale-up techniques, 3500 annual tons can be produced with approximately 400 reactors in parallel; further savings can be realized through the consolidation of current generators [22]. The decrease in electrolyte concentration not only reduces costs for the educts but also facilitates further purification processes.

Another direct syntheses is described by Vallières and Matlosz [40]. A multisectioned flow-through electrode configuration (see Figure 4.14 (b)) was used to synthesize D-arabinose from sodium gluconate via electrochemical oxidation. The influence of the number of segments in the graphite anode was investigated using 1, 2, 5 or 10 segments. Gluconate was fed into the reactor as an aqueous solution with a concentration of 0.01 M and 1.5 M sodium acetate as electrolyte. The current efficiency could be increased by a factor of two to a maximum of 80% at $1\,A\,m^{-2}$. The achievable yield of D-arabinose after 1 h of cycling in the reactor was 70%, while the remaining 30% was unconverted starting material.

One of the main advantages of microreactors is that they rely on minimized electrolyte concentrations needed to conduct synthesis. This can be explained with the small electrode gaps and the smaller diffusion paths between the electrodes due to partially overlapping electrochemical double layer. Most of the published articles therefore deal with the effect of electrolyte free synthesis routes.

4.4.2 Electrolyte free synthesis

Electrolyte free synthesis is an outstanding possibility for performing electrochemical syntheses in microreactors. This process is impossible with standard chemical reactors such as CSTRs. Hence, most of the publications concerning electrochemistry in microreactors deal with the influence of remarkably decreased electrolyte concentration on the reaction on the electrode surfaces.

- The application of electrochemistry in industry is usually direct synthesis, problems are high electrolyte concentrations and difficult workup procedures.
- The use of microreactors makes it possible to severely reduce the electrolyte concentration and contemporaneous increase the yield.
- Microreactors enable the field of electrochemically activating reagents which is only possible due to the excellent reaction control.
- The cation pool method generates a reservoir of activated carbocations which can be stored over a short time period.
- The cation flow method generates the ions in flow which are directly processed with other reagents.

The first publication dealing with electrolyte free synthesis with microstructured electrodes in thin gap cells was published by Belmont and Girault in 1994 [45]. They examined the methoxylation of furan with a self-developed cell using printed alumina electrodes with platinum coating and a distance of 250 μm. The cell was placed inside a tank for 26 to 50 h and the flow-through was enabled due to rising hydrogen gas at the cathode. The optimum voltage for methoxylation was determined to be 3 V, which decreases the power consumption by 50% compared to the packed bed bipolar cell. The current efficiency is nearly equal between the two reactors at circa 87–90% while the material yield varies from 78% for the thin gap cell and 90% for the packed bed cell. The decrease in yield can be explained through the not fully accomplished screening and optimization [45].

The optimization of the process was done by Horii and Marken in 2002 with a true flow cell using a 80 μm spacer in a potentiostatic mode [23]. Platinum and glassy carbon electrodes were used at 3 mA cm^{-2}, a voltage of 6.8 V and a flow rate of 0.1 mL min^{-1}. As reference electrode a saturated calomel microscale electrode was applied. The yield obtained with this setup was under the given conditions 98%. A complete design of experiment covering the electrochemical aspects was conducted, comparing glassy carbon and platinum electrode materials as well as voltage and current density. A rise in voltage from 6.8 V to 7.6 V and a drop in yield below 40% were observed with a setup using only glassy carbon electrodes, with Pt/Pt electrodes the yield decreased further remarkably below 10%.

He and Haswell investigated the cathodic dimerization of 4-nitrobenzylbromide by electrochemical reduction of 4-nitrobenzyl bromide in a 160 μm and 320 μm gap flow cell without added supporting electrolyte [46]. It is shown that smaller electrode gaps do not necessarily provide more product and higher residence times with lower current and do not result in higher conversion. The current efficiency of 80% was found for all reactions. The electrode area for the 160 μm flow cell was 12.5 mm^2, for the 320 μm flow cell 15 mm^2. Concentration of 4-nitrobenzylbromide was fixed at 10 mM for all experiments. Table 4.2 summarizes the influence of flow rate and electrode gap of the conversion and yield. He and Haswell showed the possibility of carrying out electrolyte free synthesis with high purities and conversion. This product could be sold

Table 4.2: Parameter investigation of 4,4'dinitrodibenzyl synthesis, voltage of 4 V for 20 μL min^{-1} and 4.8 V for 40 20 μL min^{-1}, R–R equals 4,4'dinitrodibenzyl, R–H is short for nitrobenzene [87].

Electrode gap	Current (mA)	Flow rate (μm/min)	Conversion (%)	Product distribution	
				R-R	R-H
160	0.8	20	99	68	32
160	1.3	40	95	69	31
320	0.6	20	70	93	7
320	1.2	20	91	91	9
320	0.6	40	58	94	6
320	2.5	40	92	91	9
320	2.5	40	100	76	24

Table 4.3: Process data for electrolysis of dimethylfumarate with a variety of aromatic bromides [88].

R^2	Yield (R^1-R^2) (%)
Benzyl bromide	98
4-Methoxybenzyl bromide	94
4-Methylbenzyl bromide	94
1,4-Dibromobenzene	99
(1-Bromoethyl)benzene	98

directly after solvent evaporation as technical or even purum grade material without further workup.

Further investigations were done with an improved reactor of 45 mm² electrode area with platinum foil electrodes and an electrode gap of 160 µm [47]. The double channel version for a parallel operation was built by a separated spacer layer with two equal channels with 3 × 21 mm² using the same 8 × 15 mm² platinum electrodes. A later version was equipped with four electrodes in total, so that two completely different reactions could be accomplished in this reactor. The screening was done by using 5 mM dimethylfumarate solution in DMF and a 5 mM aromatic bromide solution at a flow rate of 10 µL min^{-1}. The data given in Table 4.3 show a variety of different benzyl-bromide derivates which were coupled with yields above 94%. Altering of the reactor to the double channel version with 20 µL min^{-1} flow rate and two big electrodes for both channels gives comparable results with yields mostly above 97% in both cells. Using two and four independent electrodes with flow rates of 20 and 40 µL min^{-1} respectively confirms the results for numbering up: All yields are above 93%, however using a quadruple cell with only two big electrodes as anode and cathode the yields dropped down to 20–70%. This numbering up study shows that it is possible to divide large electrodes into small separate operating reactors for the same reaction and still obtain very good yields. These electrodes can also be operated by only one current generator, which enables further savings in investments [46, 47].

4.4.3 Activation of chemicals

One of the most famous areas of application for chemistry in microreactors is the activation of cheap bulk chemicals or protecting group free synthesis through superior mixing and temperature control [48].

These activation reactions have been well investigated for stable basic chemicals with highly reactive species where harsh reaction conditions do not enable those reactions in normal CSTRs. However, investigating of reactions with two basic chemicals and electrochemical activation of one educt with a quick reaction step has only been done by the group of Yoshida [49–51]. He proposed two types of activation processes:

The first one generates a stable ion pool which can be collected and mixed with other educts for coupling [50], the second one generates a carbocation via anodic oxidation which cannot be stored and must be reacted immediately with other educts to form carbon-carbon bonds. This method was also used as a new synthesis path for cationic polymerization at low temperatures without the presence of dormant species for molecular weight control. The weight control was achieved by altering flow rates and equivalent concentrations of monomer versus cationic species [49, 51].

4.4.3.1 Cation pool method

The cation pool method describes an electrochemical operation, where a normal H-cell is employed for generating cations without nucleophiles. The H-cell type has a membrane between cathode and anode, so that reduction and oxidation can be carried out in different chambers. However, the main advantage of decreasing the supporting electrolyte concentration cannot be done with those types due to the cm-range distances between the electrodes.

The H-cell generates and accumulates the cations, which then can be pumped through reactors under low temperatures. In this case N-acylium ions were generated by oxidation of trimethylsilylated carbamates at −78 C. The main advantage of this process, compared to the industrial acid-promoted reactions, is the irreversibility of the cation generation reaction. Acid promoted generations are usually reversible and yield not only the cation but also a nucleophile which can react with the N-acylium ion [52–69]. For high concentrations and high purities of N-acylium ions this process can be used.

Later coupling of those ions with different styrenes (e.g., p-chloro-styrene) show low yields with conventional mixing techniques. Using a multilamination micromixer under the same conditions shows up to 50% higher yields compared to a conventional mixing with a stirrer. Taking this into account and developing a separated chamber electrochemical cell leads to the cation flow method.

4.4.3.2 Cation flow method

The cation flow method is a flow adaption of the cation pool method mentioned above. A microreactor with separated chambers has been built to achieve an immunity against evolving hydrogen traveling to the anode and to easily separate both electrolytes (see Figure 4.15).

For calculating equivalents before the nucleophilic reaction, a Fourier transform-infrared spectroscopy (FT-IR) cell was built in the setup after the electrochemical cell. This enables the exact dosage of nucleophile to suppress side-reactions and minimize the cation consumption. For example, a Friedel–Crafts type monoalkylation reaction of 1,3,5-trimethoxybenzene was chosen. In batch, this reaction yields about 37% monoalkylation and 32% dialkylation product, whereas the flow process yields

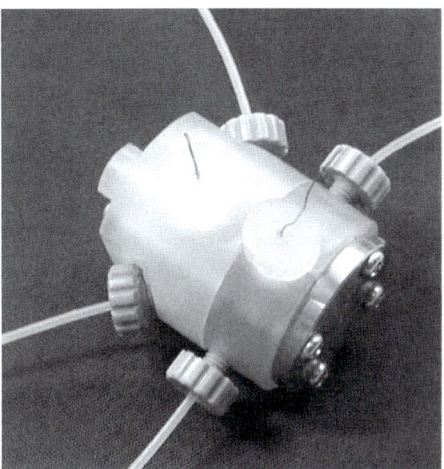

Fig. 4.15: Electrochemical microflow reactor with a PTFE membrane for separation of anodic and cathodic reaction [8] (source: J.-I. Yoshida, Department of Synthetic Chemistry and Biological Chemistry, Kyoto University).

92% monoalkylation and 4% dialkylation product. Sequential alkylation with different N-acylium ions was also tested with thiophene. N-butyl-acylium ions were coupled in the first step followed by N-cyclohexyl acylium ions with a total yield of 64%. Comparing this to batch reactions is not possible, because the monoalkylated thiophene is more reactive, yielding mostly dialkylated 2,5-di-(N-butylcarbamate) thiophenes [49].

[4+2] cycloaddition reactions and cationic polymerizations are also possible using the cation flow method. Both reactions rely on the same educts but different equivalents. For cycloaddition, a 1 : 1 mixture is most efficient, while for polymerization the ion is the initiator, suggesting low equivalents of ions versus the polymer backbone educt. This was demonstrated using the styrene derivatives under batch in cation flow reaction conditions. While batch cycloaddition reactions only yield 20% of addition product and 80% styrene polymer, the cation flow method yields 79% cycloaddition product and 20% low molecular weight by-product. The polymerization reaction using cation flow yields a very good control of the molecular weight with a polydispersity index PDI = 1.14. Compared to conventional cationic living polymerization without dormant species this is a very good value. Conventional PDI (polydispersity index) control usually is achieved with a dominant dormant species in living polymerizations, prolonging reactions to hours or even days. The polymerization using cation flow yields number averages of molecular weights up to 7000 g mol^{-1} within minutes and can simply be quenched with addition of (functional) quenchers in a second micromixer [49].

Electrochemical activation of stable educts to induce reactions which are not possible in batch reactors was demonstrated in this chapter, leaving a lot of room for further experiments for this uncommon technique. It was shown that cationic polymerizations can be controlled as well as Friedel–Crafts monoalkylation of benzene derivates with a possibility to alkylate a second time with different alkylation reagents resulting in heterogeneously dialkylated benzene derivates, which are also not available in batch reactors in high yields.

4.5 Ionic liquids in electrochemistry

Further optimizations in continuous electrochemistry address the limited potential window of common solvents.

- Usable solvents are a big problem, conductivity of organic solvents is low and adding electrolyte is expensive, furthermore the solvent itself decomposes at certain voltages.
- Ionic liquids (ILs) can solve most organic compounds while obtaining a very good electric conductivity.
- ILs can be designed to the actual need (melting point, potential window, toxicity, etc.) and are a promising alternative for organic solvents with electrolytes.
- However, workup of ILs is difficult and the solvent is very expensive.

Organic reactions are usually accomplished in organic solvents like tetrahydrofuran or toluene because the solubility of the educts and products in water is small. The solubility is only one of the core problems with aqueous organic reactions; the potential window is limited through the low decomposition potential of water at the electrodes and the amphoteric character which could hydrolyze the reactants. On the other hand, the disadvantage of organic solvents compared to water is the low conductivity and the cost of organic soluble supporting electrolyte and the solvent itself. During the last decades, a new group of solvents has evolved called the ionic liquids (ILs). These are organic salts with a melting point below 100 °C, which can cover a variety of properties. Through adjusting the chemical formulas of anion and cation melting point, polarity, solubility and electrochemical stability as well as viscosity can be changed almost at will. Hence, another name for this group of solvents is designer solvents [70].

A subdivision of those liquids is called room temperature ionic liquids (RTILs). These kind of ILs have the same adjustable properties mentioned above with the limitation of a melting point below room temperature (25 °C). A common example of RTILs is 1-ethyl-3-methylimidazolium trifluoromethane-sulfonate ([EMIM]CF_3SO_3) with a melting point of about −12 °C.

Applying those liquids in lithium ion batteries is already in research, however applying ILs in electrochemical synthesis is not well investigated [71]. One of the disad-

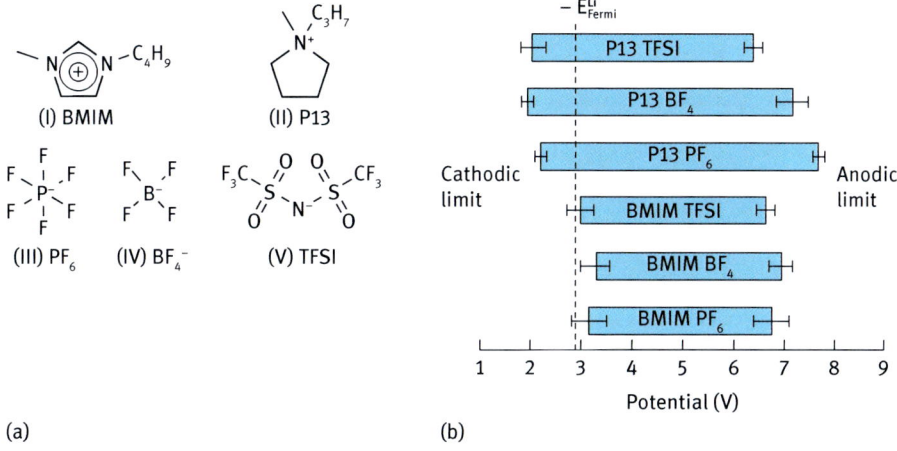

Fig. 4.16: (a) Cations and anions investigated in DFT calculations concerning potential windows. (I) 1-butyl-3-methylimidazolium (BMIM); II N,N-propylmethylpyrrolidinium (P13). Anions: III Hexafluorophosphate (PF6); IV Tetrafluoroborate (BF4); V bis-(trifluoromethylsulfonyl)imide (TFSI). (b) calculated limits of the ILs against vacuum level; the error bars denote a 95% confidence interval and the lithium Fermi level is also indicated [9]. (Reprinted with permission from Chemistry of Materials 2011;23:2979-86. Copyright 2011 American Chemical Society).

vantages of designer solvents is the design itself. Hundreds of possibilities to combine certain anions and cations result in hundreds of different properties (e.g., potential window, viscosity, acidity ...) so that one can easily get lost in the overflow of options. However, trends are usually recognizable by a chemists' eye. Ping Ong and Ceder simulated a variety of ILs concerning their potential windows using a combination of molecular dynamics (MD) density functional theory (DFT) calculations [72].

Those DFT calcucations are found to be in close agreement with other experimental findings even though the environmental effects could not be taken into account [73]. The results from Ping Ong and Ceder suggest that ILs could replace most of conventional organic solvents, at least regarding the potential window stability. Due to the nature of ionic liquids, they also exhibit a very high conductivity, also enabling the further reduction of electrolyte or, in the case of electrolyte free processes, enabling higher current rates [74].

A major disadvantage of ionic liquids is their strong hygroscopic behavior. O'Mahony and Compton investigated the effect of water on the potential limits of ILs, revealing a dramatic influence [75].

Table 4.4 shows a significant reduction in potential limits, even if the IL is used as purchased with water content in ranges of several 100 to several thousand ppm. Vacuum drying can reduce those water residues to values in the lower hundred ppm for water immiscible ILs. Those values were determined by *Karl–Fischer* titration [75].

Table 4.4: Electrochemical window limits of different ILs in vacuum dried condition, atmospheric (undried, used as received) condition and in saturated moisture environment at given temperatures. $J = 1$ equals $1\,\text{mA}\,\text{s}^{-1}$ $J = 5$ equals $5\,\text{mA}\,\text{s}^{-1}$ [89].

RTIL	Water content (ppm)			E(vac dried) (V)		E(untreated) (V)		E(wet at 298 K) (V)	
	Vac-dried	Un-treated	wet	$j = 1\,\text{ma}\,\text{cm}^{-2}$	$j = 5\,\text{ma}\,\text{cm}^{-}$	$j = 1\,\text{ma}\,\text{cm}^{-}$	$j = 5\,\text{ma}\,\text{cm}^{-}$	$j = 1\,\text{ma}\,\text{cm}^{-}$	$j = 5\,\text{ma}\,\text{cm}^{-}$
$[P_{14.6.6.6}][NTf_2]$	—	328	—	5.2	5.4	3.2	4.0	1.8	2.6
$[C_4\text{mpyrr}][NTf_2]$	133	406	11407	4.2	5.2	3.0	4.6	2.0	2.8
$[C_6\text{mim}][FAP]$	203	417	3068	4.6	4.8	3.0	3.7	2.5	3.3
$[C_4\text{mim}][NTf_2]$	144	491	5680	4.3	5.0	2.9	3.5	2.8	3.7
$[C_4\text{dmim}][NTf_2]$	295	504	9868	4.7	5.2	3.3	4.4	2.9	3.8
$[N_{6.2.2.2}][NTf_2]$	167	1150	8200	4.7	5.4	2.3	4.6	2.2	3.4
$[C_4\text{mim}][PF_6]$	268	2119	24194	4.8	5.0	3.9	4.5	2.6	3.1
$[C_2\text{mim}][NTf_2]$	105	3385	19940	4.2	4.6	2.8	2.7	2.6	3.2
$[C_4\text{mim}][BF_4]$	119	5083	∞	4.6	5.1	2.4	4.0	2.0	3.6
$[C_4\text{mim}][I]$	1050	11349	∞	2.0	2.2	1.9	2.2	1.6	1.8
$[C_4\text{mim}][OTf]$	250	15227	∞	4.2	5.1	3.3	4.8	2.7	4.2
$[C_6\text{mim}][Cl]$	2231	61049	∞	3.0	3.3	2.9	3.3	2.4	2.8

Given the advantages of greater potential windows and higher conductivities as well as the disadvantages of usually higher viscosities and a high impact of air moisture, there are only a few publications which follow up on ionic liquids in electro organic chemistry.

Sekiguchi and Fuchigami investigated the electro-polymerication of pyrrole at the anode with 1-ethyl-3-methilimidazolium trifluoromethanesulfonate ($[EMIM]CF_3SO_3$) [76]. The polymerization was done with the cyclic potential-scanning method with $-0.8\,\text{V} \sim +1.2\,\text{V}$ versus a saturated calomel electrode at a scanning rate of $0.1\,\text{V}\,\text{s}^{-1}$. Not only was synthesis in 100% $[EMIM]CF_3SO_3$ investigated, but also 0.1 M solutions of $[EMIM]CF_3SO_3$ in H_2O and in Acetonitrile were used for electro-polymerizations. The resulting film thickness, the thickness standard deviation as well as the electrochemical conductivity, were measured. Results show that in the aqueous solution the roughness is very high whereas the conductivity is very low, suggesting that most of the delocalized π-systems of the pyrroles are broken and the polymerization was not proceeding in a controlled manner. The acetonitrile solution of $[EMIM]CF_3SO_3$ yields very good values with thickness standard deviations as well as electrochemical capacities comparable to the pure $[EMIM]CF_3SO_3$. However, the electro conductivity is lowered by four orders of magnitude, again still suggesting that the delocalized π-system is broken. Polymerizing pyrrole in pure comparable to the pure $[EMIM]CF_3SO_3$ yields thinner films with highly conductive and very even manner which can be simply repelled from the anode with reduction at −0.8 V. The remaining pyrrole in the IL can be extracted with chloroform and reused without further purification [76].

Study questions
4.1. Electrochemical processes are widely used. Please provide some examples.
4.2. Please describe the principle of an electrochemical cell, indicate electrodes and explain physical and chemical processes caused by electrical forces at the electrode surfaces. Use H_2O as electro-active species in acidic and alkaline solution to form hydrogen and oxygen.
4.3. Why is it necessary to add a so-called conducting salt into an electrolyte to perform a conventional electrolysis? Is it possible to reduce the amount of conducting salt when a thin-gap cell is used?
4.4. Please provide a definition of the *Wagner*-number.
4.5. For what purpose is the so-called *Karl–Fischer* titration (KFT) necessary? Please explain briefly how this titration is performed.
4.6. Why is the application of a high current density disadvantageous?
4.7. Please make a suggestion how to avoid unwanted electrode reactions.

Further readings
- Schmidt VM. Elektrochemische Verfahrenstechnik. Weinheim: Wiley-VCH; 2003. (German)
- Sharma BK. Electro Chemistry. 5th ed. Meerut: Krishna Prakashan; 1997.
- Hamann CH, Hamnett A, Vielstich W. Electrochemistry: Wiley; 2007.
- Zoski CG. Handbook of Electrochemistry: Elsevier; 2007.
- Ref. #16: Hessel V, Hardt S, Löwe H. Chemical micro process engineering: Wiley Online Library; 2004.
- Ref. #22: Ziogas A, Kolb G, O'Connell M, et al. Electrochemical microstructured reactors: Design and application in organic synthesis. Journal of Applied Electrochemistry 2009;39.
- Ref. #49: Yoshida J-I. Flash chemistry using electrochemical method and microsystems. Chemical communications (Cambridge, England) 2005:4509-16.
- Ref #50: Suga S, Nagaki A, Tsutsui Y, Yoshida J-I. "N-acyliminium ion pool" as a heterodiene in [4 + 2] cycloaddition reaction. Organic letters 2003;5:945-7.
- Ref #56: Yoshida Ji, Suga S. Basic concepts of "cation pool" and "cation flow" methods and their applications in conventional and combinatorial organic synthesis. Chemistry-A European Journal 2002;8:2650-8.

Bibliography

[1] Venkatesh S, Tilak BV. Chlor-alkali technology. Journal of Chemical Education 1983;60:276.
[2] Janz GJ. Molten Salts Handbook: Elsevier Science; 2013.
[3] Kiehne HA. Battery Technology Handbook: Taylor & Francis; 2003.
[4] Larminie J, Dicks A, McDonald MS. Fuel cell systems explained: Wiley Chichester; 2003.
[5] Kissinger PT, Refshauge C, Dreiling R, Adams RN. An Electrochemical Detector for Liquid Chromatography with Picogram Sensitivity. Analytical Letters 1973;6:465–477.
[6] Houghton RW, Kuhn AT. Mass-transport problems and some design concepts of electrochemical reactors. Journal of Applied Electrochemistry 1974;4:173–190.

[7] Goodridge F, Scott K. Introduction to Electrochemical Engineering. Electrochemical Process Engineering: Springer US; 1995:1–16.
[8] Pickett DJ. Electrochemical reactor design. 2nd ed. New York: Elsevier; 1979.
[9] Goodrige F, Scott K. Electrochemical Process Engineering – A guide to the Design of electrolytic Plant. New York & London: Plenum Press; 1995.
[10] Pletcher D, Walsh FC. Industrial Electrochemistry: Springer; 1990.
[11] Jüttner K. Encyclopedia of electrochemistry. 1st ed. Weinheim: Wiley-Vch; 2004.
[12] Beck F. Organic electrosynthesis. Weinheim: VCH Verlagsgesellschaft mbH; 1987.
[13] Wendt H, Kreysa G. Electrochemica engineering, science and technology in chemical and other industries. Berlin: Springer-Verlag; 1999.
[14] Hamann CH, Hamnett A, Vielstich W. Electrochemistry. Weinheim: Wiley-VCH; 1998.
[15] Püttner H. Industrial electroorganic chemistry. 4th ed. New York: Marcel Dekker; 2001.
[16] Hessel V, Hardt S, Löwe H. Chemical micro process engineering: Wiley Online Library; 2004.
[17] Schultze JW, Tsakova V. Electrochemical microsystem technologies: From fundamental research to technical systems. Electrochimica Acta 1999;44.
[18] Lund H, Hammerich O. Organic electrochemistry. 4th ed. New York: Marcel Dekker; 2002.
[19] Degner D. Organic electrosynthesis in industry. Berlin: Springer-Verlag; 1988.
[20] Baizer MM, Nonaka T, Park K, Saito Y, Nobe K. Electrochemical conversion of 2,3-butanediol to 2-butanone in undivided flow cells: A paired synthesis. Journal of Applied Electrochemistry 1984;14.
[21] Küpper M, Hessel V, Löwe H, et al. Micro reactor for electroorganic synthesis in the simulated moving bed-reaction and separation environment. Electrochimica Acta 2003;48:2889–2896.
[22] Ziogas A, Kolb G, O'Connell M, et al. Electrochemical microstructured reactors: Design and application in organic synthesis. Journal of Applied Electrochemistry 2009;39.
[23] Horii D, Atobe M, Fuchigami T, Marken F. Self-supported paired electrosynthesis of 2,5-dimethoxy-2,5-dihydrofuran using a thin layer flow cell without intentionally added supporting electrolyte. Electrochemistry Communications 2005;7.
[24] He W, Wang J, Shao H, Zhang J, Cao C-n. Novel KOH electrolyte for one-step electrochemical synthesis of high purity solid K_2FeO_4: Comparison with NaOH. Electrochemistry Communications 2005;7.
[25] Paddon CA, Pritchard GJ, Thiemann T, Marken F. Paired electrosynthesis: Micro-flow cell processes with and without added electrolyte. Electrochemistry Communications 2002;4.
[26] Horii D, Atobe M, Fuchigami T, Marken F. Self-Supported Methoxylation and Acetoxylation Electrosynthesis Using a Simple Thin-Layer Flow Cell. Journal of The Electrochemical Society 2006;153.
[27] Attour A, Rode S, Bystron T, Matlosz M, Lapicque Fo. Oxidation kinetics of 4-methylanisole in methanol solution at carbon electrodes. Journal of Applied Electrochemistry 2007;37.
[28] Wendt H, Bitterlich S, Lodowicks E, Liu Z. Anodic synthesis of benzaldehydes – II. Optimization of the direct anodic oxidation of toluenes in methanol and ethanol. Electrochimica Acta 1992;37.
[29] Wendt H, Bitterlich S. Anodic synthesis of benzaldehydes – I. Voltammetry of the anodic oxidation of toluenes in non-aqueous solutions. Electrochimica Acta 1992;37.
[30] Rode S, Attour A, Lapicque Fo, Matlosz M. Thin-Gap Single-Pass High-Conversion Reactor for Organic Electrosynthesis. Journal of The Electrochemical Society 2008;155:E193.
[31] Petrii OA, Nazmutdinov RR, Bronshtein MD, Tsirlina GA. Life of the Tafel equation: Current understanding and prospects for the second century. Electrochimica Acta 2007;52.
[32] Gutman EM. Can the Tafel equation be derived from first principles? Corrosion Science 2005;47.
[33] Burstein GT. A hundred years of Tafel's Equation: 1905–2005. Corrosion Science 2005;47.

[34] Pignotti A, Shah RK. Effectiveness-number of transfer units relationships for heat exchanger complex flow arrangements. International Journal of Heat and Mass Transfer 1992;35.
[35] Bersier BM, Carlsson L, Bersier J. Electrochemistry V. Berlin: Springer-Verlag; 1994.
[36] Ziogas A, Löwe H, Küpper M, Ehrfeld W. Electrochemical microreactors: A new approach for microreaction technology. Berlin: Springer-Verlag; 2000.
[37] EU FP6 Project IMPULSE. 2005.
[38] Attour A, Dirrenberger P, Rode S, Ziogas A, Matlosz M, Lapicque F. A high pressure single-pass high-conversion electrochemical cell for intensification of organic electrosynthesis processes. Chemical Engineering Science 2011;66.
[39] Attour A, Rode S, Lapicque Fo, Ziogas A, Matlosz M. Thin-Gap Single-Pass High-Conversion Reactor for Organic Electrosynthesis. Journal of The Electrochemical Society 2008;155:E201.
[40] Vallières C, Matlosz M. A Multisectioned Porous Electrode for Synthesis of D-Arabinose. Journal of The Electrochemical Society 1999;146.
[41] Rode S, Altmeyer S, Matlosz M. Segmented Thin-Gap Flow Cells for Process Intensification in Electrosynthesis. Journal of Applied Electrochemistry 2004;34.
[42] Ming Z, Jürgen H. Electropolymerization of Pyrrole and Electrochemical Study of Polypyrrole. 2. Influence of Acidity on the Formation of Polypyrrole and the Multipathway Mechanism. The Journal of Physical Chemistry B 1999;103.
[43] Pfluger P, Street GB. Chemical, electronic, and structural properties of conducting heterocyclic polymers: A view by XPS. The Journal of Chemical Physics 1984;80.
[44] Ohno H. Electrochemical aspects of ionic liquids. New York: Wiley; 2005.
[45] Belmont C, Girault HH. Coplanar interdigitated band electrodes for electrosynthesis. Journal of Applied Electrochemistry 1994;24.
[46] He P, Watts P, Marken F, Haswell SJ. Electrolyte free electro-organic synthesis: The cathodic dimerisation of 4-nitrobenzylbromide in a micro-gap flow cell. Electrochemistry Communications 2005;7.
[47] He P, Watts P, Marken F, Haswell S. Scaling out of electrolyte free electrosynthesis in a micro-gap flow cell. Lab on a chip 2007;7:141–143.
[48] Yoshida J-I. Flash Chemistry: Fast Organic Synthesis in Microsystems. London: Wiley; 2008.
[49] Yoshida J-I. Flash chemistry using electrochemical method and microsystems. Chemical communications (Cambridge, England) 2005:4509–4516.
[50] Suga S, Nagaki A, Tsutsui Y, Yoshida J-I. "N-acyliminium ion pool" as a heterodiene in [4 + 2] cycloaddition reaction. Organic letters 2003;5:945–947.
[51] Seiji S, Masayuki O, Kazuyuki F, Jun-ichi Y. "Cation Flow" Method: A New Approach to Conventional and Combinatorial Organic Syntheses Using Electrochemical Microflow Systems. Journal of the American Chemical Society 2001;123.
[52] Naota T, Nakato T, Murahashi S. Novel method for α-substitution of amines via N-Methoxycarbonyl-α-t-butyldioxyamines. Tetrahedron Lett 1990;31:7475–7478.
[53] Moeller KD. Synthetic applications of anodic electrochemistry. Tetrahedron 2000;56:9527–9554.
[54] Shono T. Electroorganic chemistry in organic synthesis. Tetrahedron 1984;40.
[55] Yoshida J, Suga S, Suzuki S, Kinomura N, Yamamoto A, Fujiwara K. Direct Oxidative Carbon–Carbon Bond Formation Using the Cation Pool Method. 1. Generation of Iminium Cation Pools and Their Reaction with Carbon Nucleophiles. J Am Chem Soc 1999;121.
[56] Yoshida Ji, Suga S. Basic concepts of "cation pool" and "cation flow" methods and their applications in conventional and combinatorial organic synthesis. Chemistry-A European Journal 2002;8:2650–2658.
[57] Suga S, Suzuki S, Yoshida J-i. Reduction of a "Cation Pool": A New Approach to Radical Mediated CC Bond Formation. Journal of the American Chemical Society 2002;124:30–31.

[58] Suga S, Okajima M, Fujiwara K, Yoshida J-i. "Cation Flow" Method: A New Approach to Conventional and Combinatorial Organic Syntheses Using Electrochemical Microflow Systems. Journal of the American Chemical Society 2001;123:7941–7942.

[59] Suga S, Suzuki S, Yamamoto A, Yoshida J-i. Electrooxidative generation and accumulation of alkoxycarbenium ions and their reactions with carbon nucleophiles. Journal of the American Chemical Society 2000;122:10244–10245.

[60] Shimizu T, Tanino K, Kuwajima I. An intramolecular hetero Diels–Alder reaction of α-(alkynylsiloxy)aldimine derivatives. Tetrahedron Lett 2000;41:5715–5718.

[61] Veerman JN, Klein J, Aben RWM, et al. Solid-phase synthesis of piperidines by N-acyliminium ion chemistry Eur J Org Chem 2002.

[62] Sun P, Sun C, Weinreb SM. Stereoselective Total Syntheses of the Racemic Form and the Natural Enantiomer of the Marine Alkaloid Lepadiformine via a Novel N-Acyliminium Ion/Allylsilane Spirocyclization Strategy. J Org Chem 2002;67.

[63] Metais E, Overman LE, Rodriguez MI, Stearns BA. Halide-Terminated N-Acyliminium Ion–Alkyne Cyclizations: A New Construction of Carbacephem Antibiotics. J Org Chem 1997;62:9210–9216.

[64] Okitsu O, Suzuki R, Kobayashi S. Efficient Synthesis of Piperidine Derivatives. Development of Metal Triflate-Catalyzed Diastereoselective Nucleophilic Substitution Reactions of 2-Methoxy- and 2-Acyloxypiperidines. J Org Chem 2001;66:809–823.

[65] Schierle-Arndt K, Kolter D, Danielmeier K, Steckkhan E. Electrogenerated Chiral 4-Methoxy-2-oxazolidinones as Diastereoselective Amidoalkylation Reagents for the Synthesis of β-Amino Alcohol Precursors. European Journal of Organic Chemistry 2001:2425–2433.

[66] Zhang J, Wei C, Li C. Cu(I)Br mediated coupling of alkynes with N-acylimine and N-acyliminium ions in water. J Tetrahedron Lett 2002;43:5731–5733.

[67] Vanier C, Wagner A, Mioskowski C. Preparation of Resin-Bound N-(α-Methoxyalkyl)amides: An Advantageous Use of Solid-Phase Chemistry for the Handling of Unstable Precursors of the Versatile N-Acyliminium Ions. Chem Eur J 2001;7:2318–2323.

[68] Zaugg HE. α-Amidoalkylation at Carbon: Recent Advances – Part II. Synthesis 1984;85:181–212.

[69] Zaugg HE, Martin WB. Organic Reactions. New York: John Wiley & Sons Inc; 1965.

[70] Freemantle M. DESIGNER SOLVENTS. Chemical & Engineering News 1998;76.

[71] Taige M, Hilbert D, Schubert TJS. Mixtures of Ionic Liquids as Possible Electrolytes for Lithium Ion Batteries. Zeitschrift für Physikalische Chemie 2012;226.

[72] Ping Ong S, Andreussi O, Wu Y, Marzari N, Ceder G. Electrochemical Windows of Room-Temperature Ionic Liquids from Molecular Dynamics and Density Functional Theory Calculations. Chemistry of Materials 2011;23.

[73] Howlett PC, Izgorodina EI, Forsyth M, MacFarlane DR. Electrochemistry at Negative Potentials in Bis(trifluoromethanesulfonyl)amide Ionic Liquids. Zeitschrift für Physikalische Chemie 2006;220.

[74] Calado MS, Diogo JCF, Correia da Mata JL, Caetano FJP, Visak ZP, Fareleira JMNA. Electrolytic Conductivity of Four Imidazolium-Based Ionic Liquids. Int J Thermophys 2013;34:1265–1279.

[75] O'Mahony AM, Silvester DS, Aldous L, Hardacre C, Compton RG. Effect of Water on the Electrochemical Window and Potential Limits of Room-Temperature Ionic Liquids. Journal of Chemical & Engineering Data 2008;53:2884–2891.

[76] Sekiguchi K, Atobe M, Fuchigami T. Electropolymerization of pyrrole in 1-ethyl-3-methylimidazolium trifluoromethanesulfonate roomtemperature ionic liquid. Electrochemistry Communications 2002;4.

[77] Schmidt VM. Elektrochemische Verfahrenstechnik. Weinheim: Wiley-VCH; 2003.

[78] Rode S, Altmeyer S, Matlosz M. Segmented Thin-Gap Flow Cells for Process Intensification in Electrosynthesis. Journal of Applied Electrochemistry 2004;34.

[79] Ziogas A, Kolb G, O'Connell M, et al. Electrochemical microstructured reactors: Design and application in organic synthesis. Journal of Applied Electrochemistry 2009;39.
[80] Rode S, Attour A, Lapicque Fo, Matlosz M. Thin-Gap Single-Pass High-Conversion Reactor for Organic Electrosynthesis. Journal of The Electrochemical Society 2008;155:E193.
[81] Attour A, Rode S, Lapicque Fo, Ziogas A, Matlosz M. Thin-Gap Single-Pass High-Conversion Reactor for Organic Electrosynthesis. Journal of The Electrochemical Society 2008;155:E201.
[82] Bouzek K, Jiřičný V, Kodým R, Křišťál J, Bystroň T. Microstructured reactor for electroorganic synthesis. Electrochimica Acta 2010;55:8172–8181.
[83] Ziogas A, inventor Verfahren zur Durchführung elektrochemischer Reaktionen in einem Mikroreaktor. DE patent EP1217099. 2001.
[84] Yoshida J-I. Flash chemistry using electrochemical method and microsystems. Chemical communications (Cambridge, England) 2005:4509–4516.
[85] Ong SP, Andreussi O, Wu Y, Marzari N, Ceder G. Electrochemical Windows of Room-Temperature Ionic Liquids from Molecular Dynamics and Density Functional Theory Calculations. Chemistry of Materials 2011;23:2979–2986.
[86] Püttner H. Industrial electroorganic chemistry. 4th ed. New York: Marcel Dekker; 2001.
[87] He P, Watts P, Marken F, Haswell SJ. Electrolyte free electro-organic synthesis: The cathodic dimerisation of 4-nitrobenzylbromide in a micro-gap flow cell. Electrochemistry Communications 2005;7:918–924.
[88] He P, Watts P, Marken F, Haswell S. Scaling out of electrolyte free electrosynthesis in a micro-gap flow cell. Lab on a chip 2007;7:141–143.
[89] O'Mahony AM, Silvester DS, Aldous L, Hardacre C, Compton RG. Effect of Water on the Electrochemical Window and Potential Limits of Room-Temperature Ionic Liquids. Journal of Chemical & Engineering Data 2008;53:2884–2891.

Part II: **Cutting-edge applications in advanced and functional materials**

L. Zane Miller, Jeremy L. Steinbacher, and D. Tyler McQuade

5 Synthesis of materials in flow – principles and practice

5.1 Introduction

Continuous materials synthesis began over a century ago in the fiber industry. Despite this early start, the excellent control over mass transfer, heat transfer, and fluid alignment observed in flow has only recently begun to impact material synthesis broadly. The use of flow methods by academic chemists and materials scientists has emerged from the efforts of Whitesides and Weitz with their use of poly(dimethylsiloxane) (PDMS) for device fabrication and the microreactor community led by IMM in Germany and Yoshida in Japan. In this chapter, we will highlight the fluidic properties and examples demonstrating how flow reactors have advanced materials synthesis.

5.2 Unique properties of microreactors

5.2.1 Mixing

Rapid mixing can often have a very positive impact on reaction rate and yield [1]. Mixing in batch reactors is slow relative to microreactors (i.e., round bottom flasks and larger batch reactors) due to inhomogeneities in the flow fields resulting from stirring (active mixing). As fluid approaches the stirrer, convection is induced, yielding turbulence and chaotic mixing. The forces generating convection are lessened with increasing distance from the stirrer. As a result, the majority of a batch reactor remains poorly mixed leading to idle zones. These areas of poor mixing can be sites of side-reactions or unproductive chemistry.

Flow reactors achieve mixing via multiple strategies and enable mixing to occur on timescales ranging from seconds to microseconds [2]. The most common and simplest mixing approach uses a simple T-mixer (Figure 5.1) where two fluid streams are forced through a single channel. The rate of mixing in a T-mixer is controlled by flow rate. At slow flow rates, where interfacial tension and viscosity control fluid behavior, mixing is controlled solely by diffusion. As flow rate increases, convection becomes possible, and at high flow velocities, mixing is proportional to flow rate. While T-mixers are most common, many other strategies exist and are generally divided into devices that induce turbulence through inclusion of complex paths or those that increase diffusional mixing by maximizing surface area between co-flowing fluids. Complex paths are achieved by impinging flows or by flowing fluids through spiraled,

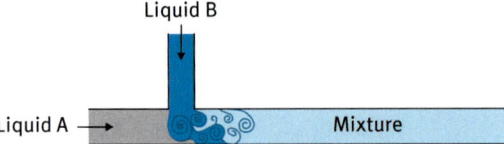

Fig. 5.1: Two liquids (*A* and *B*) colliding and mixing in a T-junction.

coiled, or wound tubing [3, 4]. Increased surface area is achieved in multilamellar devices where different fluids are separated into thin layers enabling rapid diffusional mixing. The speed of mixing is directly related to the surface-to-volume ratios of these devices which is typically much larger (30,000 $m^2\ m^{-3}$) than that found in conventional batch reactors (100 $m^2\ m^{-3}$). In these high surface-to-volume devices, mixing can occur in microseconds. These surface-to-volume ratios also have an effect on thermal and mass transport, which is discussed below.

5.2.2 Thermal and pressure control

Flow reactors provide enhanced control over reaction temperature and pressure relative to batch reactors. Heat transfer is faster and more uniform due to the smaller dimensions. Also, a much wider range of pressures can be achieved through use of flow-based back-pressure regulators. These two features enable reactions to be run at either constant temperature or at temperatures above the boiling point of the solvent used to perform the reaction. Precise control over temperature allows optimization of reactions that have multiple pathways (i.e., kinetic or thermodynamic), a feature that is significant for polymerizations where rates of propagation are similar to rates of termination. Rapid temperature changes [5] and heat exchange coefficients up to 25 $kW\ m^{-2}K^{-1}$ are possible depending on the materials and heat exchanger used [6–8]. Capacity to run reactions at high pressures in microreactors provides the opportunity to perform polymerization reactions that would otherwise be unsafe in batch [9]. For example, past explosions occurring in polyethylene research might have been avoided if performed under continuous conditions [10]. While mixing, heat transfer, and pressure attributes are significant advantages of using continuous reactors, the truly revolutionary advantage that microreactor systems offer materials chemists is the control over fluid behavior and structure.

5.2.3 Fluid behavior

Courses describing dimensionless parameters are well-covered in undergraduate engineering curricula. For most synthetic chemists, on the other hand, solvents are viewed as a medium that is not well understood beyond polarity effects, density,

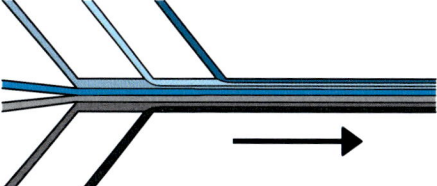

Fig. 5.2: Seven fluids co-flowing within a microfluidic channel illustrating low Reynolds number conditions or laminar flow.

and occasionally viscosity. Many synthetic chemists are surprised to learn that fluids behave differently as dimensions shrink. Because this wide knowledge gap exists between engineers and chemists, the following sections try to summarize a semester's worth of fluid dynamics in language that is approachable to chemists.

The unique properties of solvents in small dimensions was demonstrated by Whitesides' in 1999 where seven dyed solutions were co-flowed parallel to each other in a PDMS microreactor (Figure 5.2).

In this case, convective mixing of the solutions was avoided because the solutions were in a low Reynolds number regime. The only mixing that could occur under these carefully selected conditions was diffusive mixing. While many dimensionless parameters are known, Reynolds, Capillary, and Weber numbers are three of the most important when performing materials synthesis in flow. In the following sections we will provide a primer of fluid behavior, and then progress into various synthetic applications.

5.2.3.1 Reynolds number, Re

Turbulent and chaotic mixing dominates in the round bottom flask; however, moving to channels with length scales on the order of nm to mm can negate the turbulent and chaotic mixing completely. The transition from turbulent to smooth flow takes place when viscous effects begin to dominate over inertial effects in the fluid. Representing this ratio of inertial to viscous forces, the Reynolds number is defined as:

$$\mathrm{Re} = \frac{\rho v l}{\eta} \quad (5.1)$$

where ρ is density, v is mean velocity, l is a characteristic length scale (the diameter of a tube, for example), and η is dynamic viscosity. Fluids with Re below ~ 2000 exhibit a laminar flow (without turbulence) profile. The exact point of the transition from turbulent to laminar flow depends strongly on the channel geometry. By thinking of fluid as a series of arrows moving in a channel, one can obtain a feel for fluid movement on different length scales (we assume that the fluid and its velocity are held constant). Figure 5.3 shows arrows moving in a large channel where, though still inevitably constrained by the channel walls, there is enough free space to travel in any direction.

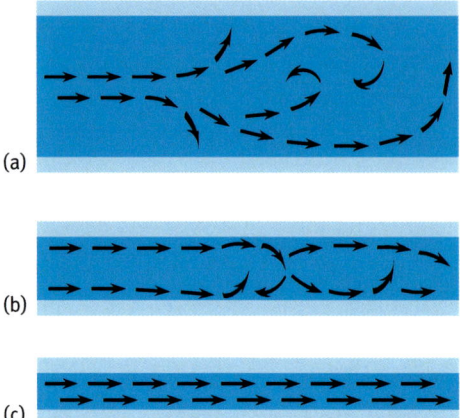

Fig. 5.3: An illustration of the effects of channel diameter on fluid flow. (a) With a large channel, the arrows, representing fluid streamlines, can move in any direction without feeling the constraint of the walls. (b) Shrinking the channel causes the arrows to move mostly in the same direction, but the channel is still large enough to turbulently mix. (c) Condensing the channel down further, the arrows are now forced to move in one direction without deviating (laminar flow).

Constricting the channel causes the arrows to align in the same direction, but the channel is still sufficiently large enough to enable turbulent mixing. Shrinking the channel further, the arrows now co-flow in one direction without signs of turbulence. The arrows represent the smallest turbulent motions available to the fluid based on the values of the variables in the Reynolds number equation (Equation (5.1)). When the channel dimensions become sufficiently smaller, the turbulence is completely quelled and the fluid moves in parallel streamlines.

An interface between two fluids, created in the laminar flow profile, offers the opportunity to perform polymerization at the interface where the two fluids come together. Figure 5.4 displays a general example similar to an early report by Whitesides and coworkers where a polymer membrane was generated by combining oppositely charged polyelectrolytes [11]. Several strategies exist, ranging from those in which the polymerization occurs with the polymer stuck to the floor and ceiling of the channel, to axisymmetric or coaxial systems that prevent one phase from touching the walls of the device so the polymer may be removed later.

An extensional flow field capable of aligning large molecules can be created using a fluid jet (Figure 5.5). While use of elongated polymers in flow has yet to see wide use in materials synthesis, Maeda has shown that elongated DNA can impact primer hybridization, suggesting that elongation impacts intermolecular interactions [12].

The consequences of fluids at low Reynolds number are significant from a materials synthesis perspective. For example, a liquid jet can be changed from a thin finger of fluid to discrete droplets depending on the parameters of the system (Figure 5.6). The process responsible for breaking the stream into droplets is called a Rayleigh–

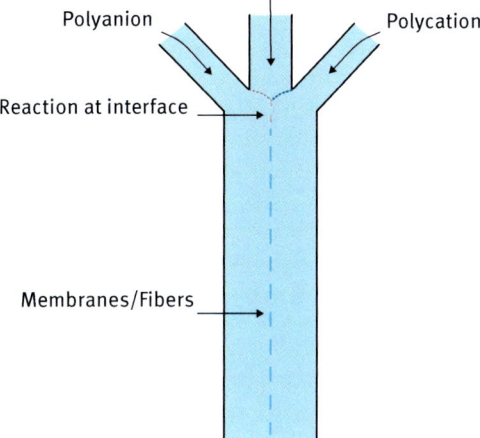

Fig. 5.4: An illustration of using the interface between two laminarly flowing liquids to template polymeric materials. Here, oppositely-charged polyelectrolytes precipitate at the interface producing membranes and fibers.

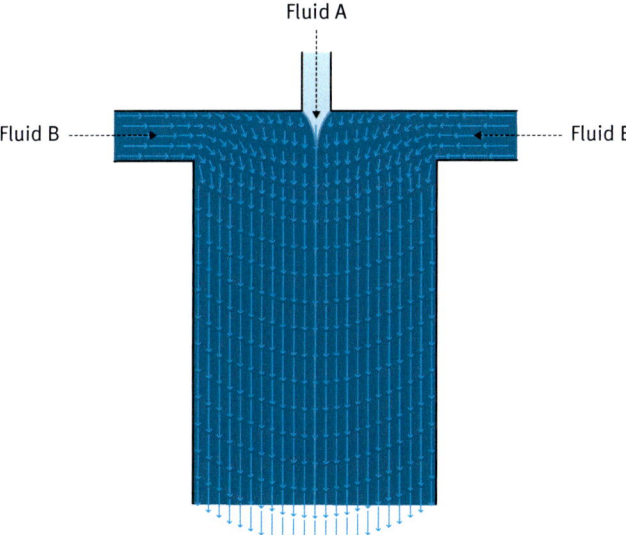

Fig. 5.5: An illustration of the flow fields created as two fluids are forced through a mixer (note the jet at the center of the large channel).

Plateau instability (when a jet thins, interfacial forces start to dominate) [13, 14]. These flow-generated emulsions are typically monodisperse [15]. Downstream from where the droplets are produced, the emulsions can be organized into various structures.

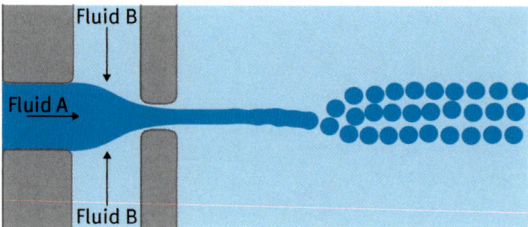

Fig. 5.6: An illustration of droplets forming during break up of a fluid jet caused by Rayleigh–Plateau instability. Perturbations in the jet become larger than the circumference of the fluid jet leading to droplet formation as the surface area is minimized.

5.2.3.2 Capillary number, Ca

Materials synthesis in flow often involves the combination of two solutions to form droplets or to perform reactions at the interface. When two fluids combine, one of the major considerations is the surface tension between the two fluids and the viscosity of the two fluids. The ratio of surface tension to viscous forces is represented as the capillary number:

$$\mathrm{Ca} = \frac{v\eta}{\gamma} \tag{5.2}$$

where v is mean velocity, η is dynamic viscosity, and γ is the interfacial surface tension. Changing the capillary number can have predictable impact on the types of droplets that form within a microfluidic device [16]. Figure 5.7 illustrates how changing capillary number can alter how two fluid streams interact. In this case, two channels containing dyed water enter a third channel containing a fluorinated solvent carrier phase.

In Figure 5.7 (a) the capillary number is low – surface tension dominates the behavior. Here, the aqueous phases entering the channel from top and bottom are stiff and do not bend under pressure from the orthogonally flowing carrier fluid. As a result, the aqueous solutions collide, mix, and snap off as a single plug that undergoes convective mixing as the plug moves along the channel. In Figure 5.7 (b), the flow rate has been increased relative to Figure 5.7 (a) resulting in an increase in the $v\eta$ term of the capillary number equation. The result is that the carrier fluid can now bend the aqueous phases such that alternating plugs of each aqueous phase forms. Further increases in flow rate of the carrier solution result in elongation of the aqueous flows along the walls of the channel resulting in smaller droplets snapping off at the termini of the aqueous flows. This example shows that with small variations in fluid properties and flow rates, one can create a range of monodisperse emulsions with different sizes. These emulsions provide an opportunity to create monodisperse beads or capsules more easily than trying to optimize a batch emulsion polymerization to be monodisperse.

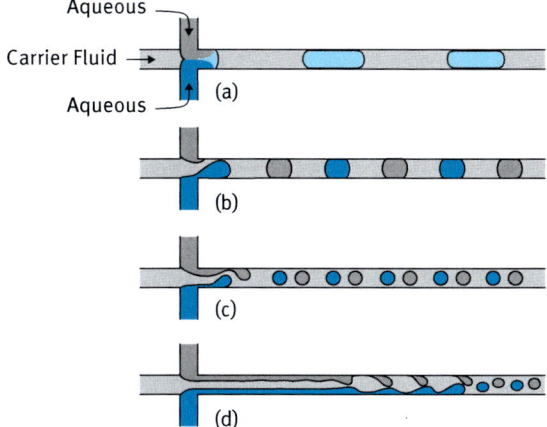

Fig. 5.7: Illustration of the effect of capillary number on the observed flow regime (Adapted from [16]). (a) At low flow rates, surface tension dominates over viscous forces, and single plug is formed from the combination of both water phases (red phase + blue phase yields the purple plug). (b) Increasing the flow rate relative to (a) leads to an increase in the $v\eta$ term of the capillary number, the water streams take turns producing plugs. (c, d) Further increases in flow rate of the carrier solution result in elongation of the aqueous flows along the walls of the channel resulting in smaller droplets snapping off the termini of the aqueous flows.

Changing capillary number not only enables droplet size and mixing variations, but also enables droplets to be organized into a wide range of geometries. A wonderful example of this organization was demonstrated when water and oil collide using a simple T-junction (Figure 5.8) [17]. By varying the ratio of water to oil pressure and total pressure, droplets could be formed and then assembled into a wide array of structures. These structures range from single monodisperse droplets to beautiful patterns of connected droplet strings. This is a treasure trove for any polymer scientist trying to create new fibers, beads, capsules, or particles of novel shape. As we will describe below and has recently been reviewed, outstanding new materials are now being realized [18, 19].

5.2.3.3 Weber number, We

Weber number is commonly used to describe the formation of bubbles and droplets. Though Ca is the most important dimensionless descriptor of droplet formation, We is also valuable for predicting droplet formation and designing experiments which deal with dripping and jetting regimes [20, 21]. The Weber number represents the ratio of inertial forces to surface tension forces:

$$\text{We} = \frac{\rho v^2 l}{\sigma} \qquad (5.3)$$

(a) Aqueous (inside)

(b) Oil (middle)

(c) Aqueous (outside)

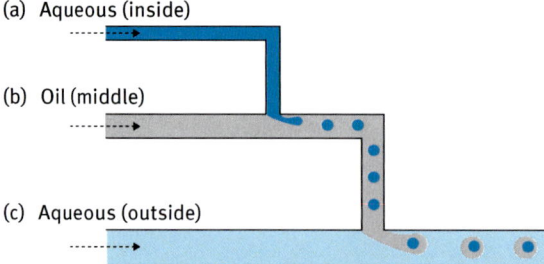

Fig. 5.8: Illustration of the formation of double emulsions (droplets within droplets) from a microfluidic T-junction device.

where ρ is the density (kg/m^3), v is the mean velocity (m/s), l is the characteristic length of the droplet (m), and σ is the interfacial surface tension (N/m). The Capillary number of the outer fluid and Weber number of the inner fluid can not only be used to predict the flow regime within the multiphase system (i.e., jetting, widened jetting, and droplet formation) but also the droplet diameter and production rates [20].

5.3 Synthesis of materials in flow

Flow synthesis can be a powerful tool for producing novel materials. Different device geometries and techniques have harnessed the physical properties of fluids flowing in narrowed dimensions (micro-/mesofluidics). This opportunity for enhanced control has been seized by many researchers to produce exciting materials that are unobtainable or very difficult to realize using batch methods. In the following sections, we provide discussion of instances where microreactors were applied to synthesize many types of materials including: linear polymers, beads, disks, other solid polymeric materials, capsules, fibers, concluding with nanoparticles and other inorganic materials. Our attention is focused on the benefits provided by flow-based platforms in each instance.

5.3.1 Linear polymers

Microreactors have emerged as useful tools to achieve greater control of synthesis of known polymers and to discover new polymers [22, 23]. Cationic, anionic [24, 25], ring-opening, and free radicals are polymerization mechanisms that have benefited from the rapid mixing and precise temperature control found in a microreactor. Yoshida was one of the first to show that microreactors can yield improved polymerizations and began by demonstrating that N-methoxycarbonyl-N-(trimethylsilylmethyl)butylamine decomposed to cation **2**. The most exciting attribute of this work is that unstable **2**

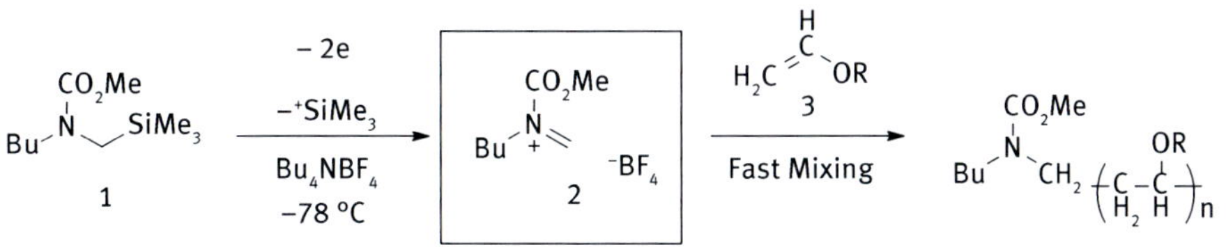

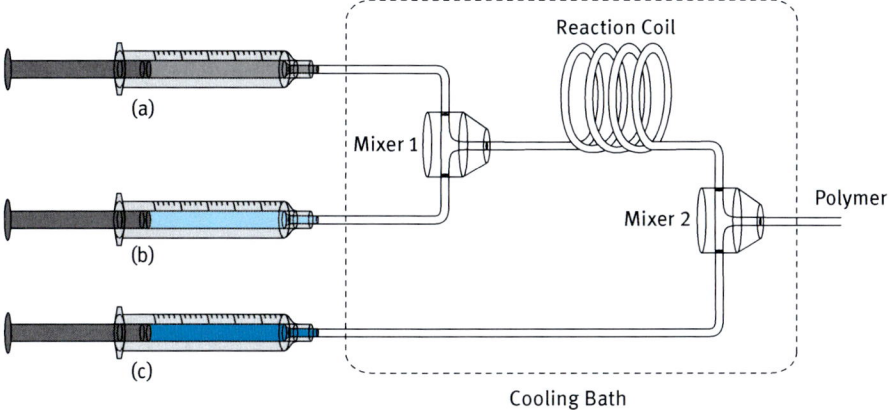

Fig. 5.9: Reaction scheme of a cationic polymerization demonstrating generation and subsequent use of an unstable intermediate.

Fig. 5.10: Illustration of a microfluidic system for exothermic polymerization reactions. The initiator (a) and monomer (b) mix at Mixer 1, react through within the reactor coil submerged in a cooling bath, and are quenched by the addition of solvent (c) at Mixer 2.

could be formed and subsequently used without isolation within a microreactor (Figure 5.9). By rapidly mixing **2** with a vinyl ether, a cationic polymerization results with superior Poly Dispersity Indexes (PDIs) relative to an analogous batch reaction (Figure 5.10) [26]. The authors also demonstrated that PDIs were dependent on flow rate. At lower flow rates, mixing is slower, providing slow initiation rates and resulting in larger PDIs. Yoshida *et al.* proceeded to demonstrate that radical polymerizations provided similar results [27]. The authors note that highly exothermic polymerizations show improved control in the microreactors relative to batch because microreactors dissipate heat much more rapidly, preventing unwanted side-reactions and/or termination reactions.

Flow reactors are not perfect and some reactor configurations do not yield improved control over radical reactions. Experiments with reversible addition-fragmentation chain transfer (RAFT) polymerization and nitroxide-mediated radical polymerizations yielded higher PDIs in tubular reactors compared to batch reactors [28–30]. In both of these cases, however, mixing was performed prior to introduction into the microreactor, and the reactions were performed within microemulsions flowed through the microreactors. In one case, PDI erosion was attributed to axial dispersion of the plugs and thus a distribution of residence times. As the plugs became large, the PDIs approached the batch values, supporting the axial dispersion model. The conclusions that should be drawn from radical polymerization lessons are that microreactors are not a silver bullet for all synthetic chemistry. Reactions that are sensitive to mass and heat transfer will benefit from being run in microreactors, and reactions that are not sensitive might be worse in microreactors because one introduces the complexity of fluid behavior. That being said, living radical polymerizations can benefit from being run in flow if the rate of heating is used as the mechanism of control. For example, RAFT polymerizations to yield Poly(N-isopropylacrylamide) polymers were faster compared to batch while exhibiting similar polydispersities and enabled immediate chain end functionalization with sugars for use in glycopolymers [31].

The benefits of microreactors are not limited to cationic and radical polymerizations as homo-polymers of N^ε-benzyloxycarbonyl-L-lysine, alanine, leucine, or glutamic acid (NCA) were formed with high control as well (Figure 5.11) [32]. In this example, an NCA was combined with triethylamine in a PDMS microreactor similar to the one observed in Figure 5.12 [33]. In all cases, the microreactor provided slightly higher molecular weights and much better PDIs. NCA polymerizations are known to dependent on mixing. The authors note that their device can produce ~ 15 g day^{-1}. Those unfamiliar with microreactors often worry that these devices cannot produce significant volumes of material. Unlike a batch reactor, a continuous reactor can run for minutes, hours, days or even weeks so even a device that produces 500 mg hour^{-1} will yield 12 g day^{-1}. Furthermore, two major strategies can be used to increase scale: (1) simply increase the channel size of the reactor. Doubling the channel diameter has a nonlinear impact in reaction scale; or (2) increase the number of reactors running in parallel. This approach is referred to as "numbering up". For example, if the aforementioned authors placed 1000 microreactors in parallel they could produce 5 Mg year^{-1}. While these volumes are small compared to the industrial production of polyethylene, they are large enough to meet the demand of many specialty polymers.

Microreactors are also useful for systematically varying the molecular weight of both homopolymers and block copolymers. Beers *et al.* carried out controlled radical polymerizations where monomer ratios and monomer-initiator ratios were altered to produce gradients of molecular weights or monomer ratios in copolymers [34]. Homopolymer synthesis was achieved using a two channel microreactor with the two channels meeting in an actively mixed reservoir. This example was one of the first examples of a microreactor with an active mixer (magnetic stir bar). The mixing reservoir empties into a "residence time" or "reactor module" channel. Polymerizations were performed by filling one inlet channel with monomer and catalyst premixed and the other inlet with the initiator. Both were dissolved in 50 : 50 water/methanol. Initiator:monomer ratios were varied by changing relative concentrations, and the poly-

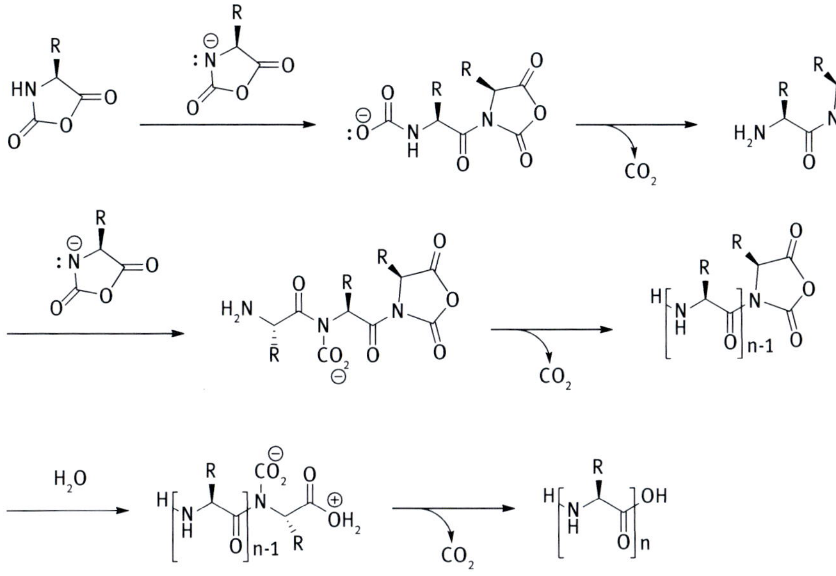

Fig. 5.11: Reaction scheme for the formation of homo-polymers of NCA.

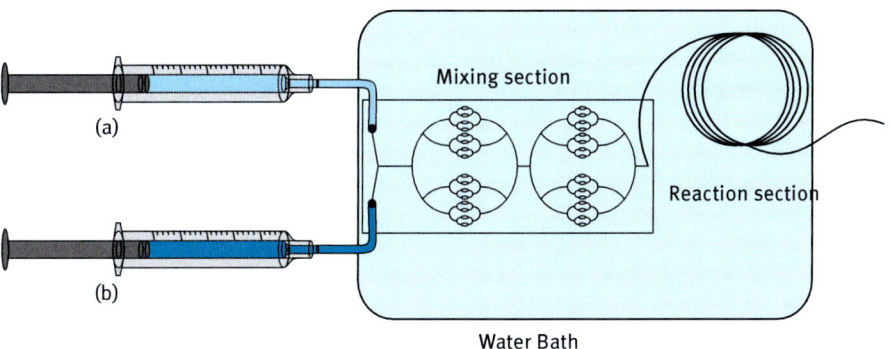

Fig. 5.12: Illustration of microreactor system composed of a PDMS micromixer and PTFE microtubes used in the polymerization reactions. The temperature of the mixer and reaction regions is controlled using a water bath.

merization at a fixed initiator:monomer ratio was then monitored as a function of flow rate. It was observed that the molecular weight decreased as a function of flow rate, which is inversely proportional to residence time. The PDIs under all conditions were very good, ranging from 1.19–1.32. Using a similar approach, but with three channels instead of two, various PEO-*b*-PHPMA were prepared with excellent control over conversion, PDI, and block ratios [35]. It should be noted that microreactors can often yield unintuitive results. Here, flow rate and reaction time were correlated which fits the normal model that longer reaction times produce higher yields. Often in microreactor systems, the opposite is observed. One must constantly be vigilant to remember that mixing and flow rate is also correlated. In many cases, slowing the flow rate in a microreactor-based reaction lowers the yield.

Beers et al., has also made surface-grafted polymer brushes with thickness gradients using a microreactor-based appraoch [36]. Gradients were produced by flowing monomer slowly over a surface activated for atom-transfer radical polymerization (ATRP). The surface at the inlet received more monomer than that near the outlet producing a linear gradient of polymer graft heights moving from the inlet end to the outlet end. The gradient forms because the dimensions of the system held the fluids in a laminar flow regime (low Reynolds number) resulting in little diffusive mixing on the time scale of polymerization. This example illustrates that microreactors offer the opportunity to create structures that would be impossible to produce under batch conditions.

5.3.2 Beads, disks, and other solid polymeric materials

Using emulsions to template polymeric structures has been a productive field for decades. Despite the many advances made in this community, producing monodisperse emulsions was often more art than science. The demonstration that microreactors could be used to produce monodisperse emulsions from any two fluids with the appropriate viscosities/interfacial tensions opened up many opportunities to prepare polymeric materials. One of the first examples prepared divinylbenzene-containing droplets in a novel microchannel emulsification device using a continuous phase of aqueous polyvinylalcohol [37, 38]. The droplets were thermally polymerized to yield cross-linked beads. The key to this system's success was that monomer droplets were prepared prior to polymerization thus avoiding any chance that the polymeric material would clog the device. Particle sizes were micrometers in diameter with coefficients of variation (CV) less than 5%. In this case, the dominant system properties controlling droplet size were interfacial tension and the size of the channel. These two properties are significantly harder to vary compared to flow rate, leaving room for future innovations.

Later efforts to produce polymeric beads in flow used a T-junction approach to droplet formation [39]. As illustrated in Figure 5.13, two fluids collide at a T-junction and the resulting droplets can be photopolymerized downstream. The monomer used by the authors was 1,6-hexanediol diacrylate and was initiated using DAROCUR 1173. In this system, flow rate was used to tune the droplet sizes (30–120 µm) and thus the final particle size post-polymerization. The method produced beads with diameter CVs less than 2%, though micron-sized satellite droplets were also produced at the T-junction. While these seminal examples were exciting and pushed the field forward, neither produced materials that could not be realized via classic emulsion polymerizations. However, these efforts paved the way for the revolutions discussed below.

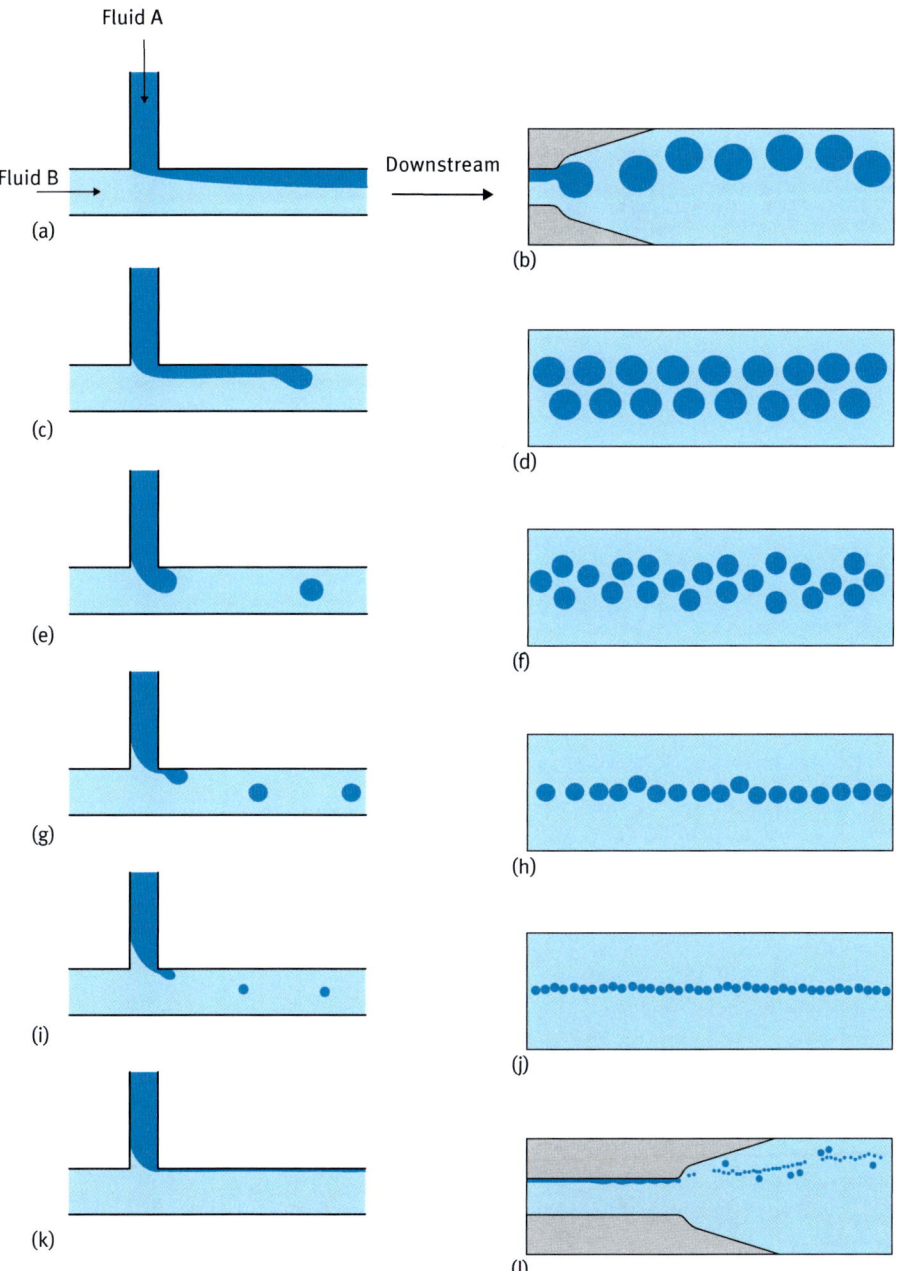

Fig. 5.13: Patterns of droplet formation observed at the T-junction (a, c, e, g, i, and k) and downstream (b, d, f, h, j, l) when the flow rate of the continuous phase was varied at a fixed dispersed phase flow rate. The droplets are observed to decrease in diameter as the continuous phase flow rate is increased.

146 — 5 Synthesis of materials in flow – principles and practice

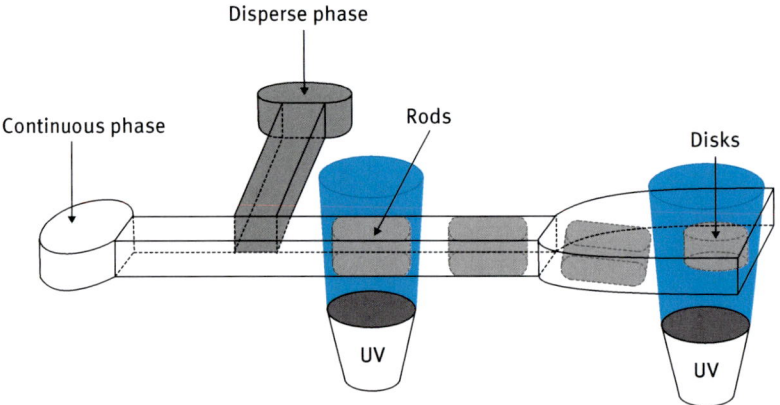

Fig. 5.14: Illustration of a microchannel geometry that can be used to create plugs and disks. The two phases are forced through the channels, where the continuous phase snaps off plugs of the disperse phase containing a monomer. The plugs can be subsequently polymerized using UV light. The channel dimensions are changed downstream enabling the capture of disks as the plugs expand laterally.

The first example of fundamentally novel materials enabled by using a flow reactor was the synthesis of solid disk- and plug-shaped particles. The system used a T-junction to create droplets of photopolymerizable monomer (Figure 5.14). The key innovation was the use of the channel dimensions to produce different structures. When a UV beam was shone on the plug in the narrow channel, a rectangular structure was captured, but if the plug was allowed to expand laterally and then polymerized, a disk was formed. Due to the small dimensions of the plugs and disks, the photopolymerization occurred in less than 1 µs using only 3 J cm^{-2} of energy [40].

In the same time frame, a flow focusing method was used to make a variety of shapes out of photopolymerized tripropyleneglycol diacrylate, dimethacrylate oxypropyldimethylsiloxane, divinylbenzene, ethyleneglycol diacylate, and pentaerythritol triacrylate [41]. Figure 5.15 shows a schematic of a flow focusing device and the various shapes realized by using channels of different geometries. Flow focusing is conceptually similar to a T-junction except that three fluids are forced through a narrow orifice. In this case, liquid A (Figure 5.15 (a)) contains the photopolymerizable monomer and liquid B is aqueous phase free of monomer (aka a continuous phase) Figure 5.15 (b) also illustrates the wavy channel used to increase residence time while the droplets and other structures are photopolymerized.

The combination of flow-focusing and channel dimension alteration allowed capture of microspheres, rods, disks, and ellipsoids. Channel clogging was prevented because the aqueous continuous phase prevented the droplets from contacting the walls. The authors note that up to 250 particles could be formed per minute and 5–10% shrinkage was observed after polymerization. Particle size was dependent on flow rate

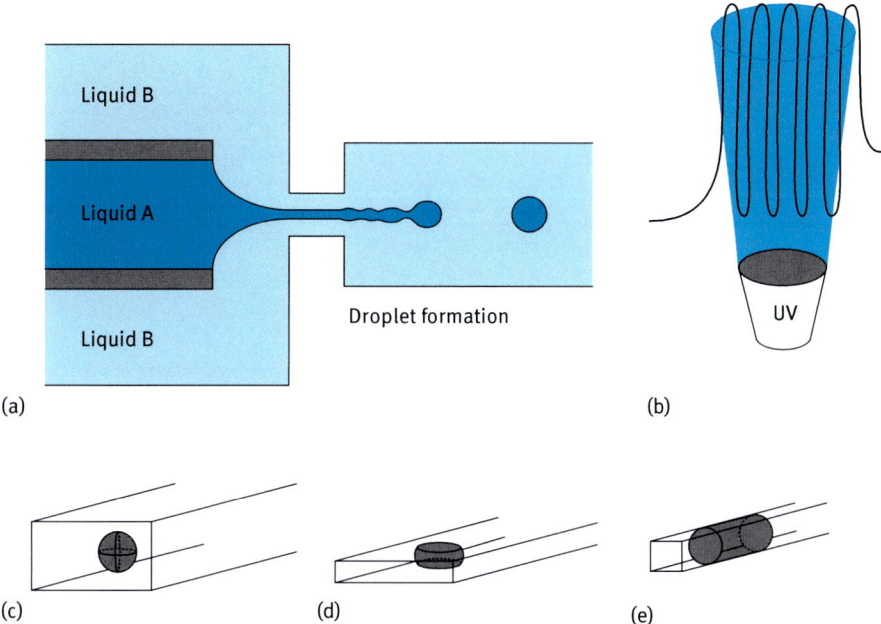

Fig. 5.15: An illustration of a flow-focusing microfluidic droplet generator. The two immiscible liquids (A) and (B) are forced into the narrow orifice where the inner liquid core breaks to release monodisperse droplets in the outlet channel (a). The device channels can be elongated, allowing for longer exposure to UV or thermal effects, for use in producing photochemically and thermally solidified particles (b). Representations of the shapes of droplets in the microfluidic channels (c–e). If the volume of the droplet produced by the device exceeds that of the largest sphere which could be accommodated in the channel, the droplet is deformed into a disk (d) or an ellipsoid or rod (e).

so that at high flow rates smaller particles were realized. At all flow rates examined, the particle CVs were less than 1.5% – monodisperse. In addition to the excellent range of shapes, complex particles were realized by incorporation of dyes, quantum dots, and liquid crystals into the microparticles. These two examples demonstrate that channel dimensions are a tunable parameter enabling the formation of particles possessing shapes that could not be realized using batch techniques. This approach is now the basis of a growing field of complex particle synthesis [18, 42, 43].

Building off this concept, it was also demonstrated that channel dimensions could be used to produce two-dimensional droplet lattices that could be captured by photopolymerization (Figure 5.16) [44]. As shown in Figure 5.16, the polymerized disks shrink. The shrinkage is important because if the discs increased in size the channels would clog. The notion that microreactors can not only enable production of monodisperse particles but also the self-assembly of these materials into complex well-ordered systems is powerful and represents a significant expanding area of research [45].

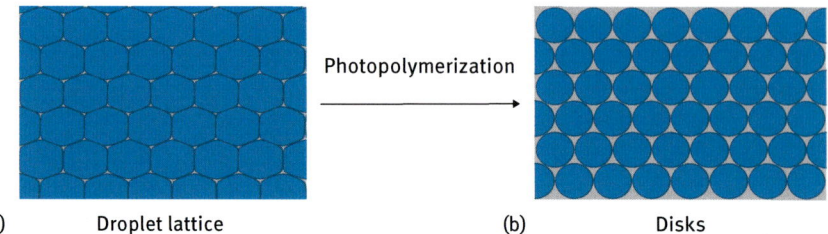

(a) Droplet lattice   (b) Disks

Fig. 5.16: An illustration of a lattice of polymer disks before and after *in situ* photopolymerization which shows the shrinking of the final polymer product.

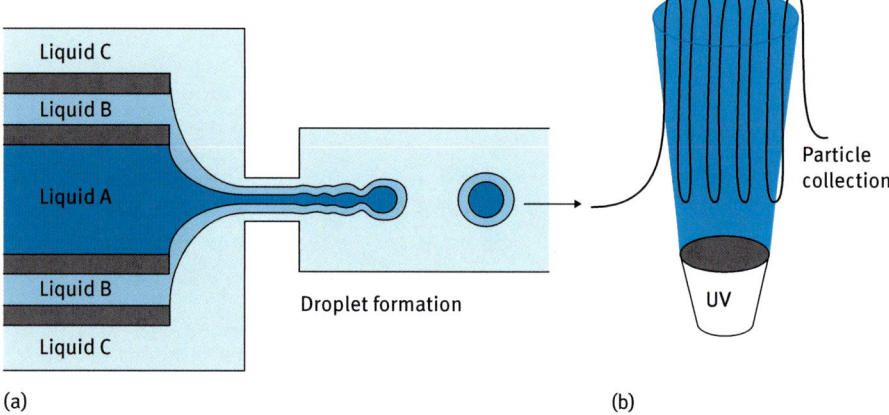

(a)   (b)

Fig. 5.17: An illustration of the production of droplets in a microfluidic flow-focusing device by laminar co-flow of silicone oil (liquid a), monomer (liquid b), and aqueous (liquid C) phases (a). Depiction of the wavy channel used for the polymerization of monomer in core-shell droplets by UV light exposure (b).

The advantages of the flow-focusing technique were further exploited by using a three fluid system to realize droplet-in-droplet structures [46]. In this seminal example, aqueous SDS (liquid C), monomers (liquid B), and silicon oil (liquid A) were combined to form droplets with liquid A encapsulated by liquid B traveling in liquid C as the disperse phase (see Figure 5.17 for an example of such a system). After photopolymerizing the two monomers, tripropyleneglycol diacrylate (TPGDA) or ethyleneglycol dimethacrylate (EGDMA), the droplet-in-droplet structures were captured, yielding novel capsule structures. Subtle changes in the ratios of the three solution flow rates facilitated the formation of a range of novel structures.

Using this flow-focusing device strategy, another group created hydrogel beads impregnated with glucose oxidase and horseradish peroxidase [47]. The enzyme containing beads were produced at rates up to 2600 per minute. A fluorescence-based activity assay established that both enzymes retained activity. Yet another group synthesized biodegradable microgels using the flow focusing strategy [48]. The rapid application of the flow-focusing methodology illustrates the power of the technique and how easily the technique can be transitioned from laboratory to laboratory.

The aforementioned approaches produced novel particle and capsule shapes by taking advantage of channel dimensions or multiphase emulsions. In each case, the shape was captured by photochemical polymerization. Doyle and coworkers recognized that the light source initiating the photochemical polymerization could be used to control particle shape if the light was first passed through a photolithographic mask [49]. In this embodiment, a photopolymerizable monomer solution is passed through a light source shaped by a mask. Because the polymerization is faster than 100 ms, the particles do not move appreciably during the patterning, minimizing distortion of the shapes. Using this approach, triangles, cuboids, cylinders, and various irregular forms with a spatial resolution of around 3 μm were produced. Oxygen diffusing through the sides of the PDMS prevented polymerization from occurring outside of the light field. Biphasic particles were also produced by co-flowing laminar streams of different monomers. The method promises to yield many exciting applications and recently this work has been reviewed in the context of drug-delivery applications [50].

5.3.3 Janus materials

Janus materials are anisotropic where, one side, or half, exhibits one property and the other side another property [51, 52]. The term "Janus" is derived from the Roman God Janus who is depicted as a single head with two faces. Janus materials are not new and have many applications including the key material in electronic paper [53, 54]. Using microreactors to form Janus particles is an ideal use of the low Reynolds number conditions that channels of small dimension impose on fluids. Figure 5.18 illustrates the formation of Janus droplets by flowing two solutions laminarly into a flow-focusing device. In one example, black and white droplets are realized by using carbon black and titanium dioxide, respectively [39]. The droplets were captured into particles by photopolymerizing the monomers included in the disperse phases. The Janus particles were of such high quality that they could be used as active elements in electroactive, switchable surfaces [19]. The switching is due to the different charging properties exhibited by carbon black and titanium dioxide. Depending on the bias of the surface, the Janus-faced particles display their white or black sides. The size control and low dispersity of these particles make this type of chromic display more sharp than displays made using other methods. The authors also created a device for making these particles in quantities of tons per year [55]. This example demonstrates that microfluidic devices can be used to produce commercially useful materials.

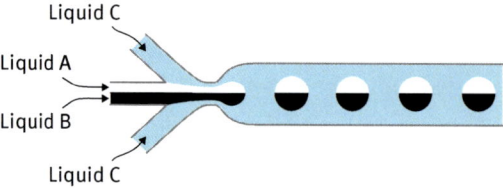

Fig. 5.18: An illustration of Janus droplets being formed in a flow-focusing geometry. Liquid A forms the white portion of the droplet, and liquid B forms the black portion. Each side of the flow-formed droplets has unique properties associated with the composition of the liquids from which they are formed.

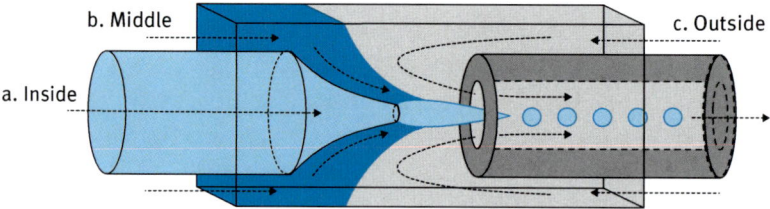

Fig. 5.19: Illustration of a microcapillary geometry/device for generating double emulsions from coaxial jets. The geometry requires the outer fluid (c) to be immiscible with the middle fluid (b) and the middle fluid to be in turn immiscible with the inner fluid (a).

5.3.4 Capsules

Complex shapes and Janus materials represent cutting edge materials. While these materials are beginning to have commercial applications, microcapsules represent a material that has had commercial applications since their first use in carbonless paper decades ago. Microcapsules are now used in applications ranging from drug delivery to agrochemical encapsulation. Although previous examples demonstrated that microcapsules can be produced using multiphase droplets and photopolymerization [56], the most common and economical strategies to produce microcapsules are via coacervation and interfacial polymerization. Coacervation is essentially precipitation of a polymer solution at the interface between two liquids. One example of this approach was accomplished by using water-in-oil-in-water emulsions formed in a flow-focusing device (see Figure 5.19 for an example of a coaxial flow-focusing device) where the diblock copolymer poly(n-butyl acrylate)-b-poly(acrylic acid) was dissolved in the middle oil phase. Upon drying, a flexible, permeable polymer shell was formed [57]. Other examples include coacervates formed from poly(d,l-lactic acid-co-glycolic acid) (PLGA) [58]. In this case, microcapsules containing encapsulated gentamycin, an antibiotic, were formed. Nonblock co-polymer coacervates have also been reported such as monodisperse polystyrene microcapsules [59].

The coacervate approach has been used to create responsive microcapsules – capsules that can encapsulate, deliver, and release payloads. One strategy involved the formation of a double emulsion followed by freezing of the middle shell phase. Temperature was then used as the mechanism to controllably release the additives [60]. A continuous phase of water with glycerol and poly(vinylalcohol) (PVA), a middle phase consisting of molten fatty glycerides, and an inner phase of water, glycerol (a viscomodifier), and FITC-Dextran as an additive. The droplets were collected in a cooled vial to speed up the solidification of the shell before being heated to 37 °C (just above the melting point of the fatty glycerides) to release added dye. The phases can be tuned to perform at different temperatures allowing for integration into a variety of applications. In addition to the heat sensitive materials, pH responsive capsules were generated by creating water-oil-water (W/O/W) emulsions [61].

Microcapsule formation via interfacial polymerization, while more chemically complex than coacervation, enables the creation of more complex materials because the tools of synthetic chemistry can be brought in to expand the attributes of the capsules. The first example using interfacial polymerization in flow to produce microcapsules was achieved using 1,6-diaminohexane in an aqueous disperse phase carried by a continuous phase of hexadecane followed by addition of third phase containing adipoyl chloride to create nylon-6,6 shells [62]. The emulsions were produced using an axisymmetric PDMS-based flow focusing device. This device enabled the acid chloride to enter the system after the droplets were formed (aq. diaminohexane surrounded by hexadecane). The key challenge in any continuous interfacial polymerization is to generate the droplets at a rate that is faster than the polymerization. If this condition is not met, the orifice where the droplets are formed becomes clogged due to premature polymerization. By adding the adipoyl chloride downstream of droplet formation this problem is avoided. This seminal example yielded capsules with monodisperse diameter CVs and allowed the formation of capsules with interesting features like encapsulated magnetic particles.

Following the proof-of-principle demonstration using the complex axisymmetric system, a series of examples demonstrated that significantly simpler devices could be used to facilitate interfacial polymerizations. One of the simplest systems used simple laboratory tubing and flat-tipped needles where the needle is inserted into the tubing, creating a T-junction [63].

This cheaper alternative to etched glass and PDMS or polymer based devices has become a valuable tool for quickly generating monodisperse emulsions. As shown in Figure 5.20, a continuous phase is pumped through 1/16″ PVC (polyvinyl chloride) tubing and a 154 µm (ID) needle introduces a disperse phase to the continuous flow. Polyamide capsules were formed by including polyethyleneimine in the continuous phase and adding sebacoyl chloride and 1,3,5-benzene tricarboxylic acid chloride in a chloroform/cyclohexane disperse phase. The interfacial polymerization begins immediately as the two phases meet, yielding capsules with diameter CVs ranging from 3.3–8.6%. As in the flow-focusing cases, the capsule sizes were easily varied from 865 µm to 313 µm by altering the flow rate. This simple approach has been used to create hierarchical capsules composed of oligomeric and crystalline diphenylsilanediol [64] as

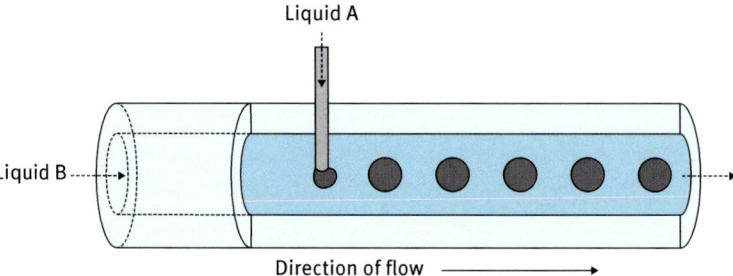

Fig. 5.20: An illustration of a simplified microfluidic device composed of laboratory tubing and flat-tipped needles. The continuous phase flows through the tubing and a disperse phase is introduced through small gauge needles (each phase is syringe pump driven).

well as silica capsules produced using highly reactive $SiCl_4$ [65]. While the tubing and needle system is inexpensive and simple to construct, some have observed that placement of the needle into the tubing can yield experiment to experiment variation and have reported intermediate solutions that provided the robustness of the axisymmetric system but are as simple to build as the needle-in-tubing systems [43, 66].

5.3.5 Membranes and fibers

The literature covering materials synthesis using microfluidic-based emulsion templates is large and overshadows some of the other possible materials that could be realized. For example, co-flowing laminar streams offer a continuous interface whereby thin membranes or fibers could be produced. The first synthesis of such a polymeric membrane was disclosed in the landmark Whitesides publication that has defined most of the materials in flow field [11]. In this case, a polymer membrane was prepared by co-flowing a solution containing a polyanion and a polycation. The two oppositely-charged polymers precipitate at the interface to yield a membrane whose properties and structure is unknown beyond the optical micrograph provided in the report.

Later examples demonstrated deposition of a nylon membrane by colliding a diamine solution with a solution containing a diacyl chloride [67]. In this example, a four-way fluidic intersection was created whereby an aqueous phase entering from one inlet ran along a path defined by hydrophilic surface treatment so that the aqueous phase exited through an outlet orthogonal from the entry point (see Figure 5.21 for device illustration). Conversely, the organic phase was guided through a diagonal path guided by hydrophobic surface treatment. The membrane exhibited a pore size smaller than 200 nm. These membrane demonstrations have led others to create reactive membranes within microreactors [68].

Building on this theme, microfibers and hollow tubes have been produced using coaxial microfluidic devices [69]. A coaxial geometry forces one fluid down the center

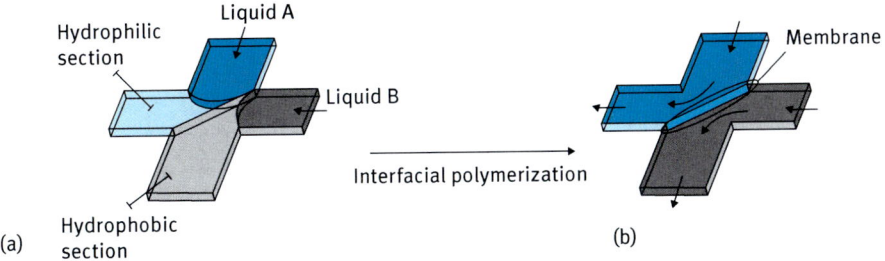

Fig. 5.21: An illustration of the surface-patterned channel (with both hydrophilic and hydrophobic halves). The two liquids flow together to form a polymer membrane at the interface before exiting the same side of the channel.

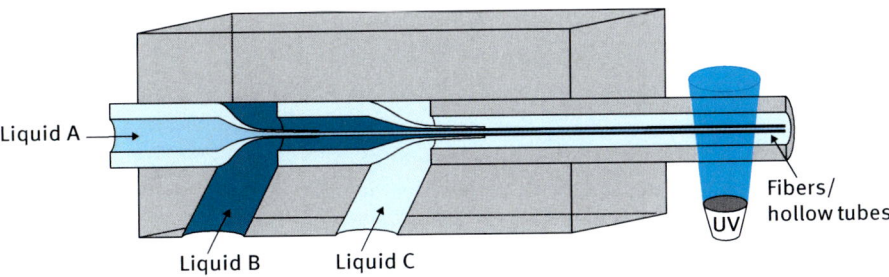

Fig. 5.22: A basic illustration of a device used for the production of hollow tubes. The design of the device can be altered to generate fibers instead of tubes by simply removing one stage (i.e., remove the pipette supplying Liquid A).

of a co-flowing stream and ensures that no part of the inner fluid touches the device wall. The result enables polymerizations to occur without clogging of the device. Figure 5.22 illustrates the device in which a sample fluid is flowed through a capillary surrounded by a sheath fluid. This approach provides a thin sample stream whose diameter can be varied by changing the sheath flow rate. By filling the sample stream with a photopolymerizable cocktail (4-hydroxybutyl acrylate [85%], acrylic acid [11%], ethyleneglycol diacrylate [1%] and 2,2′-dimethyoxy-2-phenyl-acetonephenone [3%]), the thin stream can be cured by a UV beam downstream, resulting in the formation of a dense fiber. Creating hollow microtubes was also achieved but required a more complex reactor. In this case, a second core phase was required as shown in Figure 5.22. The inert inner solution was surrounded with the same photocurable solution used to prepare the hollow fibers. Using this three phase system, photopolymerization resulted in a dense hollow fiber. This approach enabled the formation of hollow tubes impregnated with enzymes and biphasic fibers that could be woven into a responsive swatch. Although this approach to forming fibers and tubes has not yielded the same volume of publications compared to droplet/particles, the area has witnessed interesting developments [21].

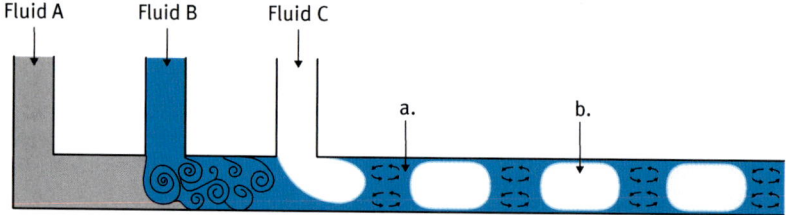

Fig. 5.23: An illustration of a channel geometry used for segmenting fluid flow. Fluid A is introduced in the first inlet followed by introduction of fluid B. Here, the two liquids are observed to rapidly mix. In the last inlet, fluid (immiscible with A, and B) is used to generate plugs. Within the plug composed of fluids A and B (a) internal circulation can be observed to increase the mixing and mass transfer. The plug composed of fluid C (b) limits the dispersion along the length of the channel.

5.3.6 Nanoparticles and inorganic nonpolymeric materials

Nanoscale materials have potential for use in many applications from solar energy conversion, LED technology, therapeutics, imaging, and drug delivery due to their size related photophysical properties. All of these applications require precise control over the size of particles. Widespread use of these particles in these innovative applications can only be realized if they can be produced by inexpensive, environmentally benign, rapid, and reproducible methods. Semiconductor nanoparticles, such as well-studied CdSe, are synthesized in batch using known hydrothermal/pyrolytic techniques [70]. Batch synthetic methods often suffer from poor control and reproducibility, for reasons we have outlined in the previous sections. Recently, attention has been devoted to microreactor synthesis of nanocrystals and nanoparticles due to the benefits associated with flow chemistry and moving to reactors with smaller dimensions (enhanced control: excellent heat and mass transfer, rapid mixing, continuous production). Early works demonstrated the feasibility of producing CdS and CdSe nanocrystals via a continuous flow approach [71, 72]. More recent reports detail controlled synthesis of high-quality materials from droplet and segmented flow microfluidics that were comparable to their batch counterparts [73, 74]. As an excellent example, PbS colloidal quantum dots (CQD) of photovoltaic quality were recently prepared by Pan *et al.* using an automated dual stage segmented flow reactor [75]. Segmenting the flow of the reacting phase (Figure 5.23 (a)) with an immiscible inert liquid or gas (Figure 5.23 (b)), the reagent becomes confined to the volume of the isolated droplets rather than dispersing along the entire length of the channel (axial dispersion) which would lead to concentration gradients. Prior to this work, the best performing photovoltaics relied on manual batch synthesis of the CQD component. The reactor allowed for separation of nucleation and growth stages of the particles within the device (dual stage = two separate stages). This enabled growth to take place at a lower temperature than nucleation, ultimately leading to narrow size distributions and high photoluminescence quantum yield and fluorescence full-width half max values.

Jensen and coworkers have prepared gold nanoparticles using a segmented-flow device [76]. Flow segmentation, or use of slugs, was predicted to produce nanoparticles with low size dispersity because of the rapid mixing and efficient mass transfer by internal circulation. The rapid reduction of $HAuCl_4$ with $NaBH_4$ in the device affords gold nuclei that subsequently grow into nanoparticles. The extent of particle growth within the device was determined by the physical properties of the continuous and disperse phases and their flow rates. Longer residence times in the device led to particles with larger size distributions.

Gold nanorods (GNRs) were prepared using a coaxial flow device coupled to a coiled aging reactor [77]. The inner phase consisted of the gold source ($HAuCl_4$), and lysine (used instead of a surfactant as a colloidal stabilizer) was introduced to an outer phase of the reductant tetramethylammonium hydroxide (TMAOH). This reaction mixture was directly injected into an aging loop, where the gold nuclei were grown into GNRs. The small volumes of the microreactor allowed for fast screening of reactants and further optimization, leading to high-quality materials in a shorter amount of time, as demonstrated by the short residence time in the aging portion of the reactor (only 16 s).

Others have discovered the utility of microreactors for synthesizing magnetic cobalt ferrite nanoparticles. The composite particles were prepared via multistep synthesis in a continuous-flow device [78]. These particles are of significant importance for applications entailing magnetic imaging and therapies. The microreactor was composed of a coaxial flow device molded into PDMS that was coupled to an aging loop made from polytetrafluoroethylene (PTFE) laboratory tubing. The first step consisted of room temperature precipitation of the hydroxides. $FeCl_3$ and $CoCl_2$ were dissolved in water (2 equiv. Fe^{3+} : 1 equiv. Co^{2+}) and then injected at a certain flow rate, Q_{in}, into the device as the innermost fluid. An outer fluid of tetramethylammonium hydroxide (TMAOH) was then injected at a certain flow rate, Q_{out}, causing Fe^{3+} and Co^{2+} hydroxides to precipitate out. The best mixing was achieved with a Q_{out}/Q_{in} ratio of 400 (this allowed for 80 ms mixing in their device) which is crucial to obtain the coprecipitated cations. This solution was then directly injected into the aging coil, which was raised to 98 °C using a heating bath, for particle growth. Crystalline $CoFe_2O_4$ was obtained with 16 min residence time. This material was then compared to $CoFe_2O_4$ nanoparticles prepared in batch. The materials produced in flow were similar in size and magnetic properties to their batch counterparts, but were produced much more efficiently (in less time using less energy; 2 hrs. for batch vs. 16 min. for flow). The key innovation is the ability of the microreactor technology to separate the formation of the hydroxides from the subsequent growth step leading to enhanced control and efficiency in comparison to the batch preparation.

5.4 Conclusions

Materials synthesis has been transformed through the use of fluids flowing in channels and tubes and which has recently enabled new materials and methods to be developed stemming from the early works of IMM, Whitesides, Weitz, and Yoshida. Precise control of fluid dynamics using micro-/mesofluidic devices has provided synthetic opportunities that could not be realized using traditional batch techniques. Exothermic polymerizations can be performed more safely in the small dimensions of a microreactor and yield high quality materials due to the exceptional heat transfer and mixing. Monodisperse emulsions can be formed, controlled, and captured by photopolymerizations within microfluidic devices. Beads and capsules can also be synthesized from reactions taking place within the droplet (beads) or at the interface (capsules). Fibers and membranes can be created conveniently at the interface generated by two co-flowing liquids. Janus materials have also been realized due to the ability to control the viscosity and miscibility of fluids flowing in narrow channels. Nanoparticle synthesis has also benefitted from the use of dual-stage and segmented microreactors as control of heat and mass transfer (and even separation of mixing and heating stages) is crucial to the formation of nuclei and growth into particles. The field of materials synthesis in flow continues to gain popularity and promise as new discoveries pertaining to devices, techniques, and materials are reported.

> **Capillary number:** The ratio of surface tension to viscous forces is represented as the capillary number:
>
> $$Ca = \frac{v\eta}{\gamma}$$
>
> where v is mean velocity, η is dynamic viscosity, and γ is the interfacial surface tension.
>
> **Weber number:** Weber number represents the ratio of inertial forces to surface tension forces:
>
> $$We = \frac{\rho v^2 l}{\sigma}$$
>
> where ρ is the density (kg/m^3), v is the mean velocity (m/s), l is the characteristic length of the droplet (m), and σ is the interfacial surface tension (N/m).

Study questions

5.1. What does a low Reynolds number imply about the flow of a fluid through a tube? When two fluids are combined what impact does Reynolds number have?

5.2. What are the major advantages of performing materials synthesis in flow?

5.3. Describe two strategies to prepare droplets in flow? Which strategy yields "satellite" or daughter droplets?

5.4. Draw an oil-in-water-in-oil droplet and describe the type of device that can produce such droplets. What are emulsions of this type used to prepare?

5.5. What is a Janus-faced particle and what forces enable such particles to be prepared in flow? What commercial products utilize Janus-faced particles?

Further readings
- Ref #11 Kenis PJA, Ismagilov RF, Whitesides GM. "Microfabrication inside capillaries using multiphase laminar flow patterning" *Science* **1999**, 285, 83.
- Ref #17 Thorsen T, Roberts RW, Arnold FH, Quake SR. Dynamic pattern formation in a vesicle-generating microfluidic device *Phys. Rev. Lett.* **2001**, 86, 4163.
- Ref #26 Nagaki A, Kawamura K, Suga S, Ando T, Sawamoto M, Yoshida J. Cation pool-initiated controlled/living polymerization using microsystems. *J. Am. Chem. Soc.* **2004**, 126, 14702.
- Ref #40 Dendukuri D, Tsoi K, Hatton TA, Doyle PS. Controlled synthesis of nonspherical microparticles using microfluidics *Langmuir* **2005**, 21, 2113.
- Ref #56 Utada AS, Lorenceau E, Link DR, Kaplan PD, Weitz DA. Monodisperse double emulsions generated from a microcapillary device *Science* **2005**, 308, 537.
- Ref #71 Edel JB, Fortt R, deMello JC, deMello AJ. Microfluidic routes to the controlled production of nanoparticles *Chem. Comm.* **2002**, 1136–1137.

Bibliography

[1] Marre, S.; Jensen, K. F. Chem. Soc. Rev. 2010, 39, 1183–1202.
[2] Jahnisch, K.; Hessel, V.; Lowe, H.; Baerns, M. Angew. Chem. Int. Edit. 2004, 43, 406.
[3] Bayer, T.; Himmler, K. Chem. Eng. Technol. 2005, 28, 285.
[4] Hessel, V.; Lowe, H.; Stange, T. Lab Chip 2002, 2, 14N.
[5] Sue, K.; Murata, K.; Kimura, K.; Arai, K. Green. Chem. 2003, 5, 659.
[6] Schwalbe, T.; Autze, V.; Hohmann, M.; Stirner, W. Org. Process. Res. Dev. 2004, 8, 440.
[7] Jahnisch, K.; Hessel, V.; Lowe, H.; Baerns, M. Angew. Chem. Int. Edit. 2004, 43, 406.
[8] Wegner, J.; Ceylan, S.; Kirchning, A. Adv. Synt. and Cat. 2012, 354, 17–57.
[9] Cortese, B.; de Croon, M.; Hessel, V. Ind. Eng. Chem. Res. 2012, 51, 1680–1689.
[10] Kendall, D. P. J. Press. Vessel. Technol-Trans. ASME 2000, 122, 229.
[11] Kenis, P. J. A.; Ismagilov, R. F.; Whitesides, G. M. Science 1999, 285, 83.
[12] Yamashita, K.; Yamaguchi, Y.; Miyazaki, M.; Nakamura, H.; Shimizu, H.; Maeda, H. Anal. Biochem. 2004, 332, 274.
[13] Rayleigh, L. Proc. R. Soc. London. 1879, 29, 71.
[14] Plateau, J. Acad. Sci. Bruxelles. Mem. 1849, 23, 5.
[15] Jillavenkatesa, A.; Dapkunas, S. J.; Lum, L. H. In National Institute of Standards and Technology Special Publication 960-1; U.S. Government Printing Office: Washington, 2001, p 149.
[16] Zheng, B.; Tice, J. D.; Ismagilov, R. F. Anal. Chem. 2004, 76, 4977.
[17] Thorsen, T.; Roberts, R. W.; Arnold, F. H.; Quake, S. R. Phys. Rev. Lett. 2001, 86, 4163.
[18] Wang, J.-T.; Wang, J.; Han, J.-J. Small 2011, 7, 1728–1754.
[19] Nisisako, T.; Torii, T.; Takahashi, T.; Takizawa, Y. Adv. Mater. 2006, 18, 1152–1156.
[20] Chen, Y.; et al., Phys. Rev. E 2013, 013002.
[21] Nunes, J. K.; Tsai, S. S. H.; Wan, J.; Stone, H. A. J. Phys. D: Appl. Phys. 2013, 46, 114002.
[22] Wilms, D.; Klos, J.; Frey, H. Macromolec. Chem. Phys. 2008, 209, 343–356.
[23] Tonhauser, C.; Natallelo, A.; Löwe, H.; Frey, H. Macromolecules 2012, 45, 9551–9570.
[24] Nagaki, A.; Tomida, Y.; Miyazaki, A.; Yoshida, J. Macromolecules 2009, 42, 4384–4387.
[25] Nagaki, A.; Takahashi, Y.; Akahori, K.; Yoshida, J. Macromolec. React. Eng. 2012, 6, 467–472.
[26] Nagaki, A.; Kawamura, K.; Suga, S.; Ando, T.; Sawamoto, M.; Yoshida, J. J. Am. Chem. Soc. 2004, 126, 14702.
[27] Iwasaki, T.; Yoshida, J. Macromolecules 2005, 38, 1159.
[28] Russum, J. P.; Jones, C. W.; Schork, F. J. Ind. Eng. Chem. Res. 2005, 44, 2484.

[29] Russum, J. P.; Jones, C. W.; Schork, F. J. Aiche J. 2006, 52, 1566.
[30] Enright, T. E.; Cunningham, M. F.; Keoshkerian, B. Macromol Rapid Commun 2005, 26, 221.
[31] Sugiura, S.; Nakajima, M.; Seki, M. Ind Eng Chem Res 2002, 41, 4043.
[32] Honda, T.; Miyazaki, M.; Nakamura, H.; Maeda, H. Lab Chip 2005, 5, 812.
[33] Yamaguchi, Y.; Ogino, K.; Yamashita, K.; Maeda, H. J. Chem. Eng. Jpn. 2004, 37, 1265.
[34] Wu, T.; Mei, Y.; Cabral, J. T.; Xu, C.; Beers, K. L. J. Am. Chem. Soc. 2004, 126, 9880.
[35] Wu, T.; Mei, Y.; Xu, C.; Byrd, H. C. M.; Beers, K. L. Macromol Rapid Commun 2005, 26, 1037.
[36] Xu, C.; Wu, T.; Drain, C. M.; Batteas, J. D.; Beers, K. L. Macromolecules 2005, 38, 6.
[37] Sugiura, S.; Nakajima, M.; Itou, H.; Seki, M. Macromol Rapid Commun 2001, 22, 773.
[38] Sugiura, S.; Nakajima, M.; Seki, M. Ind. Eng. Chem. Re.s 2002, 41, 4043.
[39] Nisisako, T.; Torii, T.; Higuchi, T. Chem. Eng. J. 2004, 101, 23.
[40] Dendukuri, D.; Tsoi, K.; Hatton, T. A.; Doyle, P. S. Langmuir 2005, 21, 2113.
[41] Xu, S. Q.; Nie, Z. H.; Seo, M.; Lewis, P.; Kumacheva, E.; Stone, H. A.; Garstecki, P.; Weibel, D. B.; Gitlin, I.; Whitesides, G. M. Angew. Chem. Int. Edit. 2005, 44, 724.
[42] Yi, G. R.; Pine, D. J.; Sacanna, S. J. Phys.-Cond. Matter 2013, 25, 193101.
[43] Serra, C. A.; Khan, I. U.; Chang, C. Q.; Bouquey, M.; Muller, R.; Kraus, I.; Schmutz, M.; Vandamme, T.; Anton, N.; Ohm, C.; Zentel, R.; Knauer, A.; Kohler, M. J. Flow Chem. 2013, 3, 66–75.
[44] Seo, M.; Nie, Z. H.; Xu, S. Q.; Lewis, P. C.; Kumacheva, E. Langmuir 2005, 21, 4773.
[45] Dendukuri, D.; Doyle, P. S. Adv. Mater. 2009, 21, 4071–4086.
[46] Nie, Z. H.; Xu, S. Q.; Seo, M.; Lewis, P. C.; Kumacheva, E. J. Am. Chem. Soc. 2005, 127, 8058.
[47] Jeong, W. J.; Kim, J. Y.; Choo, J.; Lee, E. K.; Han, C. S.; Beebe, D. J.; Seong, G. H.; Lee, S. H. Langmuir 2005, 21, 3738.
[48] De Geest, B. G.; Urbanski, J. P.; Thorsen, T.; Demeester, J.; De Smedt, S. C. Langmuir 2005, 21, 10275.
[49] Dendukuri, D.; Pregibon, D. C.; Collins, J.; Hatton, T. A.; Doyle, P. S. Nat. Mater. 2006, 5, 365.
[50] Zhang, Y.; Chan, H. F.; Leong, K. W. Adv. Drug Deliv. Rev. 2013, 65, 104–120.
[51] Walther, A.; Muller, A. H. E. Chem. Revs. 2013, 113, 5194–5261.
[52] Hu, J.; Zhou, S.; Sun, Y.; Fang, X.; Wu, L. Chem. Soc. Revs. 2012, 41, 4356–4378.
[53] Cho, I.; Lee, K. W. J. Appl. Polym. Sci. 1985, 30, 1903–1926.
[54] de Gennes, P. G. Rev. Mod.Phys. 1992, 64, 645–648.
[55] Nisisako, T.; Torii, T. Lab Chip 2008, 8, 287–293.
[56] Utada, A. S.; Lorenceau, E.; Link, D. R.; Kaplan, P. D.; Weitz, D. A. Science 2005, 308, 537.
[57] Lorenceau, E.; Utada, A. S.; Link, D. R.; Cristobal, G.; Joanicot, M.; Weitz, D. A. Langmuir 2005, 21, 9183.
[58] Martin-Banderas, L.; Flores-Mosquera, M.; Riesco-Chueca, P.; Rodriguez-Gil, A.; Cebolla, A.; Chavez, S.; Ganan-Calvo, A. M. Small 2005, 1, 688.
[59] Martin-Banderas, L.; Rodriguez-Gil, A.; Cebolla, A.; Chavez, S.; Berdun-Alvarez, T.; Garcia, J. M. F.; Flores-Mosquera, M.; Ganan-Calvo, A. M. Adv. Mater. 2006, 18, 559.
[60] Sun, B. J.; Shum, H. C.; Holtze, C.; Weitz, D. A. ACS App. Mater. & Interfac. 2010, 2, 3411–3416.
[61] Abbaspourrad, A.; Datta, S. S.; Weitz, D. A. Langmuir, 2013, 29, 12697–12702.
[62] Takeuchi, S.; Garstecki, P.; Weibel, D. B.; Whitesides, G. M. Adv. Mater. 2005, 17, 1067.
[63] Quevedo, E.; Steinbacher, J.; McQuade, D. T. J. Am. Chem. Soc. 2005, 127, 10498.
[64] Steinbacher, J. L.; Moy, R. W. Y.; Price, K. E.; Cummings, M. A.; Roychowdury, C.; Buffy, J. J.; Olbricht, W. L.; Haaf, M.; McQuade, D. T. J. Am. Chem. Soc. 2006, 128, 9442.
[65] Miller, L. Z.; Steinbacher, J. L.; Houjeiry, T. I.; Longstreet, A. R.; Woodberry, K. L.; Gupton, B. F.; Chen, B.; Clark, R.; McQuade, D. T. J. Flow Chem. 2012, 2, 92–102.
[66] Chang, C. Q.; Serra, C. A.; Bouquey, M.; Prat, L.; Haddziioannou, G. 2009, 9, 3007–3011.
[67] Zhao, B.; Viernes, N. O. L.; Moore, J. S.; Beebe, D. J. J. Am. Chem. Soc. 2002, 124, 5284.
[68] Yamada, Y. M. A.; Watanabe, T.; Ohno, A.; Uozumi, I. Chem. Sus. Chem.2012, 2, 293–299.

5.4 Conclusions

[69] Jeong, W.; Kim, J.; Kim, S.; Lee, S.; Mensing, G.; Beebe, D. J. Lab Chip 2004, 4, 576.
[70] a) Murray, C. B.; Norris, D. J.; Bawendi, M. G. J. Am. Chem. Soc. 1993, 115, 8706. b) Peng, Z. A.; Peng, X. J. Am. Chem. Soc. 2001, 123, 183; c) Qu, L.; Peng, Z. A.; Peng, X. Nano Lett. 2001, 1, 333. d) Yang, Y. A.; Wu, H.; Williams, K. R.; Cao, Y. C. Angew. Chem. 2005, 117, 6870. e) Chen, O.; Chen, X.; Yang, Y. A.; Lynch, J.; Wu, H. M.; Zhuang, J.; Cao, Y. C. Angew. Chem. 2008, 120, 8766; Angew. Chem. Int. Ed. 2008, 47, 8638. f) Dethlefsen, J. R.; Døssing, A. Nano Lett. 2011, 11, 1964. g) Kalita, M.; Cingarapu, S.; Roy, S.; Park, S. C.; Higgins, D.; Jankowiak, R.; Chikan, V.; Klabunde, K. J.; Bossmann, S. H. Inorg. Chem. 2012, 51, 4521. h) Zhang, W.; Jin, C.; Yang, Y.; Zhong, X. Inorg. Chem. 2012, 51, 531. i) Aguilera-Sigalat, J.; Rocton, S.; Sanchez- Royo, J. F.; Galian, R. E.; Perez-Prieto, J. RSC Adv. 2012, 2, 1632.
[71] Edel, J. B.; Fortt, R.; deMello J. C.; deMello, A. J. (2002) Chem. Commun. 2002, 10, 1136–1137.
[72] Nakamura, H.; Yamaguchi, Y.; Miyazaki, M.; Maeda, H.; Uehara, M.; Mulvaney, P. Chem. Commun. 2002, 23, 2844–2845; Nakamura, H.; Yamaguchi, Y.; Miyazaki, M.; Uehara, M.; Maeda, H.; Mulvaney, P. Chem. Lett. 2002, 31, 1072–1073.
[73] Shestopalov, I.; Tice, J. D.; Ismagilov, R. F. Lab Chip 2004, 4, 316.
[74] Chan, E. M.; Alivisatos, A. P.; Mathies, R. A. J. Am. Chem. Soc. 2005, 127, 13854.
[75] Pan, J.; El-Ballouli, A. O.; Rollny, L.; Voznny, O.; Burlakov, V. M.; Goriely, A.; Sargent, E. H.; Bakr, O. M. ACS Nano, 2013, 7, 10158–10166.
[76] Cabeza, V. S.; Kuhn, S.; Kulkami, A. A.; Jensen K. F. Langmuir 2012, 28, 7007–7013.
[77] Sebastián, V.; Lee, S.-K.; Zhou, C.; Kraus, M. F.; Fujimoto, J. G.; Jensen, K. F. Chem. Commun. 2012, 48, 6654–6656.
[78] Abou-Hassan, A.; Neveu, S.; Dupuis, V.; Cabuil,V. RSC Adv. 2012, 2, 11263–11266.

Genoveva Filipcsei, Zsolt Otvos, Reka Angi, and Ferenc Darvas
6 Flow chemistry for nanotechnology

6.1 Introduction to nanotechnology and graphene technology

6.1.1 Introduction

Flow chemistry has been introduced and was translated into practice during the past years to prepare nanoparticles with unique material characteristics. Nanotechnology is not a single technology; it is a complex and malleable group of diverse technologies and attracting interest for the researchers to create new materials with unexpected physicochemical properties, develop new technologies for the production of the nanomaterials creating new industries across a wide range of fields.

Advanced nanotechnology in the near future will offer new opportunities, for example, development of new or improved materials; applications within the sphere of electronics and IT; advances in health and medicine; improvements in cosmetic products and advances in food technology; developments in products for military, in-line security use, and space exploration; products and processes to improve the environmental management. Many parts of the world can not sustainably support with its attendant environmental and societal impacts. "Nano" manufacturing could provide cheap and green technologies to develop and create novel materials with enhanced characteristics, thus making sustainable twentieth century manufacturing processes.

Nanotechnology has a great potential for a wide range of industrial applications. Industries affected include automotive, electronics, packaging, aerospace, information, communications, pharmaceuticals, food and personal care. Nanotechnology is encountered in everyday materials (polymers, plastics, and rubber, agrochemicals, pharmaceuticals and cosmeceuticals, paper, foodstuffs, fabrics, textiles, and detergents) and technologies (nucleation and precipitation, liquid crystals, chromatography and ion-exchange, flotation, and heterogeneous catalysis).

This chapter gives a brief overview on the history and theoretical background of the nanoparticle production and summarizes the most important approaches of flow chemistry-based technologies currently used for fabrication of different nanomaterials.

6.1.2 Definition and concepts

Nanotechnology originates from the Greek word meaning "*dwarf*". A nanometer is one billionth (10^{-9}) of a meter, which is about one hundred thousandth of the width of a hair, only the length of ten hydrogen atoms.

> **Nanotechnology** is a multidisciplinary field of engineering science that combines the physical, chemical and biological sciences with the field of engineering conducted at the nanoscale, which is about 1 to a few 100 nanometers. It also covers the engineering of functional systems at the atomic or molecular level. Due to the wide spectrum of various technologies, such as physical, chemical and biological processes realized on nanolevel, a generally accepted and widely used definition of nanotechnology is still missing.
>
> A practical definition covering all the fields where the nanotechnology is applicable without any limitations was established by Raj Bawa and *et al.* [1] in 2005: "*The design, characterization, production, and application of structures, devices, and systems by controlled manipulation of size and shape at the nanometer scale (atomic, molecular, and macromolecular scale) that produces structures, devices, and systems with at least one novel/superior characteristic or property*". We used this definition for our work in order to show the wide diversity of the structures generated at nanoscale and demonstrate their novel size dependent material characteristics.

6.1.3 Brief history of nanotechnology

Nanotechnology is generally considered as a relatively new scientific approach to manipulate the materials at a molecular level. However, this approach is not as young a discipline as we may expect. Fine dispersion of gold (gold nanoparticles) was synthetized by Michael Faraday in 1857. The term of colloid [2] was created by Thomas Graham in 1861. John Tyndall, Hermann Helmholtz, Lord Rayleigh, James Clerk Maxwell and Albert Einstein were among the pioneers who studied the characteristics of colloids around 1900.

The "science called as colloid chemistry" dealing with the chemical and physical interactions of nanoscale particles has been re-born as a new discipline with the Nobel Laureate physicist Richard Feynman's 1959 lecture [3]. Nobel Prizes in Physics were awarded in 1986 to Gerd Binnig and Harold Rohrer for scanning tunneling microscopy. This technology revolutionized the nanotechnology and opened a new way for the generation of nanomaterials [4].

In 1991, Sumio Iijima discovered the carbon nanotubes which are fullerene-related structures consisting of rolled graphene sheets. Robert F. Curl, Jr., Richard E. Smalley and Harold W. Kroto were awarded the 1996 Nobel Prize in Chemistry for the discovery of the buckyball, a soccerball-shaped carbon molecule approximately a nanometer in diameter. The 2010 Nobel Prizes in Physics were awarded to Andre Geim and Konstantin Novoselov for the discovery and characterization of graphene, a form of carbon based on a bi-dimensional arrangement of its atoms on the nanoscale.

The first liposomal drug delivery system (Doxil®) was approved by the FDA in 1995. The first nanodrug consisting of albumin and anticancer drug, paclitaxel, was approved in the US in 2005 [5]. This formulation has proven extremely beneficial in ovarian and breast cancer and has a market size that is beginning to rival those of the most successful cancer drugs of any type. Since then, several nanocarrier-based drugs have been introduced to the market, such as drug nanocrystals (Rapamune®, Emend®, Tricor®, Megace®), limposomes (Caelyx®, Depocyt®, Daunoxome®, Adagen®), polymer-drug conjugates (Onscaspar®, Pegasys®), polymeric micelles (Genexol-PM®) and lipid colloidal dispersion (Amphotec®) [6].

6.1.4 Why nanotechnology?

Both scientific and geographical discoveries are based upon a never-ending search for a greater knowledge, wisdom and understanding of the universe. Therefore, the way natural sciences are getting extended to novel fields can be placed parallel to the history of geographical explorations. For example, the significant extension of the borderline of the Aristotelian geocentric map by Columbus, Vasco de Gama and others in the late medieval era led to great expectations and social excitement in Europe. The burgeoning European empires such as England and France followed the lead of the Spanish and Portuguese Crowns and increasingly extended their power and influence throughout the New World.

We are perhaps enjoying a similar extension of our dimensions when entering into the area of nanotechnology.

The common denominator between the late medieval geographical discoveries and excursions of the research teams to the field of nanotechnology is an opportunity to challenge the unknown. The unspoken belief of the researchers to extend the science to the nanoscale world will bring the same miracles in the development as did the discovery of the microscope by Robert Hooke or the development of the ultramicroscope by Richard Zsigmondy in the past centuries which enabled scientists to study the invisible world. Representative examples of what nanotechnology is expected to offer in the near future are given below.

In the information technology area, researchers are intensively investigating the application of nanotechnology to the development of, for example, high density/efficiency memories, computer devices, high-luminosity devices using nanomaterials such as carbon nanotubes and high-speed optical network devices using photonic crystals. In medical area, the development of nanodrugs with enhanced biological performance, smart drug delivery systems including nanorobots, nanosized biosensors (nanotechnology-on-a-chip) and nanoparticles for imaging and targeted drug delivery are in the focus. In the field of industrial organic chemistry, the researchers are working on the development of nanosized catalysts in order to improve their efficiency. At the environmental and energy industries, the utilization of nanomaterials as environmental remediation catalysts and hydrogen-loading materials is under investigation.

6.1.5 Batch and flow-chemistry based nanonization technologies

There are two main approaches to making nanomaterials: "top-down" and "bottom-up" technologies (Figure 6.1). Top-down approach basically relies on mechanical attrition to render large components into nanosized substances. The bottom-up approach relies on the arrangement of smaller components into more complex assemblies at molecular level. The combination of these two approaches is also applicable to the nanoparticle production. It relies on micro precipitation which is followed by particle size reduction.

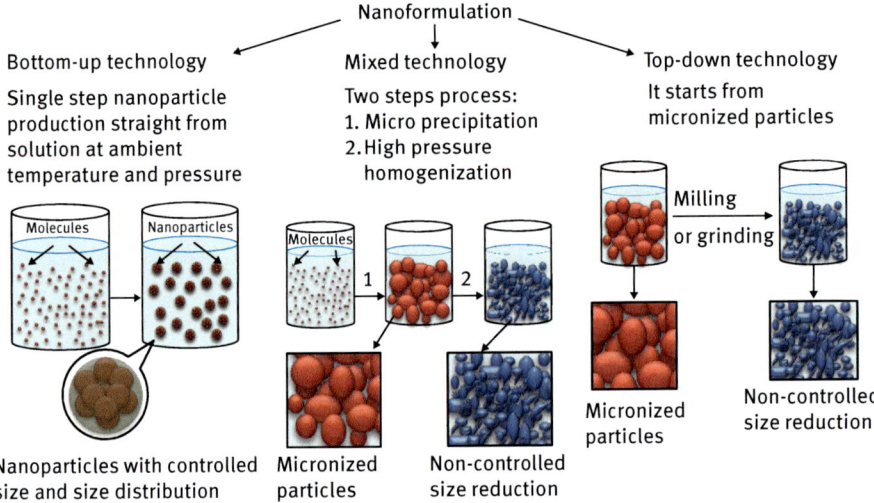

Fig. 6.1: Nanoparticle fabrication technologies.

Top-down approaches include technologies that derived from conventional solid-state methods being capable of creating materials smaller than 100 nm. For example, focused ion beams can directly remove material, or even deposit material when suitable pre-cursor gasses are applied at the same time or atomic layer deposition method allows us to fabricate thin films of the materials onto substrates. Lithography is a top-down technique where a bulk material is reduced in size to nanoscale pattern. Various techniques of nanolithography such as optical lithography, X-ray lithography, dip-pen nanolithography, electron beam lithography or nanoimprint lithography were also developed and used to create nanotubes and nanowires used in semiconductors.

Traditional top-down grinding, milling and homogenization methods are industrially accepted standard techniques for the production of nanoparticles of biologically active small molecules.

Bottom-up approaches start from atomic or molecular species and build the nanomaterials by organizing the species into covalently or physically bounded assemblies.

Beside these techniques, biomimetic approach is also available to apply biological methods and systems to design nanomaterials. Nanobiotechnology is the use of biomolecules for applications in nanotechnology, including use of viruses and lipid assemblies [7].

In summary, both top-down and bottom-up approaches are applicable to produce nanosized particles. The properties of the particles are independent of whether they were synthetized using batch methodology or flow-chemistry based technologies. However, flow technologies offer several advantages over batch technologies, such precise process control of nanoformulation parameters, better control in material properties, easy scalability, lower production costs, and less toxic loading.

Nanosized particles sometimes show increased toxicological effect on animals and environment, although their toxicological effects associated with human exposure are still under investigation. Nanoparticle toxicology, an emergent field, works toward establishing the hazard of nanoparticles, and therefore their potential risk in light of the increased use and likelihood of exposure.

Flow chemistry-based nanoparticle production approach offers a great tool to significantly reduce the toxic human exposure. This technology is a solution based method. It uses reagent solutions or suspensions of microsized particles for the production of colloid solutions of nanoparticles eliminating the potential exposure to the dust of ultrafine particle.

6.1.6 Overview and principles of microfluidic reactors

In the recent years, the importance of flow processes extensively rises, especially in the field of nanoparticle synthesis. Flow-chemistry based approaches have numerous advantages such as good mixing, improved heat and mass transfer and high space-time-yields due to short diffusion lengths and high surface-to-volume ratios of the reactors, green and sustainable production (increased safety (toxic materials) and reduced costs (expensive materials)), very homogeneous conditions (reduced turbulence, low Reynolds numbers) and fast and precise adjustment and modification of process parameters (T, p, x) over conventional batch methods.

In the context of nanomaterial synthesis, the reaction parameters such as temperature, pressure and residence time have a strong effect on the nature of the product and, in particular, effect the size, shape, and chemical composition.

Microfluidic technology can be combined with supercritical fluid technology to overcome the limitations of high boiling point solvents used for nanoparticle synthesis. The use of high pressure high temperature microreactor opens new routes for nanomaterials synthesis in microfluidic devices by enlarging the set of solvents and phases (liquid, gas, supercritical) available.

A microfluidic device has one or more channels with a width of less than 50 μm and deal with volumes of fluid on the order of nanoliters or picoliters. The flow of a fluid through a microfluidic channel can be characterized by the Reynolds number (Re), defined as

$$Re = \frac{LV_{avg}\rho}{\mu}$$

where L is the most relevant length scale, μ is the viscosity, ρ is the fluid density, and V_{avg} is the average velocity of the flow. For many microchannels, L is equal to $4\,A/P$ where A is the cross-sectional area of the channel and P is the wetted perimeter of the channel. Due to the small dimensions of microchannels, the Re is usually much less than 100, often less than 1.0. In this Reynolds number regime, flow is completely laminar and no turbulence occurs. Laminar flow ensures that the molecules are transported in a relatively predictable manner through microchannels [8, 9].

Microfluidic reactors are built from one or more inlets for the reagent feeding, network of channels in which various chemical processes are carried out, and one or more outlets where products and waste materials are extracted. For examples, a simple y-shaped reactor having two inlets and an outlet can be used to mix and react two reagents A and B. The composition and the physicochemical properties of the product and the residence time inside the microfluidic device can be controlled by varying the flow rates of the two reagent solutions, v_A and v_B. If the volume of the outlet is V and the molar concentrations of the two reagent solutions are [A] and [B], then the residence time (τ) in the reaction outlet channel is given by

$$\tau = \frac{V}{v_A + v_B}.$$

and the molar ratio (R) of A to B in the product is equal to

$$R = \frac{v_A}{v_B} * \frac{[A]}{[B]}.$$

Microfluidic devices allow us to precisely control the temperature of the inlets and reactor space using built-in heating elements. In this way, the entire reaction volume can be held at a uniform temperature to ensure the high conversion and selectivity of thermally initiated reactions. In special cases, pressure can also be applied to accelerate the reactions or run reactions which cannot be executed in other way.

6.2 Nanomaterials

6.2.1 Structure and properties: is the smaller better?

The term "nanoparticle" has become the most frequently used phrase in the last two decades. Nanoparticles can be considered as an intermediate state between macroscopic – bulk – phase and the atomic or molecular system. Therefore, their properties differ from the ones in the bulk or in the molecular state. In general, the size of a nanoparticle is in the range of 1–100 nm.

In this chapter, we consider the change of the physical properties when a transition from bulk phase to nanometer regime happens.

Morphological/structural property caused by the particle size reduction is considered to be the most important in leading to the extreme increase in the ratio of surface-area-to-volume. As a particle decreases in size, a much greater proportion of atoms are found at the surface as compared with that inside (Figure 6.2) which results in increase of dominance of the atoms' behavior on the surface over the ones in the bulk, interior phase.

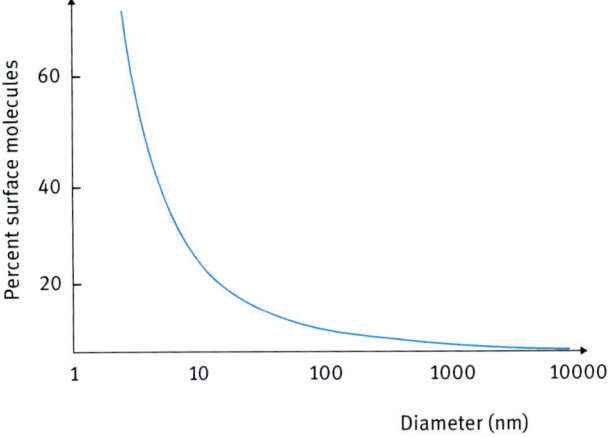

Fig. 6.2: As particles size decreases, the percentage of atoms and molecules exposed on the outside surface of the particles increases exponentially.

This high surface area has a key importance in reactivity, solubility, sintering performance; or in the field of catalysis electrodes or fuel cells.

Decreasing the particle size from the macroscopic regime to the nanoworld leads to the change of the *optical properties* of the metallic or semiconductor particle. This particle size reduction results in the emergence of discrete, quantized energy levels instead of continuous valence and conduction bands [10] (see Figure 6.3 (a)).

It is interesting to see how this phenomenon is manifested in a particular form of semiconductor nanoparticles, the quantum dots. On the other hand, semiconductor nanoparticles are different from metal particles as they do not have free conduction electrons. Instead, the electrons are contained in valence band states, and the electronic properties reflect excitation of the valence electrons into conduction band states across an energetic band gap. With the decreasing particle size of semiconductors the band gap energy increases and then more energy is needed to excite the dot, and con-

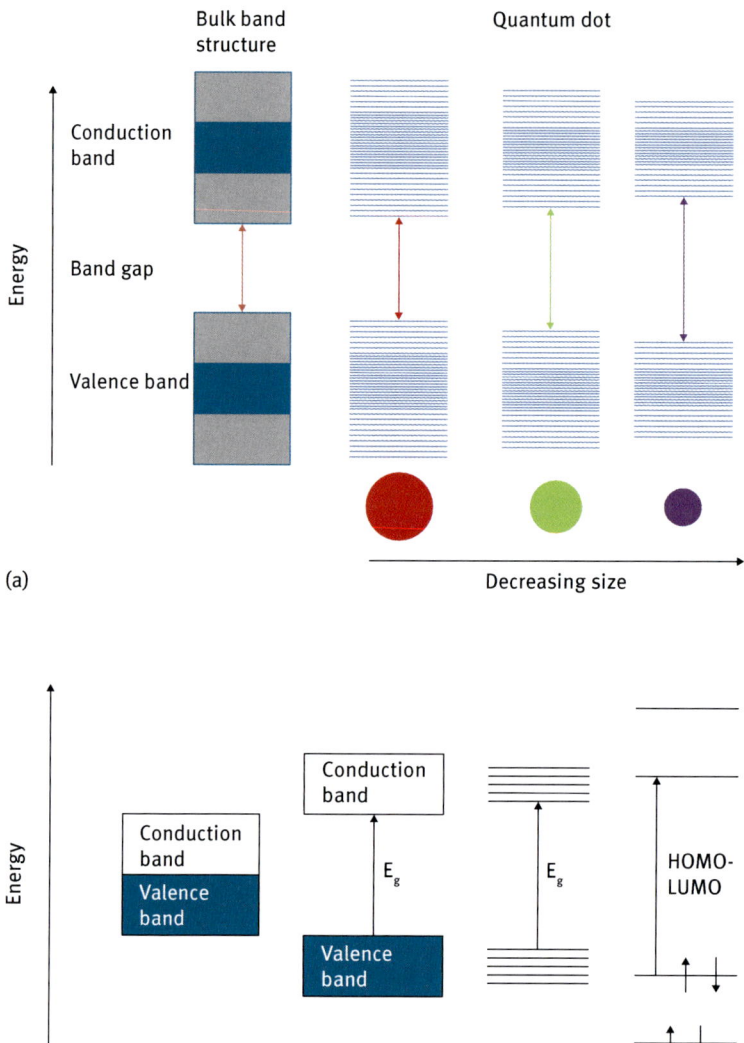

Fig. 6.3: Particle size dependence of band gap in semiconductor and metal particles.

currently, when the dot returns to its ground state, more energy is released. This phenomena result in a color shift from red to blue in the emitted light (Figure 6.3 (a)). In isolated molecules this band gap can be considered as the difference in energy levels of the highest occupied molecular orbital (HOMO) and the lowest unoccupied molecular orbital (LUMO) (Figure 6.3 (b)).

Melting temperature depression is a well-known and widely reported consequence of the increased surface-to-volume ratio evolved in nanoparticle formation [11]. As the atoms and molecules located at the particle surface become influential in the nanometer order, the melting point of the material decreases from that of the bulk material because they tend to be able to move easier at the lower temperature.

As the particle size approaches the nanometer regime, interesting size-dependent ***magnetic properties*** like enhanced coercivity, enhanced magnetization, superparamagnetism and so on are observed.

The crystalline size reduction generally leads to improved hardness of the crystalline materials. Therefore, by micronization or nanoparticle formation the mechanical strength of the materials like metals or ceramic materials can be increased.

The particle size reduction induced increased surface-to-volume ratio results in enhanced solubility and dissolution rate which plays an important role in the application of nanoparticles in medicine. Due to the increased solubility and dissolution rate, the administration of drug compositions containing nanoparticles instead of "bulk phase" active ingredient materials may cause the increase of the bioavailability of the drug [12].

6.2.2 Organic nanoparticles: biologically active small molecules

In the past few years, there has been a growing interest in the use of nanoparticles prepared from biologically active small molecules such as active pharmaceutical or agrochemical ingredients, nutraceuticals or food supplements, active ingredients of skin care products, and so on [13]. Nanoparticles of these biologically active molecules are characterized by increased solubility and dissolution rate which can lead to enhanced biological performance of the active ingredients.

Dendrimers are nanosized particles prepared from highly branched, monodisperse macromolecules such as polyamidoamine, polypropyleneimine and polyaryl ether (Figure 6.4). The particles in the 10 nm size range have a compact spherical geometry in solution. Due to the tailor-made branches, well-defined molecular weight and controlled surface functionality of the macromolecules, the dendrimers have great potential as drug delivery carriers [14].

The biologically active small molecules can be entrapped in the macromolecular globules or attached to the surface functionalities. Due to their versatility, both hydrophilic and hydrophobic drugs can be incorporated into the dendrimers. Other applications of the dendrimers include, but are not limited to, catalysis, gene and DNA delivery, and biomimetics. Some of the drug delivery applications include therapeutic and diagnostic utilization for cancer treatment; enhancement of drug solubility and permeability; and intracellular delivery [15].

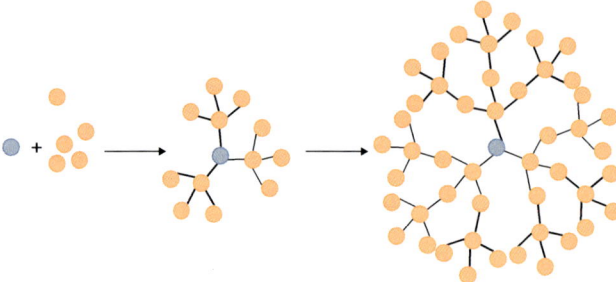

Fig. 6.4: Schematic representation of various dendrimers. These macromolecules are composed of a core molecule (blue sphere) and a known number of monomer units (orange spheres).

Polymeric nanoparticles are mainly used for the nanostructures that consist of biologically active small molecules embedded in a polymeric matrix. The polymers act as stabilizing agents to prevent the crystal growth or the aggregation of the particles. Usually, these polymers are functionalized in order to ensure the sustained release of the active substances or control the release by changing the environmental conditions, for example in pH level.

Polymeric nanoparticles having particle size between 10 and 1000 nm can be further classified into **nanocapsules** and **nanospheres**. Nanocapsules consist of biologically active small molecules in the core covered by a polymer shell. These particles have spherical hollow structures. However, in the nanospheres the active ingredients are incorporated into a polymer matrix. The active molecules are either solubilized in the polymer matrix to form an amorphous or a molecularly dispersed particle, or embedded in the polymer matrix as crystallites [16].

Polymeric micelles are formed from amphiphilic di-block or triblock copolymers having hydrophilic and hydrophobic segments (Figure 6.5). These polymers are capable of assembling into micelles in an aqueous solution achieving a polymer specific concentration, the critical micelle concentration. Poorly water-soluble substances can be solubilized in the hydrophobic segments resulting in a colloidal particle with a diameter ranging from 10 to 100 nm.

Liposomes are small artificial vesicles in the size range of 50–100 nm consisting of an aqueous core surrounded by one or more phospholipid bilayers (Figure 6.6). Liposomes made of phospholipids such as phosphatidylcholine, phosphatidylglycerol, phosphatidylethanolamine and phosphatidylserine have been intensively studied in biology, biochemistry, medicine, food and cosmetics [12].

In spite of their ability to prevent degradation of the active molecules, reduce side-effects and target drugs to site of action, low encapsulation capacity, rapid leakage of the active substance in the bloodstream, poor storage stability and high production cost strongly limit the industrial applicability of the liposomes.

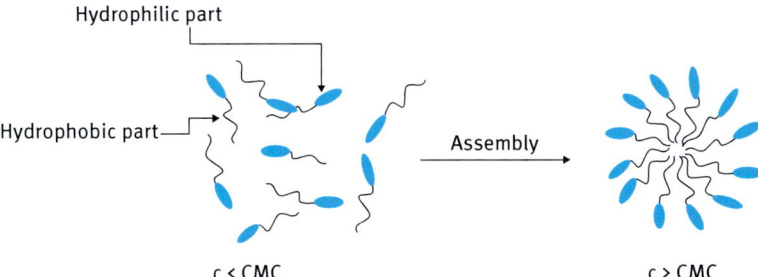

Fig. 6.5: Polymeric micelle formation by self-assembling process.

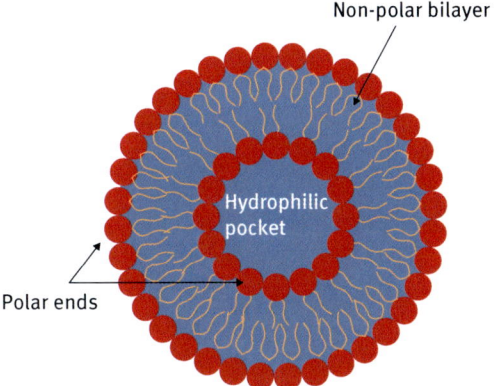

Fig. 6.6: Liposome structure formed by phospholipids.

Solid lipid nanoparticles provide an effective alternative for the liposomes due to their stability, ease of scalability and commercialisability. These nanoparticles consist of a solid lipid matrix (such as glyceryl behenat, stearic triglyceride, cetyl palmitate and glycerol tripalmitate), where the active small molecules are incorporated in it, and a surfactant or polymer layer are adsorbed on the surface to avoid aggregation and to stabilize the dispersion. The limitations of the solid lipid nanoparticles include insufficient drug loading, drug expulsion after polymorphic transition on storage and relative high water content of the dispersions.

6.2.3 Inorganic nanoparticles: metallic, bimetallic and semiconductor particles

Inorganic nanoparticles gain an intensive interest in widespread application due to their special physical properties described in Section 6.2.1. The group of inorganic nanoparticles includes the metallic and bimetallic nanoparticles (mainly from Noble metals like Au, Ag, Pt, Pd, Pt–Pd, Au–Pd etc.), oxide nanoparticles (Fe_2O_3, TiO_2, etc), semiconductor nanoparticles (or Quantum Dots, like CdSe, CdS, etc), and carbon-based nanostructures. This subchapter gives a short introduction on the various types of inorganic nanosized metallic particles – including the magnetic particles such as Fe or Co – can be generated from most metals of the periodic table. Different "bottom-up" and "top-down" techniques to obtain single- or multimetal nanoparticles such as monometallic nanoparticles like Au, Ag; Pt; Pd, Fe, Co, or bimetallic nanoparticles like FePt, AuAg or AuPt, and so on, are well known and well reported in literature (will be detailed later in Section 6.4.1). Their size-dependent physical properties listed earlier gave the opportunity to their widespread application in the field of catalysis, fuel cells, IT, electrochemical and biological sensors; and other important biological applications such as drug delivery systems, labeling, bio-sensing or bio-separations.

Semiconductors are the key units of devices used every day like computers, sensors or diodes. Semiconductors represent a special class of materials, and can be considered as material having a conductivity ranging between that of an insulator and a metal. In the ground state, the valence band is completely filled and separated from the conduction band by the "band gap". When sufficient energy is applied to a semiconductor, an excitation of electrons from the valence band into the conductive band can be observed. This excitation leaves holes in the valence band and creates electron-hole pairs. When external electric field is applied the electron and the hole will migrate in the conduction and valence bands.

The dependency of the band gap energy with the decreasing particle size has been described in Section 6.2.1. Reducing the size of semiconductors to the nanometer regime has led to materials having characteristics between the bulk and the molecular or atomic level, and that exhibits properties of quantum confinement. When the semiconductor material has a sufficiently small size – typically 10 nanometers or less, the quantum confinement effects modify the electronic and optical properties. In this case, the semiconductor nanoparticle radius is smaller than the average distance between the electron and the hole, called as the bulk exciton Bohr radius. Because of their unique size-dependent optical and electronical properties, semiconductor nanoparticles are playing an important role in the emerging new fields of applications in nanotechnology from lasers to biological fluorescence labeling.

The fabrication of semiconductor nanoparticles is, like that of metal nanoparticles, of vital relevance. Routinish synthesis of CdSe, CdS, ZnS, PbSe, GaAs semiconductor nanoparticles are well reported in the literature and will be detailed later.

6.2.4 Hybrid nanoparticles

Hybrid nanomaterials offer new opportunities for the development of novel nanostructures with improved or new synergetic properties. These materials are assembled nanosized structures consisting of inorganic and organic materials with controllable particle size, surface functionality, high drug loading, entrapment of multiple therapeutic agents, tunable drug release profile, and good serum stability [17].

Hybrid nanostructures have advantageous properties in the field of detection and biodiagnostic screening. For example, lipid-coated silica nanoparticles containing quantum dots can be used for direct visualization and quantification of the nanometer-sized structure [18].

Magnetic nanoparticles and other functional moieties like fluorescence or radiogenic tags for multimodal imaging and gene delivery can be synergistically integrated in hybrid nanoparticles to provide more accurate information in *in vitro* and *in vivo* biological systems [19].

Hybrid nanoparticles can also be utilized in cancer therapies as targeted drug delivery systems in order to modify the biodistribution of the drugs and facilitate their accumulation in the tumors. For example, magnetic nanoparticles containing immobilized drug molecules can be locally administered into the tumor. The drug release from these nanoparticles can be controlled by applying an external magnetic field at the site of the action [20].

6.3 Theoretical background of nanoparticle synthesis using flow-chemistry based approaches

The high cost associated with producing large quantities of uniform, nanosized particles limits the use of the nanomaterials in many practical applications. Flow-chemistry based approaches offer inexpensive, scalable, reproducible and safe routes for the development and production of nanosized materials. This session provides a brief overview on the principles of nanoparticle synthesis and introduces the bottom line flow technologies.

6.3.1 Principles of nanoparticle stabilization

Applying either "top-down" or "bottom-up" methods the obtained nanoparticles have to be stabilized in order to avoid the particle aggregation and the uncontrollable particle growth, and to control the particle size and the particle growth rate. The stabilizers may allow or increase the particle solubility in various solvents.

In general, the stabilization of the nanoparticles can be reached by electrostatic and steric stabilization. In electrostatic stabilization, ions are adsorbed to the surface of the particle. This creates an electrical double layer which results in a Coulombic repulsion force between the individual particles. In steric stabilization, the center of the nanoparticle is surrounded by layers of material that are sterically bulky, such as polymers or surfactants [21].

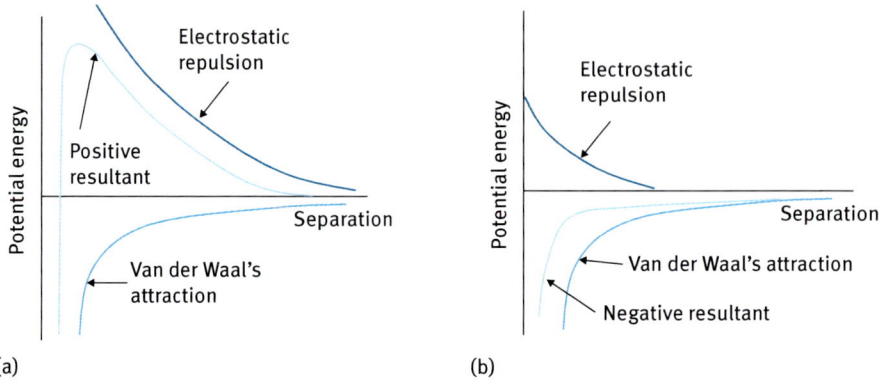

Fig. 6.7: Potential energy curves for stable (a) and unstable (b) dispersions.

> The colloid stability is the balance of various interaction forces such as van der Waals attraction, double-layer repulsion and steric interaction. These interaction forces have been described at a fundamental level by Deryaguin and Landau and Verwey and Overbeek (DLVO theory). In this theory, the van der Waals attraction is combined with the double-layer repulsion and an energy–distance curve can be established to describe the conditions of stability/instability. The electrical forces increase exponentially as particles approach one another and the attractive forces increase as an inverse power of separation. As a consequence, these additive forces may be expressed as a potential energy versus separation curve. A positive resultant corresponds to an energy barrier and repulsion, while a negative resultant corresponds to attraction and hence aggregation (Figure 6.7 (a) and (b)). It is generally considered that the basic theory and its subsequent modifications provide a sound basis for understanding colloid stability [22].

6.3.2 Classical nucleation theory

In "top-down" techniques, the large objects are modified to result in smaller species. This approach includes various litographic methods [23], condensation from a vapor using laser ablation [24], arc discharge [25], electrically heated generators, wire electrical explosion [26] or sputtering techniques [27] and mechanical methods like milling [28], or sonochemical technique [29]. For the preparation of organic nanoparticles, for example, poorly water-soluble active pharmaceutical ingredients milling techniques in various media are widely applied. Using a dry- or wet-milling process, different nanoparticulated active molecules were fabricated having enhanced pharmacological and/or physicochemical properties, such as increased bioavailability or solubility [30, 31].

The "bottom-up"-type chemical synthesis of metallic nanoparticles carried out in liquid phase involves a number of steps. First, the generation of metal atoms can be accomplished by the reduction of the metal precursor, generally a metal salt, or

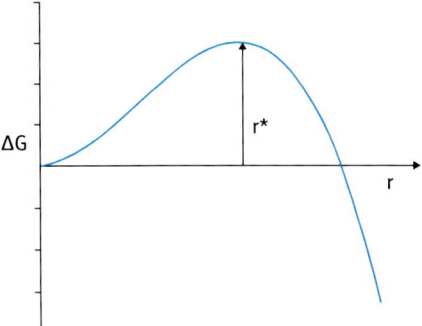

Fig. 6.8: Particle's Gibbs energy dependency on the particle size.

by the decomposition of an organometallic complex. The formed metal atoms then undergo elementary nucleation followed by a particle growth process which leads to the nanoparticle formation.

The particle formation and growth can be described by the Classical Nucleation Theory [32]. The theory describes the nucleation process in terms of the change in Gibbs free energy of the system upon transfer of molecules from the liquid phase to a solid cluster with the following equation:

$$\Delta G = -\frac{4}{V}\pi r^3 k_B T \ln(S) + 4\pi r^2 \gamma$$

where V is the molecular volume of the precipitated species; r is the radius of the nuclei; k_B is the Boltzmann constant; T is the absolute temperature; S is the saturation ratio and γ is the surface free energy per surface area.

It is clearly seen that the particle formation is strongly dependent on the saturation ratio. The particles can only be formed if $S > 1$, the system is supersaturated, and the free energy term is negative, favoring generation of solid particles and their growth.

The maximum value of ΔG corresponds to the nucleus with critical radius (r^*). A thermodynamically stable nucleus exists when the radius of the nucleus reaches to r^*. Therefore, the slope of ΔG at critical radius of nucleus will be zero [33]:

$$\frac{d\Delta G}{dr} = 0.$$

In this situation, the critical radius (r^*) of the spherical nucleus can be obtained by:

$$r^* = \frac{2V\gamma}{3k_B T \ln(S)}.$$

The behavior of formed solid particles in supersaturation solution depends on their size. Particles with the radius smaller than r^* will dissolve because this is the only way that leads to the reduction of the particle's free energy. Similarly, if $r > r^*$ particle growth will occur (see Figure 6.8).

Though both "top-down" and "bottom-up" methods are suitable for inorganic nanoparticle synthesis, the "bottom-up" approach has become favored, due to the controllability of the process. The solution phase synthesis of metal, semiconductor or organic nanoparticles can be carried out at elevated or room temperature and in batch or continuous mode. Main challenges regarding nanoparticle synthesis are to tune the size of the nanoparticle and control its dispersity. To reach these requirements, the precise control of reaction parameters and the careful selection of the ingredients are essential.

6.4 Application of flow technology in nanoparticle synthesis

6.4.1 Synthesis of metal nanoparticles

Synthesis of nanoparticles by chemical reduction of metal ions to their 0 oxidation states is a commonly used method, whose advantage is the ability of fabrication particles having different shapes like nanorods, nanowires and hollow nanoparticles. With the chemical reduction method, the shape and size of the nanoparticles can be tuned by changing the reducing agent, the dispersing agent, the reaction time and the temperature. The process uses noncomplicated equipment or instruments, and can yield large quantities of nanoparticles at a low cost in a short time. Based on this method, a wide range of metallic and bimetallic nanoparticles was synthesized such as gold [34], silver [35] platinum [36], palladium [37], iron/gold [38], iron/copper [39], cobalt/platinum [40] and so on.

Metallic nanoparticle fabrication by decomposition of an organometallic complex is a widely used method which is a suitable route for the preparation of Fe [41], Co [42], Pt [43], Au [44], FeMo [45] and so on.

For bimetallic nanoparticle preparation, the two above-mentioned synthesis routes are often used mixed, resulting in, for example, monodisperse iron-platinum (FePt) nanoparticles by reduction of platinum acetylacetonate and decomposition of iron pentacarbonyl in the presence of oleic acid and oleyl amine stabilizers [46] or FePd nanoparticles [47].

Works dealing with synthesis of gold [48] or semiconductor [49], nanoparticles in microfluidic reactors are well reported in literature, however flow-type synthesis of other nanoparticles such as silver [50], titania [51], silica [52] or ZnO [53] can also be found.

Gold nanorods (GNRs) were prepared using a coaxial flow device coupled to a coiled aging reactor[54]. The inner phase consisted of the gold source ($HAuCl_4$), and lysine (used instead of a surfactant as a colloidal stabilizer) was introduced to an outer phase of the reductant tetramethylammonium hydroxide (TMAOH). This reaction mixture was directly injected into an aging loop, where the gold nuclei were grown into GNRs. The small volumes of the microreactor allowed for fast screening of reactants and further optimization, leading to high-quality materials in a shorter amount of time, as demonstrated by the short residence time in the aging portion of the reactor (only 16 s).

Others have discovered the utility of microreactors for synthesizing magnetic cobalt ferrite nanoparticles. The composite particles were prepared via multistep synthesis in a continuous-flow device[55]. These particles are of significant importance for applications entailing magnetic imaging and therapies.

The size dependency of platinum nanoparticles on the applied flow rate in a microfuidic reactor is demonstrated in Figure 6.9. The Pt nanoparticles were synthesized by the reduction of $H_2PtCl_6 \times 6H_2O$ using methanol as reducing agent at 150 °C and 50 bar. It is clearly seen that the increasing flow rate induced a reduced residence time resulted in the formation of smaller particles.

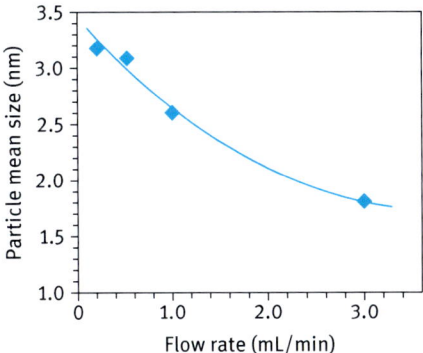

Fig. 6.9: Size dependency of platinum nanoparticles on the applied flow rate.

The successful application of metal nanoparticles in catalysis is based on the high activity and selectivity of nanoparticle-containing homogeneous or heterogeneous catalysts. Supported Au nanoparticle catalysts were found to be effective in hydrogenation or oxidation of alkenes, aldehydes and alcohols, oxidative decomposition of dioxines, C–C coupling reactions, hydroamination of terminal alkynes, water gas shift reaction, and so on. Pt nanocatalysts were used in selective hydrogenations, oxidation reactions [56], catalytic decomposition of alcohols [57], NO reduction [58], fuel cell technology, and so on. Various C–C coupling-, CO oxidation-, methane oxidation [59] and NO-reduction [60] reactions were performed over catalysts containing Pd nanoparticles.

6.4.2 Synthesis of semiconductor nanoparticles

Flow-chemistry based methods can be effectively utilized for the continuous production of semiconductor nanoparticles. For example, CdSe nanoparticles were prepared by thermal decomposition of the precursor materials (selenium in trioctylphosphine solution, cadmium acetate and trioctylphosphine) using microfluidic device. The precise reaction parameter control allowed us to tailor both the size and the surface morphology of the particles. To avoid precipitation and blockages from the evaporation of the solvent the applied pressure was chosen to be 70 bar in each case. Optical spectra of the colloid samples were recorded on-line using a flow through cell lined with a xenon light source and a spectrophotometer (Ocean Optics). Increasing the temperature of the reaction from 180 °C to 295 °C and keeping the flow rates constant, the emission wavelength of the colloid solution containing the semiconductor nanoparticles shifted to higher values indicating the formation of larger particles (Figure 6.10).

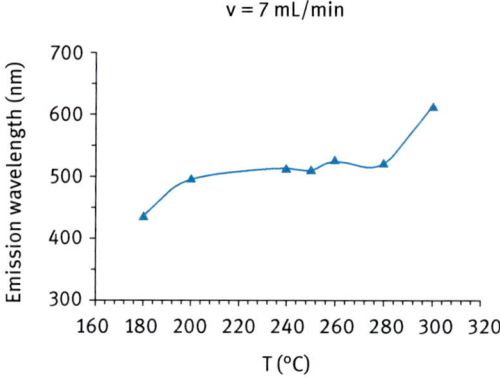

Fig. 6.10: Dependency of the optical properties on the synthesis temperature at a constant flow rate.

Varying the flow rate between 3 and 7 mL/min (decreasing the residence time from 27 to 12 seconds) resulted in blue-shift in the emission wavelength from 624 nm to 526 nm indicating the formation of smaller particles at 280 °C (Figure 6.11). Transmission electron microscopy (TEM) measurements showed that the smallest particle size was 2–3 nm.

With this reactor setup and the applied high throughput methodology it was found to be easy to follow the influence of each parameter change made on the nanocrystal structure.

6.4.3 Synthesis of biologically active organic nanoparticles

Traditional organic nanoparticle fabrication approaches rely on particle size reduction using dry- or wet-milling [61] processes. The size reduction occurs by collision of particles with the surfaces of the equipment as well as with each other. The process itself consumes a lot of energy and is characterized by the wear of the milling media

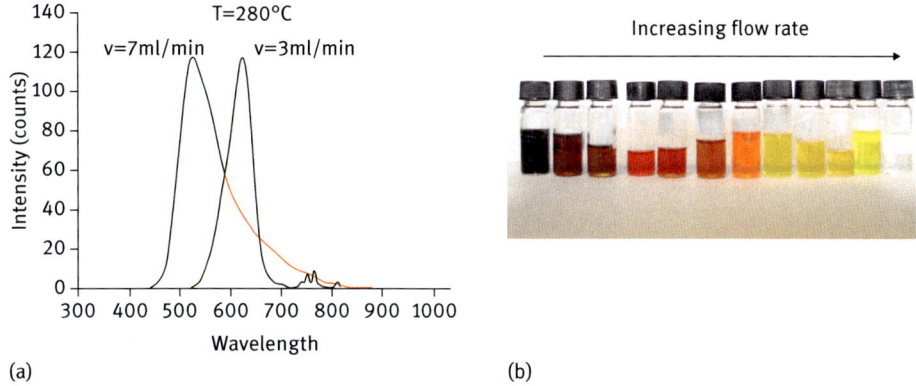

Fig. 6.11: Emission spectra recorded on samples prepared at constant temperature and different flow rates.

and with that product contamination. Due to heat-induced active form conversion, a large number of active compounds cannot be nanoformulated with this approach, for example, salt or active compounds with low melting point cannot be milled.

Recently, there has been a growing interest in exploiting the benefits of the existing flow-chemistry based technologies to overcome the shortcomings and drawbacks of the conventional batch processes.

Homogenization is a process by which the particle size of a suspension or an emulsion can be reduced down to the nanometer range. This approach combines high shear, turbulence, impact, as well as cavitation in the homogenizer. The dispersion of an active compound containing the stabilizer is passed through a very narrow gap under high pressure [62]. Number of cycles, pressure, percentage of solid content, type and combination of surfactants, hardness of active compound, and temperature of the homogenization process have strong influence on the particle size.

High-pressure homogenizers can be microfluidizers and piston gap homogenizers. In piston gap homogenizers, the powder of the active substance is dispersed in an aqueous solution and passed through a narrow homogenization gap (25 mm) at a high streaming velocity, under high pressure (500–1500 bar). The particles are broken by cavitation, high-shear forces and collision. Microfluidizers are jet stream homogenizers of two fluid streams collied frontally with high velocity (up to 1000 m/sec) under pressures up to 4000 bar [63].

Bottom-up technologies [64] provide more control over particle size, shape, and morphology as compared to mechanical processes, such as milling and homogenization. Precipitation is driven by a deviation from phase equilibrium conditions, where typical supersaturation driving forces are gradients in concentration or temperature. In general, the active compound is typically dissolved in a solvent and precipitated by addition of an antisolvent or solvent evaporation. During the precipitation process, the solubility of compound decreases drastically resulting in supersaturation, which drives nucleation. Once nucleation occurs, the particles grow by condensation [65].

Flow-chemistry based bottom-up precipitation method was successfully utilized to produce nanoparticles of biologically active small organic molecules. For example, nanoparticles of a non-steroidal inflammatory drug, Ibuprofen, was prepared in a microfluidic device. In this case, Ibuprofen sodium salt was dissolved in dextran (stabilizer) solution at 25 °C, the nanoparticles were continuously produced at atmospheric pressure due to the precipitating effect of the added hydrochloric acid solution. The particle size of the produced colloid solution was monitored by on-line dynamic light scattering (DLS) measurements. The particle size of the synthesized nanoparticles could be controlled by the flow rate and the amount of dextran in a wide range. In Figure 6.12, the effect of the antisolvent flow rate on the particle size and size distribution are shown. Changing the flow rates, the particle size could be varied from 10 up to 18 nm (see Table 6.1).

Table 6.1: Reproducibility of Ibuprofen nanoparticle production.

Transformation No	1	2	3	4	5	6	7	**Average**	**Deviation**
Particle size (nm)	16.25	15.75	16.15	16	16	15.75	16	**15.99**	**0.15**

The robustness of the flow process was also investigated. Experimental results showed that the Ibuprofen nanoparticles can be reproduced with ±0.15 nm deviation (Table 6.1).

Precipitation processes can also utilize compressed or supercritical fluids as antisolvents.

Bottom-up processes are often easier to scale-up and require less particle handling than milling and homogenization operations, resulting in higher process yields and lower impurity risks, as well as simplified cleaning and sterilization procedures. Additionally, precipitation technologies may be operated as continuous or semi continuous processes, whereas milling and homogenization operations are usually batch processes.

Nanoparticles of biologically active small molecules can be prepared by using combination technologies such as spray drying or lyophilization and high-pressure homogenization. The spray-dried or lyophilized product is dispersed in a surfactant solution and passed through a homogenizer [66]. Microprecipitation process can also be combined with high pressure homogenization with piston-gap [18] in order to produce nanoparticles. Treatment of a precipitated suspension with energy (e.g., high shear forces) avoids particle growth in precipitated suspensions (= annealing process).

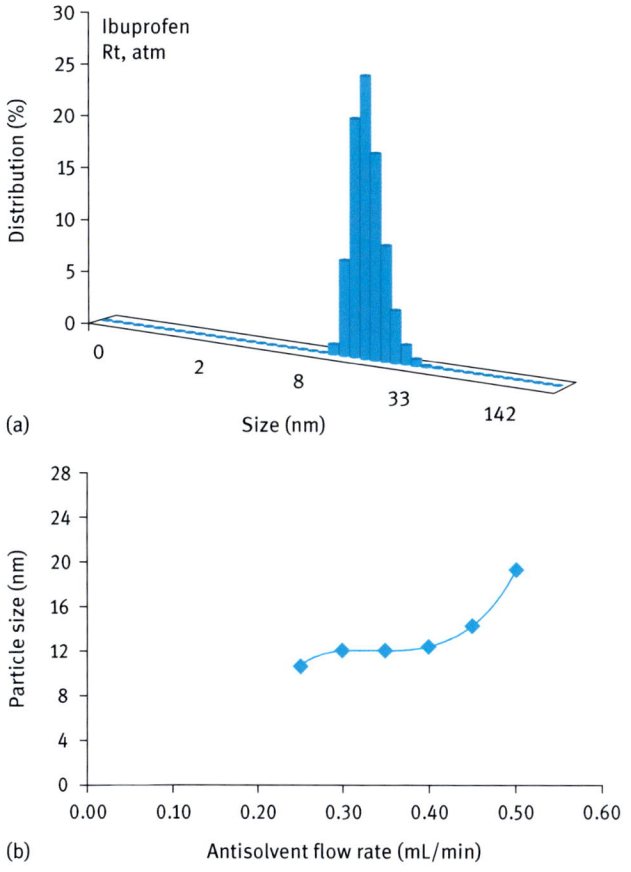

Fig. 6.12: Particle size (a) and size distribution (b) of Ibuprofen nanoparticles using different flow rates.

6.4.3.1 Applications of nanosized biologically active small molecules

The application of nanomaterials in the pharmaceutical science and medical fields is rapidly growing and has drawn increased attention recently. They can be used as fluorescent biological labels, drug and gene delivery systems and MRI contrast agents. Nanostructures can also be utilized in protein detection, DNA structure probing, tissue engineering, tumor destruction via heating (hyperthermia) and separation and purification of biological molecules and cells [67].

Recently, nanotechnology has an emerging interest in the field of cosmetics and dermal products as it offers a more effective treatment of several skin diseases than the traditional therapy. Nanocarriers used for the targeted delivery of active medicament as well as cosmetic ingredients have added advantage of improved skin penetration and depot effect with sustained release drug action. Nanomaterials in cosmetics include but are not limited to, solid lipid nanoparticles, lipid carriers, vitamin and gold loaded nanofiber facial mask, and so on [68].

Nanoparticles also have a great potential in food-related applications, such as packaging, quality monitoring, safety controlling. Nanomaterials can improve certain properties of the food products by encapsulating food components (e.g., control the release of flavors) or increasing the bioavailability of nutritional components [69].

In packaging, nanoparticles are incorporated into a traditional packaging material (e.g., metal, glass, paper, plastic) in order to slow down the leakage of gases through the packaging.

Nanostructures can also be used to monitor food quality or in detection of gases. For example, TiO_2 nanoparticles with methylene blue turn white after UV exposure.

Nanotechnology in the agrochemical industry is an enabling tool to significantly improve leaf/cuticle penetration, killing-capacity of the chemical (herbicide, insecticide or fungicide), reduce application rates of chemical per hectare (reduced environmental impact) and reduce worker exposure (improves safety by removing inflammable solvents) [70].

Novel nanopesticides are designed for sustained release applications enabling the continuous and long-term crop treatment or their controlled released forms allow the periodic treatment triggered by environmental stimuli. Smart delivery systems of herbicides could be applied seasonally in a controlled manner resulting in greater production of crops and less injury to agricultural workers.

Nanofertilizers can be used to reduce nitrogen loss caused by leaching, emissions, and incorporation in soil microorganisms. Nanomaterials allow selective release of fertilizers linked to time or environmental condition. Slow-controlled-release fertilizers may also improve the soil by decreasing toxic effects associated with fertilizer over application.

Nanosensors can detect contaminants, pests, nutrient content, and plant stress due to drought, temperature, or pressure. They could also potentially help farmers increase efficiency by applying inputs only when necessary.

6.5 Impact of nanotechnology: an outlook

The unique material properties of nanosized particles, such as different conductivity, improved reactivity, and enhanced bioavailability attributed to the small size, surface structure, chemical composition, shape, and solubility make these nanostructures attractive for different applications from catalysis to medicaments. Moreover, nanotechnology presents a unique opportunity to offer a more sustainable approach to protect public health and the environment.

Nowadays, nanotechnology is increasingly being referred to in connection with green chemistry and green engineering and manufacturing. The principles of green chemistry can be applied to produce safer and more sustainable nanomaterials and use more efficient and sustainable nano manufacturing processes.

Main elements of the future vision concerning green nanotechnology are nanostructured photovoltaics (organic, inorganic), artificial photosynthesis for fuel production, nanostructures for energy storage (batteries), solid state lighting, thermoelectrics, water treatment, medical nanorobots and smart drug delivery systems [71].

Acknowledgement: We would like to give special thanks to Professor D. Tyler McQuade for his valuable help in writing the section of synthesis of metal nanoparticles.

Study questions
6.1. What is the definition of nanomaterials?
6.2. What are the advantages of flow technologies for preparing nanomaterials?
6.3. What are the theoretical principles adaptable to prepare nanoparticles?
6.4. What are the main technologies to prepare nanoparticles?
6.5. Give a few examples of metal nanoparticles.
6.6. Give a few examples of organic nanoparticles.

Further readings
1. Zhen Guo, Li Tan. Fundamentals and Applications of Nanomaterials, Artech House nanoscale science and engineering series, Artech House, 2009.
2. Wilde G. Nanostructured Materials, 1st Edition, Oxford UK, Elsevier Science, 2008.
3. Vajtai R. Springer Handbook of Nanomaterials, Springer Berlin Heidelberg, 2013.
4. Vollath D. Nanomaterials: An Introduction to Synthesis, Properties and Applications, Weinheim, Germany, WILEY-VCH Verlag GmbH&Co. KGaA, 2008.
5. Cao G. Nanostructures & Nanomaterials: Synthesis, Properties & Applications, 1st Edition. London UK. Imperial College Press, 2004.
6. Hiemenz PC. Principles of Colloid and Surface Chemistry, Third Edition, Revised and Expanded (Undergraduate Chemistry Series), NY, USA, CRC Press, 1997.
7. Antonietti M. Colloid Chemistry I. Germany, Springer-Verlag Berlin Heidelberg, 2003.
8. Ahmed W, Jacksom MJ. Emerging Nanotechnologies for Manufacturing, 1st Edition, Oxford, UK, Elsevier, 2009.
9. Bhushan B. Springer Handbook of Nanotechnology, Springer-Verlag Berlin Heidelberg, 2010.
10. Rotello V. Nanoparticles: Building Blocks for Nanotechnology, New York, USA, Springer Science and Business Media Inc., 2004.
11. Feldheim DL, Foss CA Jr. Metal Nanoparticles: Synthesis, Characterization and Applications, New York, USA, Marcel Dekker Inc., 2002.
12. Kumar CSSR. Semiconductor Nanomaterials, Weinheim, Germany, WILEY-VCH Verlag GmbH & Co. KGaA, 2010.

Bibliography

[1] Bawa R, Bawa SR, Maebius SB, Flynn T, Wei C. Protecting new ideas and inventions in nanomedicine with patents. Nanomedicine: Nanotechnology, Biology and Medicine 2005, 1(2), 150–158.
[2] Hayashi C. Ultrafine particles. Physics Today 1987, 40(12), 44–51.
[3] Richard W, Siegel RW. Exploring mesoscopia: The bold new world of nanostructures. Physics Today 1993, 46(10), 64–68.
[4] Baird D, Shew A. Probing the History of Scanning Tunneling Microscopy. Discovering the Nanoscale, Amsterdam: IOS Press, 2004.
[5] FDA Approval for Paclitaxel Albumin-stabilized Nanoparticle Formulation. National Cancer Institute at the National Institutes of Health, 2013. (accessed September 6, 2013 at http://www.cancer.gov/cancertopics/druginfo/fda-nanoparticle-paclitaxel)
[6] Bamrungsap S et al. Nanotechnology in Therapeutics: A Focus on Nanoparticles as a Drug Delivery System. Nanomedicine 2012, 7(8), 1253–1271.
[7] Niemeyer CM, Mirkin CA. Nanobiotechnology: Concept, Applications and perspectives. Wiley-VCH Verlag GmbH&Co KGaA, Weinheim, 2004.
[8] Lee CY, Chang CL, Wang YN, Fu LM. Microfluidic mixing: A review. International Journal of Molecular Science 2011, 12, 3263–3287.
[9] Stone HA, Stroock AD, Ajdari A. Engineering flows in small devices: Microfluidics toward a lab-on-a-chip. Annual Review of Fluid Mechanics 2004, 36, 381–411.
[10] Baskoutas S, Terzis AF. Size-dependent band gap of colloidal quantum dots. Journal of Applied Physics 2006, 99 (1), 013708.
[11] Olson EA, Efremov MY, Zhang M, Zhang Z, Allen LH. Size-dependent melting of Bi nanoparticle. Journal of Applied Physics 2005, 97(3), 034304-034309.
[12] Muller RH, Bohm BHL, Grau J. Nanosuspensions: A formulation approach for poorly soluble and poorly bioavailable drugs. In Donald L. Wise (Ed.) Handbook of pharmaceutical controlled release technology, 2000, 345–357.
[13] Professor Steve Rannard. Opportunities for organic nanoparticles; Speciality Chemicals Magazine; June 2008 (Accessed September 4, 2013 at http://www.iotanano.com/newsdocs/Opportunities{%}20for{%}20Organic{%}20Nanoparticles,{%}20Speciality{%}20Chemicals{%}20Magazine{%}20June{%}202008{%}20www.specchemonline.com.pdf)
[14] Boas U, Christensen JB, Heegaard PMH, Dendrimers in Medicine and Biotechnology: New Molecular Tools, Royal Society of Chemistry, 2006.
[15] Ochekpe NA, Olorunfemi PO, Ngwuluka NC. Nanotechnology and Drug Delivery Part 2: Nanostructures for Drug Delivery. Tropical Journal of Pharmaceutical Research 2009, 8 (3), 275–287.
[16] Eerikäinen H. Preparation of nanoparticles consisting of methacrylic polymers and drugs by an aerosol flow reactor method, Phd Thesis, 2005.
[17] Sailor MJ, Park JH. Hybrid Nanoparticles for detection and treatment of cancer. Advanced Materials 2012, 24(28), 3779–3802.
[18] van Schooneveld MM et al. Imaging and quantifying the morphology of an organic–inorganic nanoparticle at the sub-nanometre level. Nature Nanotechnology 2010, 5, 538–544.
[19] Cheong J, Lee JH. Synergistically Integrated Nanoparticles as Multimodal Probes for Nanobiotechnology. Accounts of Chemical Research 2008, 41(12),1630–1640.
[20] Hung-Wei Yang et al. Potential of magnetic nanoparticles for targeted drug delivery. Nanotechnology, Science and Applications 2012, 5, 73–86.
[21] Anand M. Synthesis and steric stabilization of silver nanoparticles in neat carbon dioxide solvent using fluorine-free compounds. J. Phys. Chem. B, 2006, 110 (30), pp 14693–14701.

[22] Tadros TF. Colloid Stability: The Role of Surface Forces, Part I. WILEY-VCH Verlag GmbH & Co. KGaA, Weinheim, 2007.
[23] Langford RM. Focused ion beam nanofabrication: a comparison with conventional processing techniques, J Nanosci Nanotechnol 2006, 6(3), 661–668.
[24] Semaltianos NG. Nanoparticles by Laser Ablation. Critical Reviews in Solid State and Materials Sciences 2010, 35(2), 105–124.
[25] Xie SY et al. Preparation and self-assembly of copper nanoparticles via discharge of copper rod electrodes in a surfactant solution: A combination of physics and chemical processes. Journal of Solid State Chemistry 2004, 177, 3743–3747.
[26] Kotov YA. Electric explosion of wires as a method for preparation of nanopowders. Journal of Nanoparticle Research 2003, 5(5–6), 539–550.
[27] Chung BX, Liu CP. The synthesis of Cobalt nanoparticles by DC magnetron sputtering and the effects of electron charging. Materials Letters 2004, 58(9), 1437–1440.
[28] Chakka VM, Altuncevahir B, Jin ZQ, Li Y, Liu JP. Magnetic nanoparticles produced by surfactant-assisted ball milling. Journal of Applied Physics 2006, 99(8) 08E912.
[29] Okitsu K, Ashokkumar M, Grieser F. Sonochemical synthesis of gold nanoparticles: effects of ultrasound frequency. Journal of Physical Chemistry B 2005, 109 (44), 20673–20675.
[30] Peng Liu et al. Nanosuspensions of poorly soluble drugs: Preparation and development by wet milling. International Journal of Pharmaceutics 2011, 411(1–2), 215–222.
[31] Ghosh I, Schenck D, Bose S, Ruegger C. Optimization of formulation and process parameters for the production of nanosuspension by wet media milling technique: Effect of Vitamin E TPGS and nanocrystal particle size on oral absorption. European Journal of Pharmaceutical Sciences 2012, 47(4), 718–728.
[32] Becker, R. Doring, W. Kinetic treatment of germ formation in supersaturated vapour. Ann. Phys.-Berlin 1935, 24, 719.
[33] Mersmann A. Crystallization Technology Handbook. Taylor & Francis Group, Basel, 2001.
[34] Tapan KS, Murphy CJ. Room temperature, high-yield synthesis of multiple shapes of gold nanoparticles in aqueous solution. Journal of the American Chemical Society 2004, 126 (28), 8648–8649.
[35] Panacek A et al. Silver colloid nanoparticles: synthesis, characterization, and their antibacterial activity. Journal of Physical Chemistry B 2006, 110(33), 16248–16253.
[36] Chen J, Herricks T, Xia Y. Polyol synthesis of platinum nanostructures: Control of morphology through the manipulation of reduction kinetics. Angewandte Chemie 2005, 117(17), 2645–2648.
[37] Cookson J. The preparation of palladium nanoparticles: Controlled particle sizes are key to producing more effective and efficient materials. Platinum Metals Review 2012, 56(2), 83–98.
[38] Lin J et al. Gold-coated iron (Fe@Au) nanoparticles: Synthesis, characterization, and magnetic field-induced self-assembly. Journal of Solid State Chemistry 2001, 159(1), 26–31.
[39] Tanori J, Duxin N, Petit C, Lisiecki I, Veillet P, Pileni MP. Synthesis of nanosize metallic and alloyed particles in ordered phases. Colloid & Polymer Science 1995, 273, 886–892.
[40] Zhang X, Chan KW. Microemulsion synthesis and electrocatalytic properties of platinum–cobalt nanoparticles. Journal of Materials Chemistry 2002, 12, 1203–1206.
[41] Park SJ et al. Synthesis and magnetic studies of uniform iron nanorods and nanospheres. Journal of the American Chemical Society 2000, 122, 8581–8582.
[42] Puntes VF, Zanchet D, Erdonmez CK, Alivisatos AP. Synthesis of hcp-Co Nanodisks. Journal of the American Chemical Society 2002, 124(43), 12874–12880.
[43] Wan Wang C, Daimon H, Lee Y, Kim J, Sun S. A general approach to the size- and shape-controlled synthesis of platinum nanoparticles and their catalytic reduction of oxygen. Journal of the American Chemical Society 2007, 129 (22), 6974–6975.

[44] Zhang J, Gao Y, Alvarez-Puebla RA, Buriak JM, Fenniri H. Synthesis and SERS properties of nanocrystalline gold octahedra generated from thermal decomposition of HAuCl4 in block copolymers. Advanced Materials 2006, 18(24), 3233–3237.

[45] Li Y, Liu J, Wang Y, Wang Z. Preparation of monodispersed Fe-Mo nanoparticles as the catalyst for CVD synthesis of carbon nanotubes. Chemistry of Materials 2001, 13, 1008–1014.

[46] Sun S, Murray CB, Weller D, Folks L, Moser A. Monodisperse FePt nanoparticles and ferromagnetic FePt nanocrystal superlattices. Science 2000, 287(5460) 1989–1992.

[47] Chen M, Nikles DE. Synthesis of spherical FePd and CoPt nanoparticles. Journal of Applied Physics 2002, 91, 8477–8479.

[48] Nishikawa H, Morita T, Sugiyama J, Kimura S. Formation of gold nanoparticles in microreactor composed of helical peptide assembly in water. Journal of Colloid and Interface Science 2004, 280 (2), 506–510.

[49] Krishnadasan S, Tovilla J, Vilar R, deMello AJ, deMello JC. On-line analysis of CdSe nanoparticle formation in a continuous flow chip-based microreactor, Journal of Materials Chemistry 2004, 14, 2655–2660.

[50] Lin XZ, Terepka AD. Yang H. Synthesis of silver nanoparticles in a continuous flow tubular microreactor. Nano Letters 2004, 4(11), 2227–2232.

[51] Takagi M, Maki T, Miyahara M, Mae K. Production of titania nanoparticles by using a new microreactor assembled with same axle dual pipe. Chemical Engineering Journal 2004, 101(1), 269–276.

[52] Khan SA, Günther A, Schmidt MA, Jensen KF. Microfluidic synthesis of colloidal silica. Langmuir 2004, 20, 8604–8611.

[53] He S, Liu Y, Uehara M, Maeda H. Continuous micro flow synthesis of ZnO nanorods with UV emissions. Materials Science and Engineering B 2007, 137, 295–298.

[54] Sebastián V, Lee SK, Zhou, C, Kraus MF, Fujimoto JG, Jensen KF. Chem. Commun. 2012, 48, 6654–6656.

[55] Abou-Hassan A, Neveu S, Dupuis V, Cabuil V. RSC Adv. 2012, 2, 11263–11266.

[56] Kim SC, Shim WG, Lee MS, Jung SC, Park YK. Preparation of platinum nanoparticle and its catalytic activity for toluene oxidation. Journal of Nanoscience and Nanotechnology 2011, 11(8), 7347–7352.

[57] Mostafa S, Croy JR, Heinrich H, Cuenya BR. Catalytic decomposition of alcohols over size-selected Pt nanoparticles supported on ZrO2: A study of activity, selectivity, and stability. Applied Catalysis A: General 2009, 366(2), 353–362.

[58] Miyazaki A, Balint I, Nakano Y. Morphology control of platinum nanoparticles and their catalytic properties. Journal of Nanoparticle Research, 2003, 5(1–2), 69–80.

[59] Yuranov I, Moeckli P, Suvorova E, Buffat P, Kiwi-Minsker L. Pd/SiO2 catalysts: synthesis of Pd nanoparticles with the controlled size in mesoporous silicas. Journal of Molecular Catalysis A: Chemical 2003, 192, 239–251.

[60] Castegnaro MV, Kilian AS, Baibich IM, Alves MCM, Morai J, On the reactivity of carbon supported Pd nanoparticles during NO reduction: Unraveling a metal–support redox interaction. Langmuir 2013, 29(23), 7125–7133.

[61] Morales JO, Watts AB, McConville JT. Mechanical Particle-Size Reduction Techniques. In: RO Williams, Watts AB, Miller DA., ed. Formulating Poorly Water Soluble Drugs, Springer 2012, 133–163.

[62] Sawant, SV, Kadam JV, Jadhav KR, Sankpal SV. Drug nanocrystals: novel technique for delivery of poorly soluble drugs. International Journal of Science Innovations and Discoveries 2011, 1, 1–15.

[63] Bruno RP, McIlwrick R. Microfluidizer processor technology for high performance particle size reduction, mixing and dispersion. European Journal of Pharmaceutics and Biopharmaceutics 1999, 56, 29–36.
[64] Rowe JM, Johnston KP. Precipitation technologies for nanoparticle production. In: RO Williams, Watts AB, Miller DA., ed. Formulating Poorly Water Soluble Drugs, Springer 2012, 501–553.
[65] Weber M, Thies M. Understanding the RESS process, In: Sun YP., ed. Supercritical fluid technology in materials science and engineering, Marcel Dekker, New York, 2002, 387–437.
[66] Shegokar R, Muller RH. Nanocrystals: industrially feasible multifunctional formulation technology for poorly soluble actives. International Journal of Pharmaceutics 2010, 399, 129–139.
[67] Salata OV. Applications of nanoparticles in biology and medicine. Journal of Nanobiotechnology 2004, 2(3), 1–6.
[68] Bangale MS, Mitkare SS, Gattani SG, Sakarkar DM. Recent nanotechnological aspects in cosmetics and dermatological preparations. International Journal of Pharmacy and Pharmaceutical Sciences 2012, 4(2).
[69] Ducan TV. Applications of nanotechnology in food packaging and food safety: Barrier materials, antimicrobials and sensors. Journal of Colloid Interface Sciences 2011, 363(1), 1–24.
[70] Gruère G, Narrod C, Abbott L. Agriculture, food, and water nanotechnologies for the poor: Opportunities and constraints 2011. (Accesed September 5, 2013, at http://www.ifpri.org/sites/default/files/publications/bp019.pdf)
[71] OECD, Nanotechnology for Green Innovation, OECD (2013), "Nanotechnology for Green Innovation", OECD Science, Technology and Industry Policy Papers, No. 5, OECD Publishing.

Samar Haroun, Paul C. H. Li

7 Continuous-flow synthesis of carbon-11 radiotracers on a microfluidic chip

7.1 Introduction to continuous-flow microreactors and carbon-11 radiolabeling

> **Positron (β^+):** An antiparticle with a +1 charge, the counter part of an electron.
>
> **Radiotracer:** A chemical compound labeled with a radioactive isotope used to study biochemical and physiological processes.
>
> **Half-life:** The amount of time it takes for an isotope to decay by half. (This parameter is denoted by $t_{1/2}$ and is depicted by the following equation, $t_{1/2} = \ln 2/\lambda$, where λ is the decay factor).
>
> **Positron Emission Tomography (PET):** An imaging technique used to study biochemical and physiological processes. It uses radiotracers labeled with positron-emitting isotopes (e.g., carbon-11, nitrogen-13, oxygen-15, fluorine-18). The positron emitted annihilates with an electron and emits a pair of gamma rays that are to be detected.

Positron emission tomography (PET) is a nuclear imaging technique used in various areas of medicine such as oncology, cardiology, neurology and pharmacology. This highly sensitive imaging technique uses various radioactive probes for *in vivo* monitoring of biochemical and physiological processes [1–7]. These detectable radioactive probes (also known as '**radiotracers**') incorporate short-lived positron emitting isotopes, such as ^{11}C (see equation below) at high specific activities [2, 3, 8–11].

$$_{6}^{11}C \rightarrow {}_{5}^{11}B + {}_{1}\beta^{+} + \nu.$$

Decay of carbon-11 (^{11}C, C-11) to boron-11 by releasing a positron (β^+) and a neutrino (ν).

Carbon-11 (C-11) is a commonly used radionuclide for PET research due to the high abundance of carbon in biomolecules. In addition, biomolecules labeled with C-11 retain their biological and chemical behaviors [3].

The challenge with C-11 radiotracer production is in developing a rapid, reliable and versatile **radiolabeling** platform [2]. Considering that the radiolabeled precursors are short-lived (e.g., C-11 has a half life of 20.4 min), most production processes completed within 2–3 half-lives can result in over 75–88% radioactivity loss [12]. Therefore, large amounts of initial activity are needed to compensate for the preparation time and in some cases the low radiochemical yields (RCY). As a result, the synthesis process is usually carried out using remote operations within a shielded hot ('radioactive') cell (similar to a fumehood but made of lead). Additionally, excess amounts of the expensive precursor are used to accelerate the radiolabeling process [13]. This and other limitations add to the high costs of the radiotracer production.

> **Radiochemical yield (RCY):** The percentage of the radionuclide in the precursor (e.g., ^{11}C in [^{11}C]methyl iodide) that is transformed into that in the product (e.g., [^{11}C]raclopride).
>
> **Product purity:** This represents the percentage of the radioactive product among all products including radioactive and nonradioactive side-products.
>
> **Relative activity:** The percentage of the radionuclide activity in the desired product (e.g., [^{11}C]raclopride) among all radioactive products including side-products.
>
> **% Conversion:** The percentage of the radionuclide in the radioactive precursor (e.g., [^{11}C]methyl iodide) that is converted to the radioactive product (e.g., [^{11}C]raclopride), while disregarding any unreacted [^{11}C]methyl iodide.
>
> **Absolute activity:** Product radioactivity measured at the time of product collection. This is usually decay-corrected.

The growing use of microfluidic reactor chips for a wide range of biological and chemical applications has resulted in more research and development of new devices and lab-on-a-chip (LOC) systems [14, 15]. Therefore, the development of microfluidic reactor chips for PET radiotracer synthesis has attracted a great deal of attention over the past decade [9, 16]. The use of these chips has the potential of reducing probe precursor amounts and reducing the reaction scale even further, and with automation, these chips can provide better fluid handling and control over reaction conditions [12]. Also, the enhanced mass and heat transfer and reduced diffusion distances in the microchannels of the chips can result in an increase of RCY and product purity [12]. In addition, the small chip dimensions can reduce workspace and facilitate more localized shielding requirements [9, 12].

> **Continuous-flow:** A process that the fluids continuously move inside the flow channels. This can be carried out by pressure or by electrokinetic effects using either a peristaltic pump or an electric field, respectively.
>
> **Plug-flow:** A flow in which the velocity of the fluid is assumed to be constant across any cross-section of the flow channel perpendicular to the channel axis.
>
> **Flow mix:** A reaction mode where multiple reagent streams are introduced into the microchannels through separate inlets to be mixed when the separate channels combined into a common channel downstream.
>
> **Premix:** A reaction mode where reagents are premixed before being introduced into the microchannels.

Residence (space) time: The reaction completion time in a reactor which also corresponds to the time the reagents take to flow through the micron-sized channels for reaction.

Many investigations have demonstrated the potential applicability and benefits of implementing microfluidic modules for radiolabeling purposes (see Table 7.1) [9, 16]. For example, Lu et al. were able to radiolabel a carboxylic ester using [^{11}C]**methyl iodide** on a continuous flow T-shaped microchip that was 200 µm wide and 60 µm deep with a total volume capacity of 0.2 µL. When the precursors, 3-(3-pyridinyl)propionic acid and [^{11}C]methyl iodide were both introduced through separate inlets at a 1-µL/min flow rate, the RCYs as high as 88% (see Table 7.1, entry 4) had been achieved [17]. This was completed using smaller reagent amounts in a residence (space) time of 12 s and total processing time of 10 min, at room temperature. In these examples, the purification process was easily carried out on an analytical high performance liquid chromatography (HPLC) system and the RCYs were calculated in a similar manner to the conventional radiolabeling methods using the initial carbon-11 activity and final product carbon-11 activity [9]. Brady et al. filed a patent for a microchip with a serpentine design that was used for carbon-11 methylation reactions for synthesizing radiotracers [18].

Lab-on-a-chip (LOC): The device that is designed to incorporate laboratory processes on a 'chip' with channels that are micrometers in size. Microfluidics is a technique where small amounts of fluids are studied on the LOC device.

Polydimethylsiloxane (PDMS): A porous flexible polymer material, which is made of repeating units of $-O-Si(CH_3)_2-$, has become a favorable material to construct a chip, especially for fast prototyping.

No loop chip: A microchip with a serpentine channel but no micromixer loops incorporated.

Abacus chip: A microchip with a combination of serpentine channels and micromixer loops incorporated in the channels.

Full Loop chip: A microchip with micromixer loops incorporated throughout the microchannels.

These carbon-11 radiolabeling examples have shown reaction times and RCYs that are similar to the conventional methods. The benefits of miniaturization of the radiolabeling processes, such as reduced precursor amounts and more controlled reaction conditions, were highlighted in various reports [9, 16, 19]. However, much research is still needed before this technology can be implemented for the routine production of PET radiotracers [9, 12, 16].

Table 7.1: List of ^{11}C-labeled PET tracers synthesized on various glass microchips [9].

Compound	^{11}C-labeling reagent	Reaction condition	RCY (%)
1	[^{11}C]CH$_3$I	NaOH, acetone, RT	Not reported
2	[^{11}C]CH$_3$I	Acetone, RT	10%
3	[^{11}C]CH$_3$I	NEt$_3$, acetone, RT	5–19%
4	[^{11}C]CH$_3$I	Bu$_4$NOH, DMF, RT	56%*, 88%**
5	[^{11}C]CH$_3$I	Bu$_4$NOH, DMF, RT	45%*, 65%**

RT = room temperature, Pd = palladium, DMF = N,N-dimethylformamide,
*10 µL/min, **1 µL/min.

7.2 Microfluidic synthesis of raclopride

The radioactive synthesis of a ^{11}C radiotracer, **[^{11}C]raclopride ([^{11}C]rac)** will be explored. This is an important compound for brain research due to its high binding affinity and selectivity as a dopamine D2/3 receptor antagonist [20]. Currently, [^{11}C]raclopride is synthesized in a methylation reaction, using [^{11}C]methyl iodide (see Figure 7.1) [17]. The reaction consumes up to 1.7 mg of expensive precursor (desmethyl

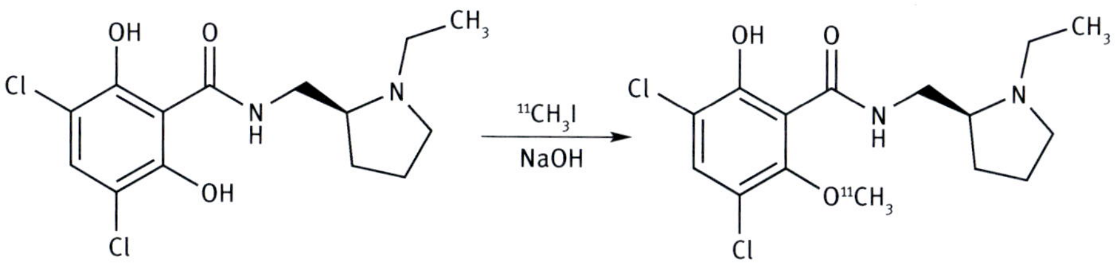

7.2 Microfluidic synthesis of raclopride

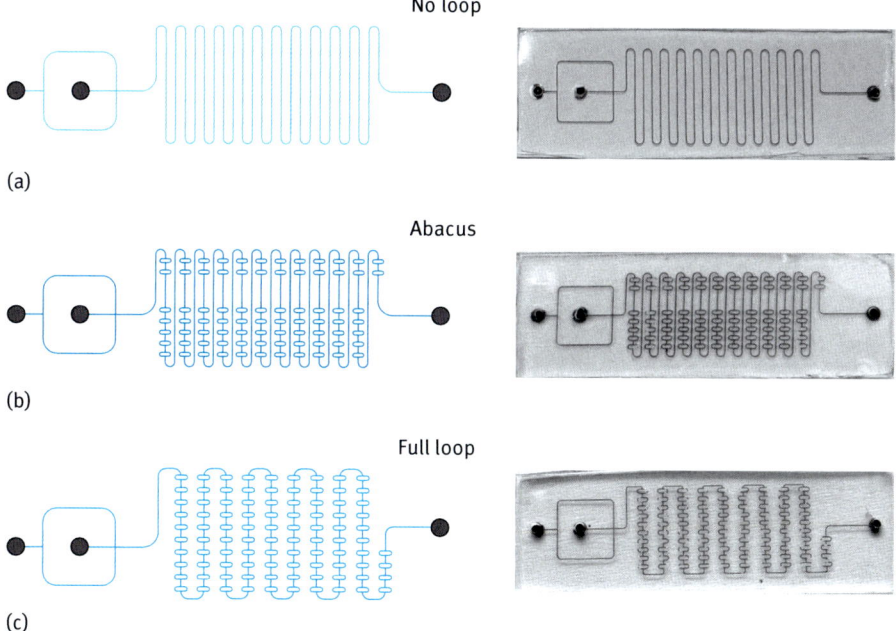

Fig. 7.1: [^{11}C]Raclopride synthesis using [^{11}C]methyl iodide [3, 13, 21].

raclopride) with a total production time of up to 45 min, limiting the commercial availability of [^{11}C]raclopride [3, 4, 13, 21–23].

A continuous-flow microreactor chip will be used for the rapid, efficient and safe radiosynthesis of [^{11}C]raclopride. The operation of using the chip involves the continuous pumping of fluids into the microchannels (plug flow) and collection of product from the outlets [24]. The use of this type of microreactor is found to shorten the reaction time, reduce the precursor consumption, and allow for safe and remote operation.

Fig. 7.2: Schematics and images of the microchips of 3 micromixer designs (45 mm × 15 mm): (a) no loop chip (total channel length = 272 mm, volume = 2.7 µL), (b) abacus chip (length = 437 mm, volume = 4.4 µL), and (c) full loop chip (length = 356 mm, volume = 3.6 µL). All channels were 100 µm wide and 100 µm deep.

The 2D passive micromixer structure in the microreactor chip was initially explored with the intention to design a portable analytical system that was required to mix and analyze samples using a hand-operated vacuum source [25]. It was anticipated that the micromixer loop design would induce more mixing, as compared to a serpentine or a T-shaped microfluidic reactor chip, previously used for radiolabeling [9, 17, 26–28]. Inspired by these, three microchip designs (shown in Figure 7.2) were made: no loop chip (or serpentine, no micromixer loop designs were incorporated), abacus chip (micromixer loop design and serpentine design were incorporated in the microchannel), and full loop chip (micromixer loop design was incorporated throughout all the microchannels). The split-and-recombine path of the microreactor loop design is expected to induce passive mixing through splitting and recombining the fluid streams. Therefore, the full loop chip that has more microreactor loops is expected to have more efficient mixing and higher yield. Polydimethylsiloxane (PDMS) was used to prototype the microreactor chip using a single-layer mold to fabricate the chip [29, 30]. Once the PDMS mold (1.2 mm thick) is created it is bonded permanently to a glass wafer (0.6 mm thick) using an O_2 plasma system. The dimensions of the PDMS-glass microchips were 45 mm (length) × 15 mm (height) × 1.8 mm (thickness), with channel depth of 100 μm. The polymer (PDMS) material was used due to its low cost of fabrication and its compatibility with the solvents (dimethyl sulfoxide and acetonitrile) used in the [^{11}C]raclopride synthetic process.

7.2.1 Microfluidic nonradioactive synthesis of raclopride

The microfluidic reactor chips were first used to perform nonradioactive synthesis of raclopride by mixing the precursor and methyl iodide. So, two solutions were made: (1) 0.8 mg of desmethyl raclopride (2.4 μmol), unless otherwise noted, was dissolved in 400 μL of dimethyl sulfoxide and mixed with 7 μL of 5 M NaOH solution and (2) 120 μmol of methyl iodide added to HPLC-grade acetonitrile (750 μl). Both solutions were loaded on two separate syringes and the experiment was carried out using a setup similar to the one shown in Figure 7.3. The microchip was mounted in a microchip holder for interfacing to the fluid delivery system. The microchip holder was then placed on a hot plate, with the temperature monitored using a calibrated thermocouple. During collection of product with a volume of 40 μL (or 80 μL, depending on the collection time), it was immediately quenched using 100 μL (or 200 μL) of 0.1 M ammonium formate (HC(O)ONH$_4$) solution. Multiple runs at various temperatures were carried out using the same batch of solutions at several intervals. The product samples were then analyzed using HPLC and the raclopride yield was then quantified using an internal standard (IS), 2,4-dihydroxybenzoic acid.

The fluidic connections were made using polyether ether ketone (PEEK) tubing (OD = 1.6 mm), with the inlet tubing ID of 175 μm and outlet tubing ID of 400 μm. The tubing with a larger ID was used at the outlet in order to avoid pressure build up [19].

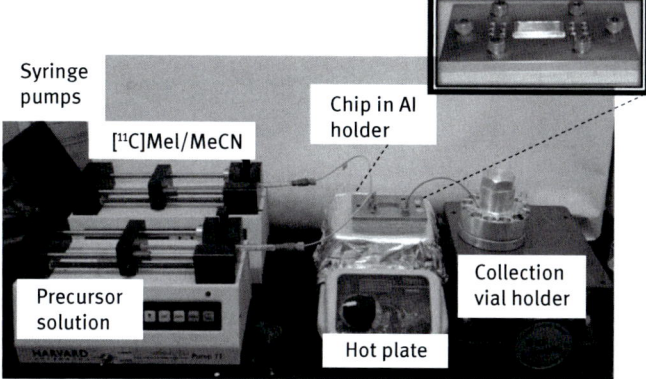

Fig. 7.3: Microchip synthesis reaction setup in which syringe pumps are used to deliver the precursor solution (desmethyl raclopride (DMR) in dimethyl sulfoxide (DMSO)) and [^{11}C]methyl iodide ([^{11}C]MeI) solution (in Acetonitrile (MeCN)). See [52] for a colored figure of this setup.

High Performance Liquid Chromatography (HPLC): A chemical separation method of components in a liquid that is based on the differential partition of the components between the stationary phase and mobile phase in the chromatographic column.

UV-HPLC: HPLC separation that uses an ultraviolet (UV) detector.

RAD-HPLC: HPLC separation that uses radioactivity (RAD) detectors/counters.

Internal standard (IS): A compound (e.g., 2,4-dihydroxybenzoic acid) added in constant amount to the standard solutions (with reagents and products) for the calibration of the product concentrations.

Table 7.2: Results of nonradioactive synthesis of raclopride on the no loop and abacus microchip designs using 2.4–3.9 μmol of desmethyl raclopride and 7.5 μL (120 μmol) methyl iodide in 750 μL of acetonitrile at 62–92 °C.

Temperature (°C)	Raclopride yield (%)	
	No Loop chip[a]	Abacus chip[b]
62	–	2.8 ± 0.7[c]
72	0.7	3.3 ± 0.1[c]
82	1.3	5.8
92	2.2	–

Note: [a] 3.9 μmol of desmethyl raclopride was used.
[b] 2.4 μmol of desmethyl raclopride was used.
[c] Errors represent the standard deviation of 2–3 experimental results. In other cases, errors are < 0.7% according to the calibration curve.

As shown in Table 7.2, at the same temperature, the abacus chip produced a higher yield than that of the no loop chip. It was believed that the micromixer loop design induces mixing through the split-recombine path of the reagent streams. Also, the split-recombine path increases the microchannel length, the reagents' exposure to high temperatures and the reaction time, possibly contributing to the higher yields.

7.2.2 Microchip radioactive synthesis of [^{11}C]raclopride

Then, the radioactive synthesis of [^{11}C]raclopride was conducted in a setup shown in Figure 7.3. In this case, an aluminum rotating collection vial holder capable of holding 12 vials, was used to facilitate vial switching during consecutive collections and to decrease radiation exposure due to reduced contact with vials by the user. Desmethyl raclopride (0.8 mg or 2.4 µmol, unless otherwise noted) was dissolved in 250 µL of extra dry dimethyl sulfoxide in a reaction vial and mixed with 7 µL of 5 M NaOH. On the other hand, 50–80 mCi of [^{11}C]methyliodide, which varied by the different amounts of [^{11}C]CH$_4$ supplied by the cyclotron, was trapped in 250 µL of extra dry acetonitrile. With a specific activity of 10 Ci/µmol, the radioactivity of [^{11}C]methyl iodide translated to 5–8 nmol [13]. For each batch of desmethyl raclopride, multiple runs were conducted using various flow rates and collection times. A 100-µL (or 200 µL) of 0.1 M ammonium formate (HC(O)ONH$_4$) solution was added to each collection vial so that the product sample of 40 µL (or 80 µL) can be quenched immediately after collection.

Based on the findings of nonradioactive work, the abacus chip was selected to conduct radioactive synthesis. Figure 7.4 shows the results of the UV-HPLC and RAD-HPLC analysis of the product mixture. On the lower trace, a nonradioactive (^{12}C) raclopride peak was not observed, indicating that the concentration was below its detectable level in the product sample. Meanwhile, on the RAD-HPLC trace (inverted) the corresponding radiolabeled products, [^{11}C]raclopride and unreacted [^{11}C]methyl iodide were observed, among substantial ^{11}C side-products formed. Since the yield (now based on the limiting reagent [^{11}C]methyl iodide) was not optimized, the relative activity of [^{11}C]raclopride (as defined in Equation (7.1)), but not the radiochemical yield (RCY), was adopted for comparison purpose. The [^{11}C]raclopride relative activity represents the percentage of C-11 activity (from [^{11}C]methyl iodide) used to make [^{11}C]raclopride.

$$[^{11}C]\text{Rac relative activity (\%)} = \frac{[^{11}C]\text{Rac RAD HPLC pk area} \times 100\%}{\text{Total RAD HPLC pk areas}}. \quad (7.1)$$

As shown in Table 7.3, three consecutive experiments were conducted at the flow rate of 10 µL/min (with residence time of 13 s). The relative activities of [^{11}C]raclopride in the three runs are quite reproducible, giving rise to 2.0%. This value is low which is in part due to the under-utilization of [^{11}C]methyl iodide, resulting in substantial

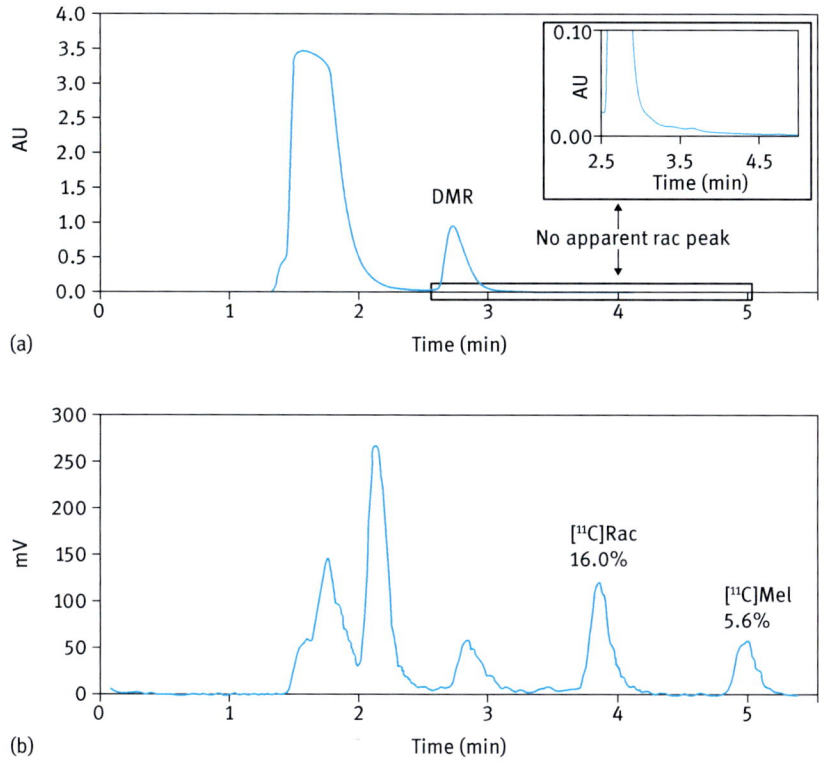

Fig. 7.4: (a) UV-HPLC and (b) RAD-HPLC for [^{11}C]raclopride ([^{11}C]Rac) synthesis on the abacus microchip at 82 °C and a flow rate of 2 µL/min in premix mode. Desmethyl raclopride (DMR) was detected at ~ 2.7 min on the UV-HPLC and [^{11}C]raclopride and [^{11}C]methyl iodide ([^{11}C]MeI) were detected at ~ 3.8 and 5.0 min on the RAD-HPLC, respectively.

relative activity of unreacted [^{11}C]methyl iodide. However, because of its low UV absorbance, unreacted methyl iodide was undetected in UV-HPLC.

In an attempt to improve the production of [^{11}C]raclopride, the rate of fluid delivery into the microchannels or the liquid flow rate was decreased. It was believed that the decrease in flow rate had increased the residence time, leading to more methylation products and less unreacted [^{11}C]methyl iodide. For instance, a decrease in flow rate from 10 µL/min to 2 µL/min increased the residence time from 13 to 66 s and the relative activity of [^{11}C]raclopride from 2.0% to 5.1%.

Although the relative radioactivity of unreacted [^{11}C]methyl iodide decreased from 84.0% to 51.8%, that of the undesirable ^{11}C side-products increased from 14.0% to 43.1%. This suggested that the increased residence time also enhanced side-product formation, possibly through other methylation processes or the decomposition of the radiolabeled products [31]. The undesirably high side-product formation indicates that the microchip flow mixing might be efficient for other fast and simple methyla-

Table 7.3: Results of [^{11}C]raclopride synthesis using the abacus microchip where 0.8–1.0 µmol of desmethyl raclopride and [^{11}C]methyl iodide are reacted at 82 °C, at various flow rates.

Flow rate (µL/min)	Calculated residence time (s)*	[^{11}C]Raclopride relative radioactivity (%)
10	13	2.0
		2.0
		2.0
5	26	4.0
2	66	5.1
1	131	6.7

Note: *residence time is calculated using the flow rate and the microchannel length.

tion syntheses, while not performing efficiently with the kinetically slower formation of [^{11}C]raclopride, resulting in its low relative activity.

For further investigation of the residence time effects, the reactant solutions (desmethyl raclopride and [^{11}C]methyl iodide) were premixed before being introduced into the microchannels. This premix mode showed an increase in the [^{11}C]raclopride relative activity with reduced levels of unreacted [^{11}C]methyl iodide, in comparison to the flow mix mode. As shown in Table 7.4, at a flow rate of 10 µL/min, the [^{11}C]raclopride relative activities increased by a factor of ~ 5 in the premix mode as compared to the flow mix mode. The highest relative activity of 16% was observed when the reagents were introduced at a flow of 2 µL/min. Premixing the reactants increases the extent of interaction between desmethyl raclopride and [^{11}C]methyl iodide in the microchannel, even before reaching the T-junction. This operation mode may also prevent [^{11}C]methyl iodide from being too concentrated which can prevent it from reacting with the chip substrate, rather than with desmethyl raclopride.

Another factor explored was the microchip design; synthesis on the full loop chip in addition to the abacus chip was performed. Results in Table 7.3 confirmed the increased efficiency when premixing the reagents for the [^{11}C]raclopride synthesis us-

Table 7.4: [^{11}C]raclopride conversion and absolute activity at various flow rates using the abacus and full loop microchip in the premix mode, at 82 °C.

Microchip	Flow rate (µL/min)	Calculated residence time (s)	[^{11}C]Raclopride relative activity	[^{11}C]Raclopride conversion (%)	*[^{11}C]Raclopride absolute activity (µCi)
Abacus	10	13	10.0	20.5	411
	5	26	11.2	16.8	386
	2	66	16.0	16.9	330
Full loop[#]	2	53	25.6 ± 0.6	25.6 ± 0.6	441 ± 138

Note: *decay-corrected
[#]errors for the full loop microchip represent the standard deviation of 4 ($n = 4$) experimental results.

ing the full loop microchip (25.6%) over the abacus design (16.0%) at a flow rate of 2 µL/min. This improvement was attributed to a higher ratio of micromixer loops to channel length (loop number/length in mm) in the full loop microchip (0.26 mm^{-1}), as compared to the abacus microchip design (0.18 mm^{-1}). The higher relative activity obtained with the full loop chip is attributed to the better mixing and hence better utilization of [^{11}C]methyl iodide, as quantified by the parameter called [^{11}C]raclopride conversion, as defined by Equation (7.2). This parameter determines the efficiency of converting [^{11}C]methyl iodide into ^{11}C compounds, not ^{11}C side products, and disregards the unreacted [^{11}C]methyl iodide, that is, a higher conversion indicates less side-products formed. The highest conversion of 25.6 ± 0.6% was observed when the reagents were flowed in at 2 µL/min in the full loop chip. However, a conversion of 25.6% meant a side-product level of 74.4%, most likely due to the radioactivity loss through the use of the PDMS polymer as the microchip substrate (*vide infra*).

$$[^{11}\text{C}]\text{Rac conversion (\%)} = \frac{[^{11}\text{C}]\text{Rac RAD HPLC pk area} \times 100\%}{\text{Total RAD HPLC pk areas} - \text{unreacted } [^{11}\text{C}]\text{MeI RAD HPLC pk areas}}. \quad (7.2)$$

From the relative activity of [^{11}C]raclopride, the absolute activity of [^{11}C]raclopride was calculated, (see Table 7.3). After starting with 7.48 mCi of [^{11}C]methyl iodide, the actual [^{11}C]raclopride activity was as high as 164 µCi (411 µCi after decay correction) for the abacus chip's premixed experiment at a flow rate of 10 µL/min. This absolute activity was obtained at a short time (3 min), and lower precursor amounts (~1/10 [^{11}C]methyl iodide and ~1/20 of desmethyl raclopride), as compared to the conventional method. Although the specific activity cannot be evaluated due to the undetected levels of raclopride nonradioactive mass, it is noted that the final activity is comparable to the doses used in some [^{11}C]raclopride animal studies [32–34]. For example, 20 MBq (or 500 µCi) of [^{11}C]raclopride activity was used to study the affect of amphetamine on the binding affinity of dopamine D2 receptors over a 90 min period [32].

Computational fluid dynamics (CFD): A technique used to analyze the dynamic flow of fluids in confinement of different geometries using numerical methods and algorithms where calculations are carried out on a computer.

COMSOL Multiphysics: A software used for modeling and simulation of problems such as fluid flow, heat transfer, and so on, based on multiple physical principles.

Reaction engineering lab (REL) module: A modeling platform used to investigate the reaction kinetics and to generate the "ideal" microreactor conditions where it is assumed that reagents are perfectly mixed.

Microelectrochemical systems (MEMS) module: A modeling platform combines the optimized reaction conditions from the REL module with the microreactor geometry in order to investigate the reaction progress at different parts of the microreactor.

7.3 Computational fluid dynamics (CFD)

Obtaining an efficient microfluidic reactor design for a particular application can be done more accurately by implementing a computational fluid dynamics (CFD) investigation. In such an investigation, the information of the fluid dynamics and the reaction kinetics is compared with the experimental data [14, 35]. This allows the researchers to evaluate the effect of the flow and mass transfer within the microchannel on the reaction progress and product formation. Computational analysis displays many advantages, especially the ability to generate information about a particular microfluidic reactor design that can be used to compare with the experimental data to enhance a particular microreactor design with optimal conditions [35].

Extensive CFD analysis has been carried out to study micromixer designs such as the T-shaped micromixer chip. This, in many cases, has helped create new and more efficient designs to overcome slow diffusion in the microchannels [14, 36–38]. The modeling process usually involves generating a simple and realistic model that is able to incorporate various geometries while evaluating parameters such as the diffusion coefficient, reaction temperature, flow rate, residence time and so on. To study the effect of various reaction conditions on chemical selectivity and yields, it has become important to incorporate both the momentum and mass transfer for a more accurate estimation of the microfluidic reactor conditions [14]. The CFD study allows for the investigation of micromixing efficiency at different Reynold numbers (Re) while implementing the reaction kinetics and exploring the mass flux at different parameters as well [39–42]. Variations in reactor lengths, residence times, micromixer structures and reaction conditions (temperature, pressure, rate constants) and many more parameters are all possible through CFD analysis. For example, Buddoo *et al.* investigated the first-order synthesis of linalool (a component used in fragrances) on a multichannel microfluidic reactor (14 channels, 0.5 mm × 0.4 mm) in comparison to a mini-channel tubular reactor (ID = 4.6 mm) [35]. They were also able to explore the effects of temperature and residence time on the decomposition of the compound during the synthesis process. The simulation results estimated the optimal reaction parameters for the synthetic process which were then validated by experiments [35].

The microchip radiosynthesis of [^{11}C]raclopride (see Figure 7.1) was studied by CFD, using a commercial software package, COMSOL Multiphysics®. Although the flow and mass transfer in microchips have been previously studied [43], reaction kinetics has not been included, especially not for the purpose of radiotracer synthesis investigation. Initially, the COMSOL's reaction engineering lab®(REL) module was used to investigate the reaction kinetics and generate the "ideal" microreactor conditions while assuming that reagents are perfectly mixed [44]. Next, the reaction kinetics were coupled with the microelectromechanical system (MEMS) module (with flow and mass transfer models) in order to investigate the reaction progress in the microchip geometry. The simulation using these modules is conducted by varying the conditions such as the flow rate, reagent concentrations, diffusion coefficient, rate constant (k) and reaction mode. For these simulations, the rate constant and diffusion coefficient values were also varied in order to cover the estimated reaction condition ranges possibly employed during experimental investigations [45–48].

7.3.1 Reaction engineering lab®(REL) module – "ideal" flow-reactor model

Initially, a **plug-flow reactor** (**PFR**) was used to serve as the ideal flow-reactor model [36]. PFR was used for the comparison of the radiotracer synthesis in two microreactors with varying lengths independent of their microchannel geometry. Two plug-flow reactors, PFR 1 (3.10×10^{-10} m^3, width or height = 100 μm, length = 31.0 mm) and PFR 2 (4.56×10^{-10} m^3, width or height = 100 μm, length = 45.6 mm) are modeled and compared.

> **Continuity equation:** Describes the transport of a quantity such as mass that is conserved.
>
> **Incompressible Navier–Stokes equation:** Is a nonlinear differential equation used to calculate the pressure-driven flow of an incompressible liquid within the microchannels. This equation arises from applying Newton's second law to liquid motion with the assumption of a viscous flow.
>
> **Fick's law of diffusion:** A differential equation that relates diffusive flux of a diffusing species to its concentration gradient, in which the species moves from an area where concentrations are high to an area where concentrations are low.
>
> **Convection-diffusion equation:** Describes the transport of solutes inside a physical system through diffusion and convection.
>
> **Diffusion coefficient:** The proportionality constant of *Fick's law of diffusion* which is the ratio between the diffusive flux and the concentration gradient of the diffusing species.
>
> **Rate constant:** Is a constant in chemical kinetics that relates the rate of a reaction to the reagent concentrations.

The expressions for rates of change of the nonradioactive chemical species are shown in Table 7.5. In the radioactive reaction, due to the decay of carbon-11 to boron-11, it is necessary to account for the loss in [^{11}C]methyl Iodide and [^{11}C]raclopride. This is done by including the decay factor (λ) which is given by $\ln 2/t_{1/2}$, where $t_{1/2}$ is the half-life of the decay of the ^{11}C radionuclide (20.4 min.) [46, 49]. The two reagents are introduced simultaneously into the plug-flow reactors with the same volumetric flow rate. Using the results for different variables (concentrations, flow rates and k, see Table 7.6) on both PFR 1 and PFR 2, the progress along the microreactors' length was evaluated by examining the normalized yields.

Similarly, the flow rate variation study was carried out on another plug-flow reactor, PFR 2, with a longer channel ($L = 45.6$ mm). With respect to PFR 1, at a flow rate of 15 μL/min, it became apparent that a reactor with a longer microchannel (PFR 2)

Table 7.5: The rates of changes of various chemical species involved in the simulations of radioactive synthesis and nonradioactive synthesis.

Species	Nonradioactive synthesis	Radioactive synthesis
$d[\text{DMR}]/dt$	$-k[\text{DMR}][\text{MeI}]$	$-k[\text{DMR}][\text{MeI}]$
$d[\text{MeI}]/dt$	$-k[\text{DMR}][\text{MeI}]$	$-k[\text{DMR}][\text{MeI}] - \lambda[\text{MeI}]$
$d[\text{Rac}]/dt$	$k[\text{DMR}][\text{MeI}]$	$k[\text{DMR}][\text{MeI}] - \lambda[\text{Rac}]$

Note: k (m^3/mol/s) denotes the rate constant, and λ the decay factor.

had an effect of producing higher [^{11}C]raclopride yields as a result of a longer residence time (see Figure 7.5). For example, at a flow rate of 15 µL/min, increasing the microchannel length from 31.0 mm (PFR 1) to 45.6 mm (PFR 2) showed an increase in the normalized [^{11}C]raclopride yield by a factor of ~1.5, which was similar to the increase in the length ratio between the two microreactors. Furthermore, reducing the flow rate from 15 µL/min to 2 µL/min, PFR 1 showed an increase in the normalized [^{11}C]raclopride yield from 1.0% to 7.3%, while PFR 2 showed a higher increase from 1.5% to 10.6% showing a ~ 7-fold increase for both PFRs. This behavior was similar to the flow rate variations in the experimental results (see Table 7.3).

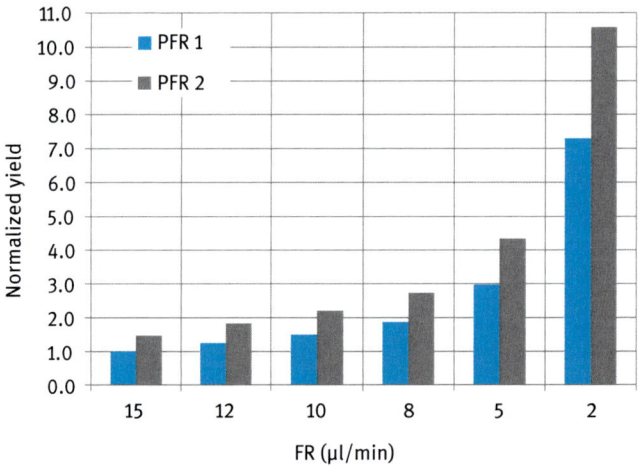

Fig. 7.5: Normalized final [^{11}C]raclopride yields in PFR 1 (31.0 mm) and PFR 2 (45.6 mm) in a flow rate variation experiment.

7.3.2 Microelectromechanical system (MEMS) module – "geometry-dependent" flow study

Next, the effect of geometry on mass transport in the synthetic process due to liquid flow was studied. This study is based on a 2D model of the repeating units (RU) of three microchip designs (see Figure 7.6).

The velocity field property of the fluid (i.e., the solvent) was represented by the continuity equation (Equation (7.3)) and incompressible Navier–Stokes equation (Equation (7.4)) [15].

$$\nabla \cdot \vec{u} = 0 \tag{7.3}$$

$$\rho(\vec{u} \cdot \nabla)\vec{u} = \eta \nabla^2 \vec{u} - \nabla p + B \tag{7.4}$$

where $\vec{u}$ denotes the velocity in the microchannels (m/s); p is the pressure (Pa); η is the dynamic viscosity (Pa · s); B is the body force per unit volume (N/m³) which is the gravitational force usually canceled by buoyancy force in microfluidic applications and hence neglected [15].

The mass transport of a chemical species in the microfluidic reactors is assumed to be based on diffusion unless micromixer structures (passive or active) are introduced to the channel design to induce convection [50]. First, the diffusive flux (J) of a species can be described using Fick's first law of diffusion,

$$J = -D\nabla c \tag{7.5}$$

where J is the diffusive flux (mol/m² · s); D is the diffusion coefficient; ∇c is the concentration gradient [15, 24, 50]. Then, the mass transfer of species i is formulated using the convection-diffusion equation as follows:

$$\frac{\partial c_i}{\partial t} - \nabla \cdot (D_i \nabla c_i) = R_i - \nabla \cdot (\vec{u} c_i) \tag{7.6}$$

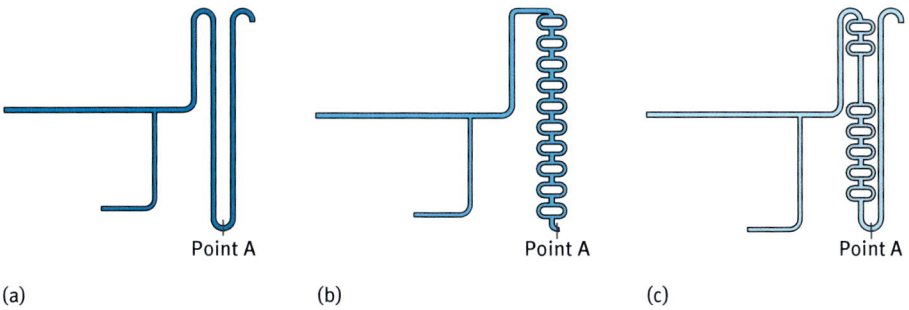

Fig. 7.6: Schematic of (a) no loop, (b) full loop and (c) abacus microchips' RUs showing the endpoint A used to determine the concentration of [^{11}C]raclopride.

where D_i (m²/s) is the diffusion coefficient of species i; c_i (mol/m³) is the concentration of species i; $\vec{u}$ is the velocity (m/s) [51]; R_i denotes the rate of reaction (mol/m³/s) identifying the rate of consumption of the reactants or production of reaction products (see Table 7.6).

To simulate the effects of variation of flow rate, reagent species concentrations, microchip design and reaction mode parameters, the MEMS module was used. The product concentrations and the yields that were calculated at the RUs' outlet were used to evaluate the reaction efficiency.

The concentrations of the reagents (desmethyl racloipride and methyl iodide) used in the model are based on the experimental values. A high methyl iodide concentration (53.4 mol/m³) was adopted as previously used in the nonradioactive synthesis of raclopride, and the low [¹¹C]methyl iodide concentrations (0.024 mol/m³) was adopted as previously used in the radiosynthesis of [¹¹C]raclopride [48, 52]. Meanwhile, the rate constant covers a range determined from a similar methylation investigation of a phenoxide site (3-nitrophenol) (see Table 7.6) [46–48]. For instance, a value of k of 5.663×10^{-4} m³/mol/s was used, which was quoted in the literature in terms of the decay factor, λ (1/min).

Furthermore, since the reagent species are dilute in our experimental process, their diffusive properties are estimated from the diffusion coefficient of the dominant species in solution, the reaction solvents (i.e., dimethyl sulfoxide and acetonitrile) [45, 52, 53]. Furthermore, the reaction mode was varied by introducing the reagents through the same inlet, indicating a premixed environment (premix mode), as compared to the flow mix (default) mode where reagents are introduced through separate inlets.

When evaluating the concentration of [¹¹C]raclopride generated at point A (see Figure 7.6) which is the outlet in the RUs of the 3 chip designs, the highest concentration is observed with the full loop design (see Table 7.7). The concentrations acquired with the full loop design were determined to be more than two-fold of those generated with the no loop design at various flow rates.

From the surface concentration plots (see Figure 7.7), an increase in the [¹¹C]raclopride concentration is observed when the flow rate is reduced, and when the mixing mode is switched from flow mix (Figure 7.7 (a–d)) to premix (Figure 7.7 (e–h)). When

Table 7.6: The range of inputted parameters for all simulations.

Parameter	Nonradioactive synthesis	Radioactive synthesis
Flow rate (µL/min)	2–15	2–15
Reagent concentration (mol/m³)	Desmethyl raclopride: 21.6 Methyl iodide: 53.4	Desmethyl raclopride: 21.6 Methyl iodide: 0.024
k value range (m³/mol/s)	5.663×10^{-4}–5.663×10^{-8}	5.663×10^{-4}–5.663×10^{-8}
D (m²/s)	10^{-8}–10^{-9}	10^{-8}–10^{-9}

calculating the concentration at the end of the 10th loop of the full-loop chip design, it is apparent that premixing the reagents before introducing them into the microchannels produced much higher product concentrations than in the flow mix mode.

In addition, other parameters such as the diffusion coefficient, reagent concentrations and reaction rate constant were varied (see Table 7.8). As expected, the results showed that an increase in the value of D (from 10^{-9} to 10^{-8}) will slightly increase the normalized yield by a factor of ~ 1.1 (from 17.6% to 18.7%) which was lower

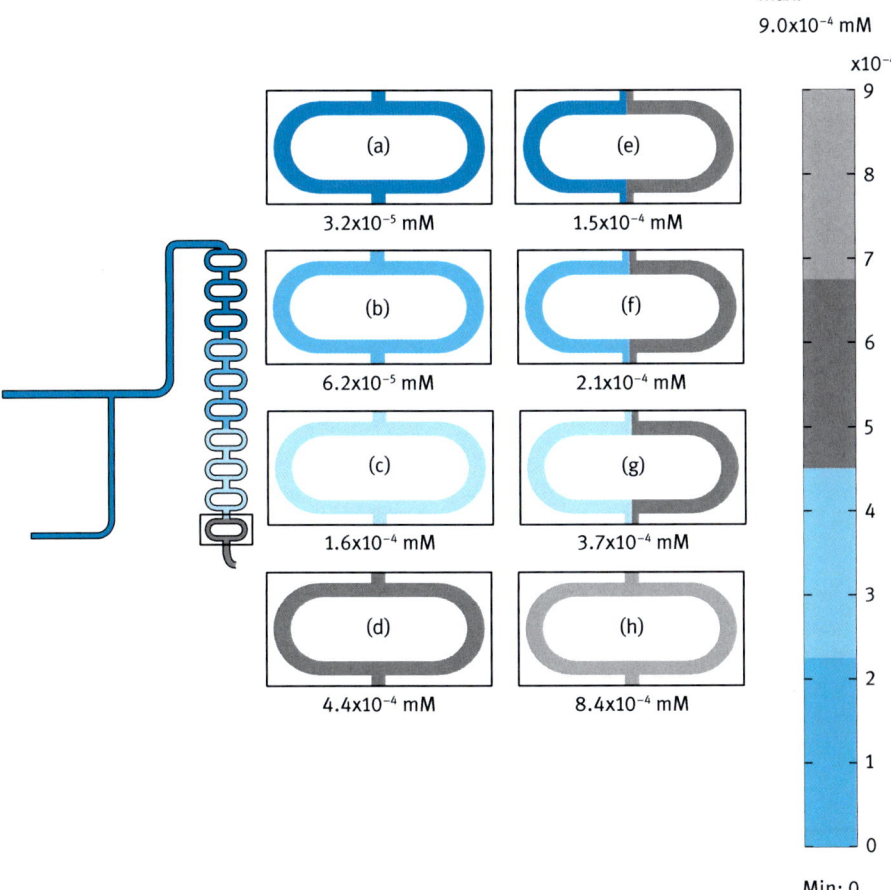

Fig. 7.7: [^{11}C]Raclopride concentration (mol/m^3) surface plot diagrams for the 10th micromixer loop in the full-loop design RU for the radioactive synthesis in flow mix mode at a flow rate of (a) 15 µL/min, (b) 10 µL/min, (c) 5 µL/min, (d) 2 µL/min, and in premix mode at flow rate of (e) 15 µL/min, (f) 10 µL/min, (g) 5 µL/min, (h) 2 µL/min, with reaction conditions of: $D = 10^{-9}$ m^2/s, $k = 5.663 \times 10^{-4}$ (m^3/mol/s) with reagent concentrations of [desmethyl raclopride] = 21.6 mol/m^3 and [^{11}C]methyl iodide = 0.024 mol/m^3. The scale shown represents the distribution of [^{11}C]raclopride concentration. See [48] for a fully colored figure of the concentration distribution.

Table 7.7: The concentration of [^{11}C]raclopride at point A (see Figure 7.6) at the end of the first loop in all three microchannel designs ([methyl iodide] = 0.024 mol/m^3).

Flow Rate (µL/min)	[^{11}C]raclopride concentration (× 10^{-5} mol/m^3)		
	No loop	Abacus	Full-loop
15	1.07	2.33	3.28
10	2.22	4.73	6.30
5	6.20	12.50	16.10
2	18.70	35.80	45.10

Table 7.8: The effect of reagent concentration, diffusion coefficient and rate constant variations on the normalized raclopride yield on the full loop RU using a flow rate of 2 µL/min in flow mix mode.

Variation	Reaction condition	Normalized raclopride
Default	*Default concentrations, $D = 10^{-9}$ m^2/s, $k = 5.663 \times 10^{-4}$ (m^3/mol/s), Flow mix	17.6
D (m^2/s)	*Default concentrations, $D = 10^{-8}$ m^2/s, $k = 5.663 \times 10^{-4}$ (m^3/mol/s), Flow mix	18.7
k (1/(mM s))	*Default concentrations, $D = 10^{-9}$ m^2/s, $k = 5.663 \times 10^{-6}$ (m^3/mol/s), Flow mix	1.80×10^{-1}
	*Default concentrations, $D = 10^{-9}$ m^2/s, $k = 5.663 \times 10^{-8}$ (m^3/mol/s), Flow mix	1.80×10^{-3}
Reagent amount (mM)	**High concentrations, $D = 10^{-9}$ m^2/s, $k = 5.663 \times 10^{-4}$ (m^3/mol/s), Flow mix	41.6
	**High concentrations, $D = 10^{-9}$ m^2/s, $k = 5.663 \times 10^{-4}$ (m^3/mol/s), Premix	75.1

Note: The final yield produced by the no loop design RU simulation with default reaction conditions at a flow rate of 15 µL/min is set to 1 and is used to normalize all the raw data.
*Default: reaction conditions include [[^{11}C]methyl iodide] = 2.4×10^{-2} mol/m^3 and [desmethyl raclopride]= 21.6 mol/m^3 which are related to the radioactive synthesis concentrations.
**High concentration: reaction conditions include [methyl iodide]= 53.4 mol/m^3 and [desmethyl raclopride]= 21.6 mol/m^3 which are related to the nonradioactive synthesis concentrations. See Table 7.5 for the expressions of the rate of changes of reaction species.

than the effects of flow rate variation and the use of the premix mode. On the other hand, increasing the concentrations of the reagents to reflect the nonradioactive synthesis experimental values ([methyl iodide] = 53.4 mol/m^3 and [desmethyl raclopride] = 21.6 mol/m^3) displayed an increase in the raclopride yield as per the rate of reaction. Furthermore, a drop in the rate constant showed a large effect on the normalized raclopride yield as expected, that is, 1/100* reaction rate constant gives 1/100* radiochemical yield.

7.4 Conclusion

It has been demonstrated that the use of the microfluidic reactor chip has improved the nonradioactive synthesis yields in comparison with the conventional process by reducing the reaction time and the amount of methyl iodide used in conventional preparation. The use of the chip with the micromixer designs was found to provide a better yield than the serpentine microchips. To evaluate the microchip method further, the reactants were premixed and it was found that the full loop chip still produced better results than the abacus chip, reinforcing the importance of a micromixer design for this process. Moreover, the fabrication of a glass chip with the micromixer design is necessary for the full investigation and optimization of the microchip radiosynthesis of [^{11}C]raclopride and other labeled compounds without the complication of side-reactions occurring with the PDMS chip material.

In addition, the behavior of reagents in the microchannels is explored in a CFD study. The simple and realistic models for the nonradioactive and radioactive synthesis of raclopride were generated. Various geometries were incorporated, while evaluating the effects of parameters such as the flow rate, residence (space) time, reaction mode, diffusion coefficient and reagent concentrations. It was demonstrated that this simulation study does provide useful information which would be compared with the experimental results in order to optimize this microchannel reaction.

Study questions
7.1. What are the definitions of radiolabeling and PET?
7.2. Why will 2–3 half-lives result in 75–88% radioactivity loss?
7.3. Why was 'microfludics' explored as technique for carbon-11 radiolabeling?
7.4. What feature is incorporated in the microchannels for the purpose of improving the synthesis process?
7.5. How did the flow rate variation affect the radioactive reactions conducted on chip?
7.6. Why was relative activity used to determine the reaction progress for the radioactive microfluidic synthesis?
7.7. How could the use of PDMS affect the loss of radioactivity and why?

Further readings
1. Ref 2. Miller, P. W., Long, N. J., Vilar, R., Gee, A. D. Synthesis of C-11, F-18, O-15, and N-13 Radiolabels for Positron Emission Tomography. Angew Chem Int Edit 2008;47:8998–9033.
2. Ref 12. Miller, P. W. Radiolabelling with short-lived PET (positron emission tomography) isotopes using microfluidic reactors. J Chem Technol Biot 2009;84:309–15.
3. Ref 13. Adam, M. J., Jivan, S., Huser, J. M., Lu, J. C-11-methylations using C-11-methyl iodide and tetrabutylammonium fluoride. Radiochimica Acta 2000;88:207–9.
4. Ref 9. Wang, M. W., Lin, W. Y., Liu, K., Masterman-Smith, M., Shen, C. K. F. Microfluidics for Positron Emission Tomography Probe Development. Molecular Imaging 2010;9:175–91.
5. Ref 16. Lu, S. Y., Pike, V. W. Micro-reactors for PET tracer labeling. Ernst Schering Res Found Workshop 2007:271–87.

6. Ref 43. Garstecki, P., Fischbach, M. A., Whitesides, G. M. Design for mixing using bubbles in branched microfluidic channels. Appl Phys Lett 2005;86:-.
7. Ref 48. Haroun, S., Wang, L., Ruth, T. J., Li, P. C. H. Computational fluid dynamics study of the synthesis process for a PET radiotracer compound, [11C]raclopride on a microfluidic chip. Chemical Engineering and Processing: Process Intensification 2013;70:140–7.
8. Ref 52. Haroun, S., Sanei, Z., Jivan, S., Schaffer, P., Ruth, T. J., Li, P. C. H. Continuous-flow synthesis of [11C]raclopride, a positron emission tomography radiotracer, on a microfluidic chip. Canadian Journal of Chemistry 2013;91:326–32.

Bibliography

[1] Blokland, J. A. K., Trindev, P., Stokkel, M. P. M., Pauwels, E. K. J. Positron emission tomography: a technical introduction for clinicians. European Journal of Radiology 2002;44:70–5.
[2] Miller, P. W., Long, N. J., Vilar, R., Gee, A. D. Synthesis of C-11, F-18, O-15, and N-13 Radiolabels for Positron Emission Tomography. Angew Chem Int Edit 2008;47:8998–9033.
[3] Welch, M. J., Redvanly, C. S. Handbook of radiopharmaceuticals: radiochemistry and applications. New York: J. Wiley; 2003.
[4] Ehmann, W. D., Vance, D. E. Radiochemistry and nuclear methods of analysis. New York: Wiley; 1991.
[5] Bailey, D. L. Positron emission tomography basic sciences. In. New York: Springer; 2005:x, 382 p. 30–32, 50.
[6] Zacheo, A., Arima, V., Pascali, G., et al. Radioactivity resistance evaluation of polymeric materials for application in radiopharmaceutical production at microscale. Microfluid Nanofluid 2011;11:35–44.
[7] Kealey, S., Plisson, C., Collier, T. L., et al. Microfluidic reactions using (11)C carbon monoxide solutions for the synthesis of a positron emission tomography radiotracer. Org Biomol Chem 2011;9:3313–9.
[8] Phelps, M. E. Molecular imaging with positron emission tomography. Annual Review of Nuclear and Particle Science 2002;52:303–38.
[9] Wang, M. W., Lin, W. Y., Liu, K., Masterman-Smith, M., Shen, C. K. F. Microfluidics for Positron Emission Tomography Probe Development. Molecular Imaging 2010;9:175–91.
[10] Paans, A. M. J., van Waarde, A., Elsinga, P. H., Willemsen, A. T. M., Vaalburg, W. Positron emission tomography: the conceptual idea using a multidisciplinary approach. Methods 2002;27:195–207.
[11] Valk, P. E. Positron emission tomography clinical practice. In. London: Springer; 2006:xiv, 475 p.3.
[12] Miller, P. W. Radiolabelling with short-lived PET (positron emission tomography) isotopes using microfluidic reactors. J Chem Technol Biot 2009;84:309–15.
[13] Adam, M. J., Jivan, S., Huser, J. M., Lu, J. C-11-methylations using C-11-methyl iodide and tetrabutylammonium fluoride. Radiochimica Acta 2000;88:207–9.
[14] Cherlo, S. K. R., Sreenath, K., Pushpavanam, S. Screening, Selecting, and Designing Microreactors. Ind Eng Chem Res 2009;48:8678–84.
[15] Bruus, H. Theoretical microfluidics. Oxford; New York: Oxford University Press; 2008.
[16] Lu, S. Y., Pike, V. W. Micro-reactors for PET tracer labeling. Ernst Schering Res Found Workshop 2007:271–87.
[17] Lu, S. Y., Watts, P., Chin, F. T., et al. Syntheses of C-11- and F-18-labeled carboxylic esters within a hydrodynamically-driven micro-reactor. Lab on a Chip 2004;4:523–5.

[18] Brady, F., Luthra, S. K., Gillies, J. M., Jeffery, N. T. Use of microfabricated devices for radiosynthesis of radiotracers for positron emission tomography. WO 03/078358, 2003.
[19] Elizarov, A. M. Microreactors for radiopharmaceutical synthesis. Lab Chip 2009;9:1326–33.
[20] Kegeles, L. S., Abi-Dargham, A., Frankle, G., et al. Increased Synaptic Dopamine Function in Associative Regions of the Striatum in Schizophrenia. Arch Gen Psychiatry 2010;67:231–9.
[21] Fei, X. S., Mock, B. H., DeGrado, T. R., et al. An improved synthesis of PET dopamine D-2 receptors radioligand C-11 raclopride. Synthetic Communications 2004;34:1897–907.
[22] Langer, O., Nagren, K., Dolle, F., et al. Precursor synthesis and radiolabelling of the dopamine D-2 receptor ligand C-11 raclopride from C-11 methyl triflate. J Labelled Compd Rad 1999;42:1183–93.
[23] Cheung, M. K., Ho, C. L. A simple, versatile, low-cost and remotely operated apparatus for [11C]acetate, [11C]choline, [11C]methionine and [11C]PIB synthesis. Appl Radiat Isot 2009;67:581–9.
[24] Wirth, T. Microreactors in organic synthesis and catalysis. Weinheim: Wiley-VCH; 2008.
[25] Garstecki, P., Fuerstman, M. J., Fischbach, M. A., Sia, S. K., Whitesides, G. M. Mixing with bubbles: a practical technology for use with portable microfluidic devices. Lab on a Chip 2006;6:207–12.
[26] Lee, C. C., Sui, G. D., Elizarov, A., et al. Multistep synthesis of a radiolabeled imaging probe using integrated microfluidics. Science 2005;310:1793–6.
[27] Gillies, J. M., Prenant, C., Chimon, G. N., Smethurst, G. J., Dekker, B. A., Zweit, J. Microfluidic technology for PET radiochemistry. Applied Radiation and Isotopes 2006;64:333–6.
[28] Gillies, J. M., Prenant, C., Chimon, G. N., et al. Microfluidic reactor for the radiosynthesis of PET radiotracers. Applied Radiation and Isotopes 2006;64:325–32.
[29] Hessel, V., Lowe, H., Schonfeld, F. Micromixers – a review on passive and active mixing principles. Chem Eng Sci 2005;60:2479–501.
[30] Nguyen, N. T., Wu, Z. G. Micromixers – a review. Journal of Micromechanics and Microengineering 2005;15:R1-R16.
[31] Pascali, G., Mazzone, G., Saccomanni, G., Manera, C., Salvadori, P. A. Microfluidic approach for fast labeling optimization and dose-on-demand implementation. Nucl Med Biol 2010;37:547–55.
[32] Kilbourn, M. R., Domino, E. F. Increased in vivo [11C]raclopride binding to brain dopamine receptors in amphetamine-treated rats. European Journal of Pharmacology 2011;654:254–7.
[33] Pedersen, K., Simonsen, M., Ostergaard, S. D., et al. Mapping the amphetamine-evoked changes in C-11 raclopride binding in living rat using small animal PET: Modulation by MAO-inhibition. Neuroimage 2007;35:38–46.
[34] Hoekzema, E., Herance, R., Rojas, S., et al. THE EFFECTS OF AGING ON DOPAMINERGIC NEUROTRANSMISSION A microPET STUDY OF 11C -raclopride BINDING IN THE AGED RODENT BRAIN. Neuroscience 2010;171:1283–6.
[35] Buddoo, S., Siyakatshana, N., Zeelie, B., Dudas, J. Study of the pyrolysis of 2-pinanol in tubular and microreactor systems with reaction kinetics and modelling. Chem Eng Process 2009;48:1419–26.
[36] Hossain, S., Ansari, M. A., Kim, K. Y. Evaluation of the mixing performance of three passive micromixers. Chem Eng J 2009;150:492–501.
[37] Hossain, S., Ansari, M. A., Husain, A., Kim, K. Y. Analysis and optimization of a micromixer with a modified Tesla structure. Chem Eng J 2010;158:305–14.
[38] Meijer, H. E. H., Singh, M. K., Kang, T. G., den Toonder, J. M. J., Anderson, P. D. Passive and Active Mixing in Microfluidic Devices. Macromol Symp 2009;279:201–9.
[39] Shih, T. R., Chung, C. K. A high-efficiency planar micromixer with convection and diffusion mixing over a wide Reynolds number range. Microfluid Nanofluid 2008;5:175–83.

[40] Kim, D. S., Lee, S. H., Kwon, T. H., Ahn, C. H. A serpentine laminating micromixer combining splitting/recombination and advection. Lab Chip 2005;5:739–47.
[41] Yang, J. T., Huang, K. J., Tung, K. Y., Hu, I. C., Lyu, P. C. A chaotic micromixer modulated by constructive vortex agitation. Journal of Micromechanics and Microengineering 2007;17:2084–92.
[42] Jeon, W., Shin, C. B. Design and simulation of passive mixing in microfluidic systems with geometric variations. Chem Eng J 2009;152:575–82.
[43] Garstecki, P., Fischbach, M. A., Whitesides, G. M. Design for mixing using bubbles in branched microfluidic channels. Appl Phys Lett 2005;86:-.
[44] Levenspiel, O. Chemical reaction engineering. 3rd ed. New York: Wiley; 1999.
[45] Frederikse, H. P. R., Lide, D. R. Handbook of chemistry and physics: a ready-reference book of chemical and physical data. In. Taylor and Francis Group. Cleveland, Ohio: Chemical Rubber Pub. Co. [etc.]; 2010:p. 6–216 – 6–7.
[46] Langstrom, B., Obenius, U., Sjoberg, S., Bergson, G. KINETIC ASPECTS OF THE SYNTHESES USING SHORT-LIVED RADIONUCLIDES. Journal of Radioanalytical Chemistry 1981;64:273–80.
[47] Langstrom, B., Bergson, G. THE DETERMINATION OF OPTIMAL YIELDS AND REACTION-TIMES IN SYNTHESES WITH SHORT-LIVED RADIONUCLIDES OF HIGH SPECIFIC ACTIVITY. Radiochem Radioa Let 1980;43:47–54.
[48] Haroun, S., Wang, L., Ruth, T. J., Li, P. C. H. Computational fluid dynamics study of the synthesis process for a PET radiotracer compound, [11C]raclopride on a microfluidic chip. Chemical Engineering and Processing: Process Intensification 2013;70:140–7.
[49] Langstrom, B., Antoni, G., Halldin, C., Svard, H., Bergson, G. SYNTHESIS OF SOME C-11-LABELED ALKALOIDS. Chemica Scripta 1982;20:46–8.
[50] Hartman, R. L., Jensen, K. F. Microchemical systems for continuous-flow synthesis. Lab on a Chip 2009;9:2495–507.
[51] Dertinger, S. K. W., Chiu, D. T., Jeon, N. L., Whitesides, G. M. Generation of gradients having complex shapes using microfluidic networks. Anal Chem 2001;73:1240–6.
[52] Haroun, S., Sanei, Z., Jivan, S., Schaffer, P., Ruth, T. J., Li, P. C. H. Continuous-flow synthesis of [11C]raclopride, a positron emission tomography radiotracer, on a microfluidic chip. Canadian Journal of Chemistry 2013;91:326–32.
[53] Johnson, C. Numerical solution of partial differential equations by the finite element method. Cambridge[Cambridgeshire]; New York: Cambridge University Press; 1987.

Part III: **Additional features of the Flow Process: in-line analytics, safety and green principles**

Ferenc Darvas, György Dormán, and Melinda Fekete

8 Lab environment: in-line separation, analytics, automation & self optimization

8.1 The role of analytics in flow applications

When optimizing a flow synthesis, it is common practice to collect samples corresponding to a range of reaction conditions or a selection of reactants, and then submit these to a communal gas chromatography (GC), liquid chromatography (LC) or liquid chromatography/ mass spectrometry (LC/MS) instrument for off-line analysis [1].

An optimum set of reaction conditions or choice of reactant can then be selected from the matrix of results obtained. It is important that the reaction is quenched at the time of sampling so that the results do not reflect any continuing reactions in the collection vial. The time delay and level of operator intervention associated with analysis in this way do not lead naturally to rapid reaction optimization, which is the desired principle of flow chemistry.

There has been some progress in the development of in-line and on-line monitoring techniques for the rapid optimization of preparative flow syntheses.

> The terms 'in-line' and 'on-line' refer to methods of analysis that do not require manual transfer of samples. All the flow is continuously analyzed by an in-line technique, whereas representative aliquots are periodically analyzed by a technique that is described as on-line. Analytical techniques that are disruptive can only be used on-line.

Systems that automatically sample the flow stream and initiate on-line analysis by GC, LC or LC/MS have been demonstrated [2–4].

Some techniques require dilution of the reaction mixture before analysis to prevent overloading of the detector. Equipment that additionally dilutes the sample prior to analysis is now commercially available from both Syrris (Syrris Ltd., UK) and Accendo (Accendo Corporation, USA). These systems use a six-port valve fitted with a sample loop to extract an aliquot from the flow stream. Sampling can occur automatically, at pre-determined time intervals.

Some of the most challenging issues concerning the optimization and synchronized control of continuous chemical processing today include:
- Monitoring (and quantization) of dispersion, worsened by diffusion.
- In-line monitoring of products and intermediates without chemically transforming them.
- Analysis within rapid screening processes.
- Detection of hazardous compounds.
- Permanent qualitative monitoring of product formation on a large scale.

In this chapter we present some of the recent developments of in-line and on-line analytical tools.

8.1.1 Applications of mass spectroscopy

There are numerous examples of the direct *in-line* coupling of mass spectrometers to flow chemistry reactors. Typically, the application involves screening of library compounds or investigations of reaction mechanisms, intermediates, and kinetics. In both cases, the purpose of the flow chemistry system is to produce sufficient material for analysis by the mass spectrometer, which by necessity consumes all the available flow material. As mass spectrometry is a destructive technique, these applications are analytical rather than preparative.

An example for *on-line* analysis with mass spectrometer is presented below [1].

In the experiments, the generation of benzyne via diazotisation of anthranilic acid, and its subsequent cycloaddition to furan was investigated. This reaction was specifically chosen as it is complex and would lead to both reactive intermediates and gaseous by-products.

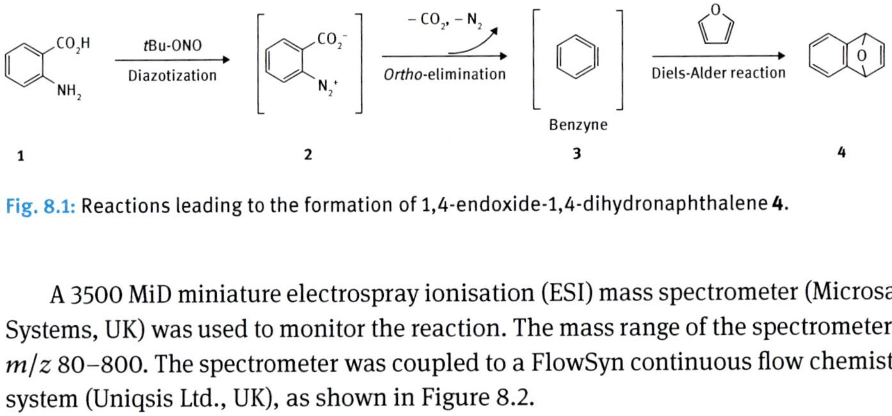

Fig. 8.1: Reactions leading to the formation of 1,4-endoxide-1,4-dihydronaphthalene **4**.

A 3500 MiD miniature electrospray ionisation (ESI) mass spectrometer (Microsaic Systems, UK) was used to monitor the reaction. The mass range of the spectrometer is m/z 80–800. The spectrometer was coupled to a FlowSyn continuous flow chemistry system (Uniqsis Ltd., UK), as shown in Figure 8.2.

A six-port switching valve was configured such that a sample of the solution leaving the flow reactor could be periodically passed to the mass spectrometer. With the switching valve in the load position, the flow from the reactor passed through a 5 mL loop (G) fitted across two of the valve ports and thereafter into a collection vessel (H).

Representative positive ion mode mass spectra were acquired at 25 °C and 50 °C reaction temperatures (Figure 8.3.) The $[M + H]^+$ ions resulting from unreacted anthranilic acid **1** and the desired product **4** can be immediately identified at m/z 138 and 145, respectively. The peak at m/z 120 is also due to anthranilic acid, which apparently undergoes collision-induced loss of water from the $[M + H]^+$ ion in the vacuum interface of the mass spectrometer.

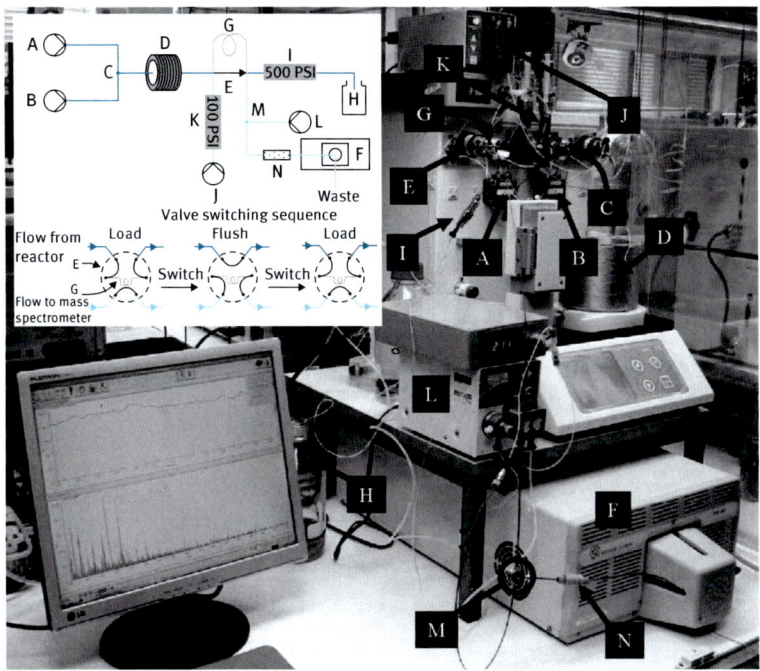

Fig. 8.2: FlowSyn flow chemistry system. The reactant solutions were pumped by high-pressure pumps (A and B) to a mixing tee (C, obscured in photo). The flow stream then flowed through the reactor coil (D) and sampling loop (G), which was fitted across two ports of the six-port valve (E). After passing through a back-pressure regulator (I), the solution was discharged into a collection vessel (H). When the six-port valve was switched to the inject position, a third high-pressure pump (J), which was stabilized by back-pressure regulator (K), pumped acetonitrile through the loop to flush the contents into mixing tee (M). A 50 : 50 (v/v) mixture of acetonitrile and water with 0.1% formic acid was pumped into the mixing tee by a fourth high-pressure pump (L) in order to further dilute the sample and modify the solution to aid the ESI process. An in-line filter (N) removed particulates before analysis of the sample by the mass spectrometer (F). Figure reproduced from [1] with permission from Wiley-VCH Verlag GmbH & Co.

In the low-temperature spectrum there is a peak at m/z 149 that can be assigned to protonated benzenediazonium-2-carboxylate (**2**). This was an unexpected result for the authors as it had been assumed that all the benzenediazonium-2-carboxylate would have been consumed during the 20 min transit of the reactor coil. The prominent peak at m/z 121 can be attributed to the loss of N_2 from benzenediazonium-2-carboxylate in the detector.

The flow system had previously been operated for some time at a coil temperature of 25 °C in the belief that all the benzenediazonium-2-carboxylate generated was being consumed. Accumulation of explosive intermediers in the collector vial can be especially dangerous. Using the ESI mass spectrometer detector, it became apparent that not all the explosive intermediate was consumed by the reaction. Such information can be critical in a preparative scale process.

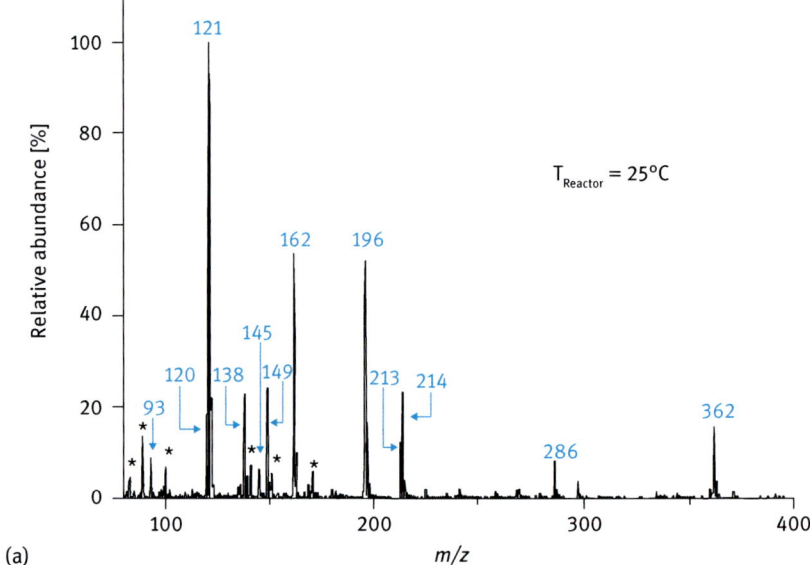

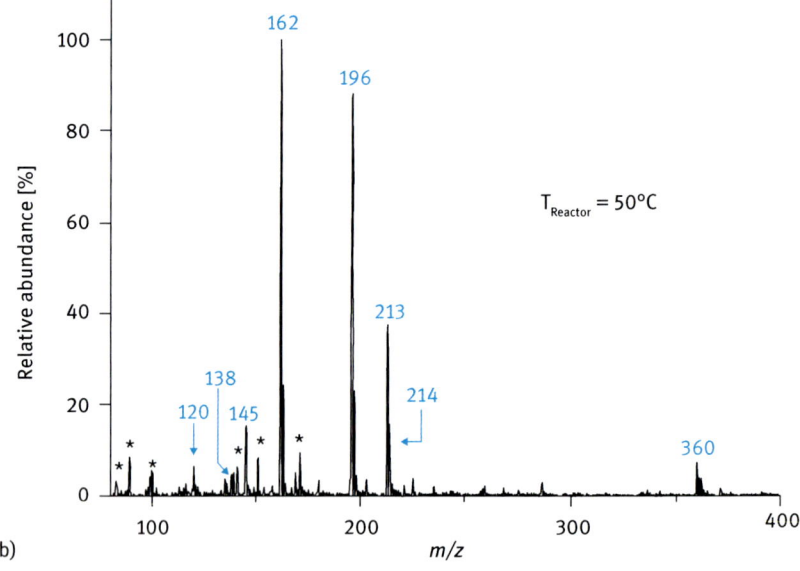

Fig. 8.3: Representative spectra corresponding to reactor temperatures of (a) 25 °C and (b) 50 °C. The reactor residence time was 20 min in both cases. Background peaks are labeled with asterisks [1]. Reproduced with permission from Wiley-VCH Verlag GmbH & Co.

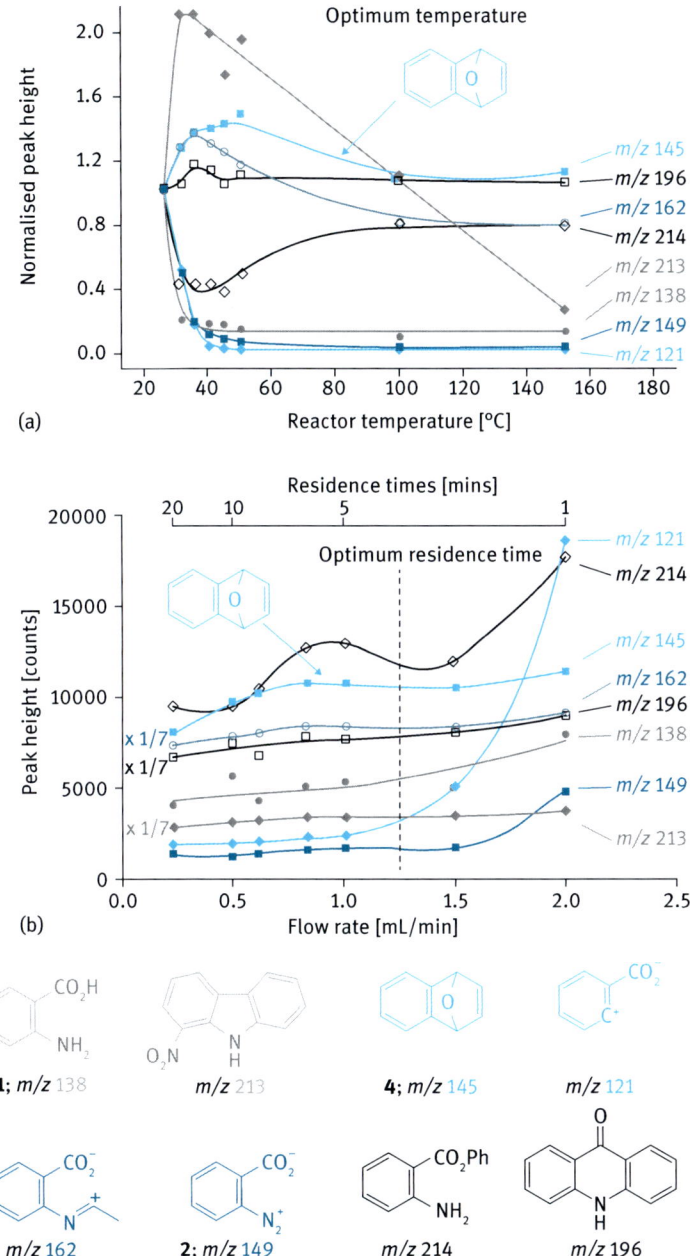

Fig. 8.4: Response of mass spectral peak intensities to (a) reactor temperature at a residence time of 20 min and (b) flow rate at a reactor temperature of 50 °C. Some data sets in (b) have been scaled to aid presentation. The compounds in the legend are labeled with the m/z value of the corresponding $[M + H]^+$ ion [1]. Reproduced with permission from Wiley-VCH Verlag GmbH & Co.

Increasing the temperature to 50 °C, the peak corresponding to this intermediate disappeared from the product mixture (Figure 8.3 (b)).

Optimization of flow parameters (Figure 8.4 (a)) shows how the intensities of selected peaks changed as the reactor temperature was increased. Each data point in Figure 8.4 (a) represents the height of the indicated peak in the cumulative mass spectrum. The data sets have been normalized such that the signal intensities are unity at 25 °C.

The optimal conditions can be determined from Figure 8.4 as 1 ml/min flow rate and 50 °C temperature. The overwhelming benefit of direct and rapid analysis of samples from the flow stream by the mass spectrometer is that unstable or reactive intermediates and products can be identified. Nevertheless, this method requires some knowledge of mass spectroscopy. Trends in peak heights as a function of residence time and reactor temperature can be used to guide the optimization of the flow conditions.

8.1.2 ReactIR flow cell

One of the most convenient and nondestructive methods for real-time in-line monitoring is infrared (IR) spectroscopy. A newly developed ReactIR flow cell (Mettler Toledo) is reported as a convenient and versatile *in-line* analytical tool for continuous flow chemical processing.

The flow cell, operated with ATR (Attenuated Total Reflectance) technology, is attached directly into a reaction flow stream using standard OmniFit (HPLC) connections and can be used in combination with both meso- and microscale flow chemistry equipment. The iC IR™ analysis software (product of Mettler Toledo as well) enables the monitoring of reagent consumption and product formation, aiding the rapid optimization of procedures.

8.1.2.1 Technical details of the instrument

The FT-IR device used in this work is a ReactIR 45 m fitted with a 24 h mercury cadmium telluride (MCT) detector. The ReactIR instrument was directly connected to a newly developed, microscale flow cell (Figure 8.5). The flow cell comprises of an integrated ATR gold sealed diamond sensor, referred to as DiComp. This allows *in situ*, real-time monitoring of a continuous flow stream [5].

The full infrared spectral region is available with this microflow cell (650–1950 cm^{-1} and 2250–4000 cm^{-1}) excluding the diamond "blind spot" which only allows very weak absorbance in this region. The IR flow cell has a removable head allowing for easy cleaning (Figure 8.5 (c)) and an internal volume of 51 µL. It can be heated up to 60 °C using an external controller, and operated up to 30 bar pressure.

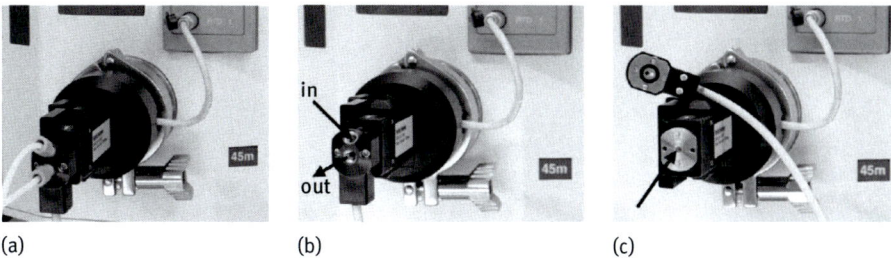

Fig. 8.5: ReactIR 45 m with prototype of IR flow cell attached. (a) Flow cell in-line, connected with OmniFit adapters; (b) flow cell with OmniFit adapters removed, with direction of flow stream through the cell indicated; (c) head unscrewed, free view on diamond window (arrow) [5]. Figure reproduced with permission from the American Chemical Society.

An integrated resistive thermal device (RTD) temperature sensor is also built into the cell in order to track its temperature, paying deference to the fact that the IR spectroscopy is a temperature-dependent spectroscopy. While the flow stream is running through the cell, scans are taken at predefined intervals. The system is controlled and the raw data are collected and analyzed by the iC IR™ reaction analysis software. The software allows real-time analysis of the spectra, such as substraction of the reference spectra or creating trend lines. The software also has a built-in knowledge for helping functional group profiling.

8.1.2.2 In-line monitoring of a hydrogenation reaction

In the here presented example, the commercially available hydrogenation flow reactor H-Cube Midi™ (ThalesNano) was used for the full saturation of nicotinate **5**. The reaction set-up was designed to incorporate the IR flow cell without exposing it to the high temperatures and pressures sometimes required in these transformations (Figure 8.6). Thus, the React IR flow cell was connected in-line before the pump and a recycling process was run as it is presented on Figure 8.6 [5].

As hydrogenations are typically performed at high dilution, the authors performed a concentration screen to ascertain the minimum concentration at which the different bands in the starting material could be monitored (Figure 8.7). As visible on Figure 8.7, measurable information could still be obtained at 0.01 M using the solvent subtraction feature in the software. (Note: Using the same technique, a concentration calibration line can be prepared, when necessary.)

Strong absorption bands (such as ester carbonyl) can be monitored at very low concentrations even without solvent subtraction. Absorptions of more weakly absorbing bands such as the C=C require higher concentrations in general.

The hydrogenation reaction of **5** was operated at 0.1 M with a 5 mL/min flow rate, and the gradual disappearance of the C=C bond was monitored (1306 cm^{-1}). The appearance of the saturated ester band (1661 cm^{-1}) was also distinguished from the dis-

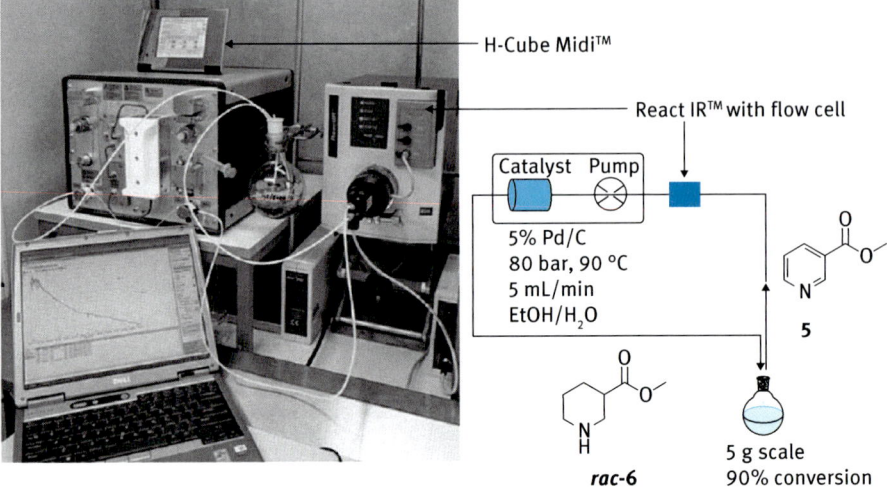

Fig. 8.6: Set-up for monitoring flow hydrogenation with the IR flow cell using a recycling process [5]. Figure reproduced with permission from the American Chemical Society.

appearance of the absorption for the conjugated ester (1719 cm^{-1}) (Figure 8.8). This demonstrates that it is beneficial to have more than one point of structural reference in the same molecule to validate the results, particularly for weak bands at high temperature. Strong absorption bands will give results with higher accuracy, since it is apparent from the baseline at a much lower concentration.

On Figure 8.8, changes in the band intensities were monitored. Both the conjugated ester bond and the C=C bond shows a decreasing trend, in a very similar manner. The saturated ester bond, representing the product, is constantly increasing during the experiment. Sum of the intensities along the whole monitored time period is 100%, which shows that the results are consistent.

8.1.2.3 Accurate addition of sequential reagents using IR spectroscopy

When performing multistep sequences in flow, the unavoidable dispersion of the reaction "plug" is a significant issue, particularly when polymer-supported reagents are used for in-line purification. Dispersion is dependent on many variables (including diffusion, reactor set-up and interaction of intermediates with polymer-supported reagents) and therefore it is difficult to predict without direct in-line measurement. The controlled addition of exact stoichiometries of further reagents to a product stream is therefore challenging [6].

Technical solutions to this problem involve using IR or UV-vis detectors to measure the dispersion of a product stream from a single step. The dispersion curve obtained is then used as a visual signal to manually start a pump (C) delivering a third stream of reagent at a predetermined concentration (Figure 8.9).

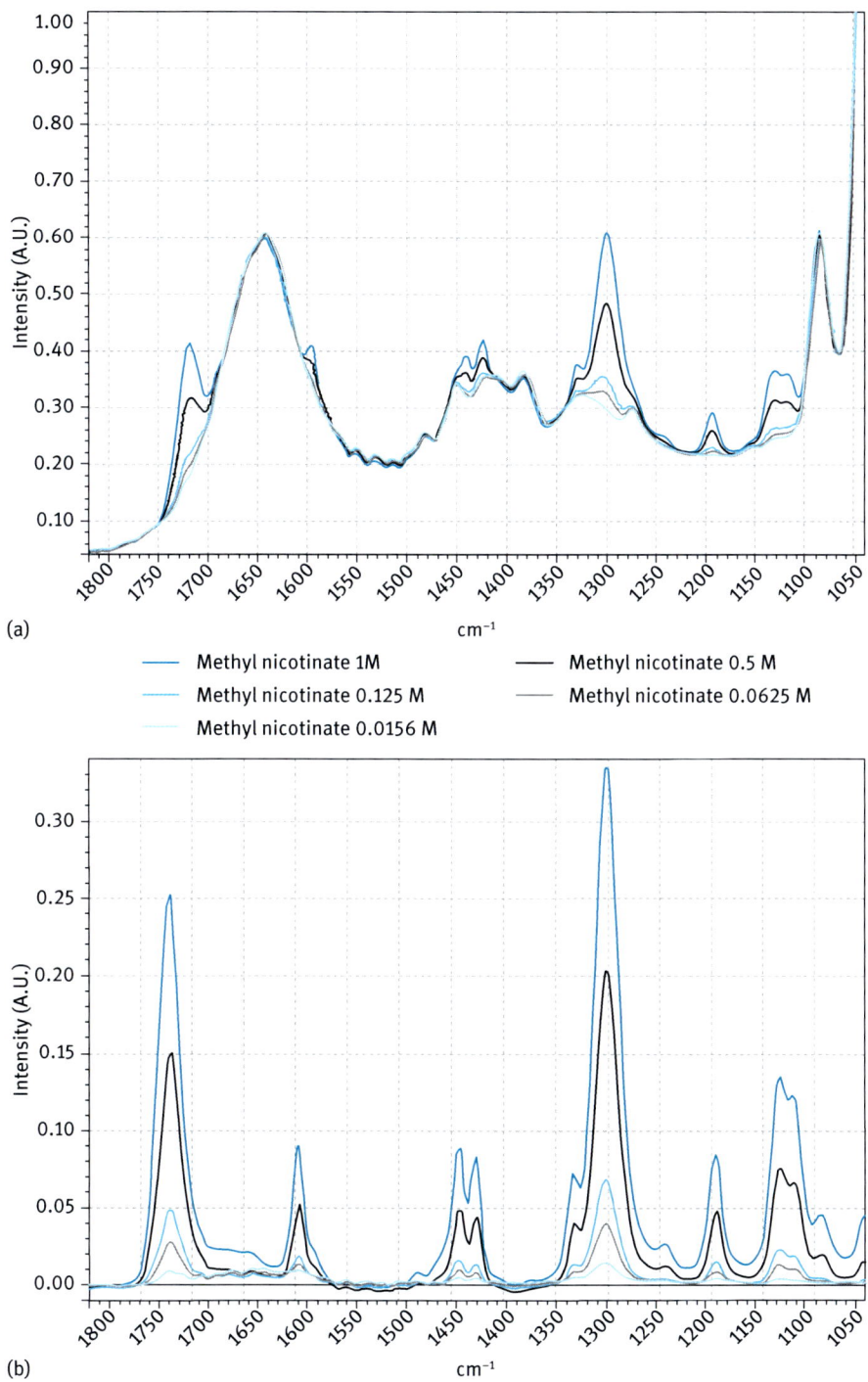

Fig. 8.7: IR flow cell detection limit screen: a) Raw data. b) Spectra obtained after solvent substraction (40% EtOH/H$_2$O). Modified after [5].

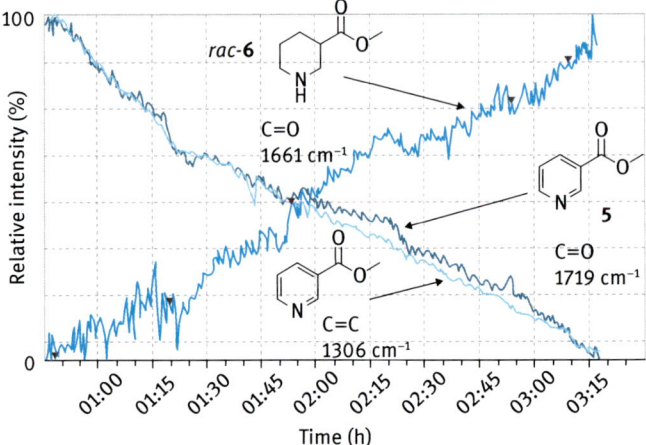

Fig. 8.8: Multiple band monitoring within the heterocycle saturation reaction. Modified after [5].

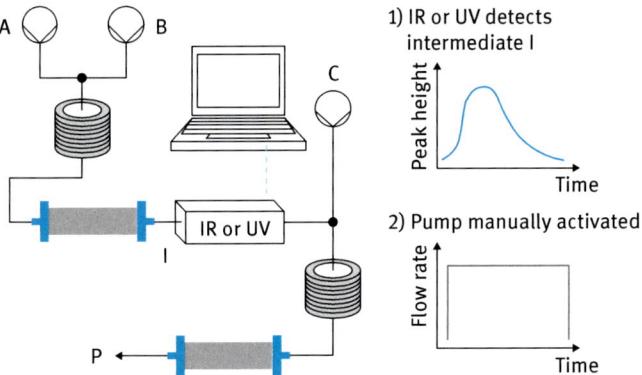

Fig. 8.9: IR/UV detection with manual pump control. A, B, C refer to reagent solutions (and the transporting pumps), I and P refer to Intermediate and Product, respectively. Modified after [6].

However, this simple on/off method results in a large amount of the third component being wasted when the reaction is performed up to appreciable scales. This is particularly problematic if the component is precious (as is likely to be the case in the synthesis of complex molecules), toxic or hazardous. Additionally, this method is ineffective for reactions requiring an accurate and maintained stoichiometry ratio throughout the dispersion curve.

As the intensity of IR radiation absorbed at a particular wavelength is defined by Beer's law to be proportional to concentration, a much better solution to this "third-stream problem" would be to control the flow rate of the third pump in real-time by using the concentration of a product-specific band measured *in situ* by an IR flow cell (Figure 8.10).

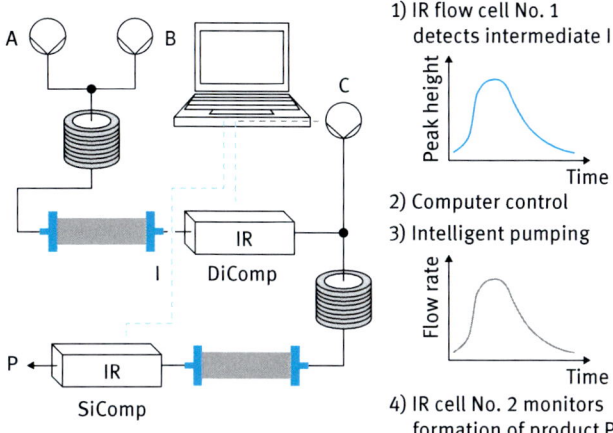

Fig. 8.10: IR-detector coupled to a computer to control a pump. Modified after [6].

The iC IR™ software is able to export the values of the absorptions measured at selected wavelengths to a Microsoft Excel data sheet via a Mettler-Toledo prototype ActiveX interface in real-time. In order to provide a means of using this data to control a pump, a simple application was developed using LabVIEW that continuously reads the Excel. This data is then converted into a flow rate (also considering in the calculation the different absorption coefficients of the different peaks).

This concept was proven using two inert compounds to determine if a 1:1 stoichiometry could be obtained along a flowing plug of material with varying concentration.

Firstly, a solution of 4-chlorobenzophenone (**7**) was pumped through a column of PS-Polyol (IRA-743) to disperse the solution. The C=O stretch (1659 cm^{-1}) was monitored in the first IR flow cell, generating the dispersion curve shown in Figure 8.11.

The peak height recorded was automatically converted in real-time to a flow rate for a second pump, which dispensed a solution of 3-methyl-4-nitroanisole (**8**). The second IR flow cell, positioned in-line at the outlet of the mixing chip, was used to measure the concentration profiles of both reagents, which appear to be accurately matched.

The absolute ratio of the two components was determined by collecting aliquots of the output of the reactor in fixed intervals and analyzing them using ^{1}H NMR spectroscopy. The results (Figure 8.12) indicated that the desired 1:1 stoichiometry was closely obtained for the majority of the eluted material.

This method would also offer a distinct advantage in the event of a reaction failure, since the lowered concentration of intermediate would be accounted for in real-time thus saving potentially precious material.

The higher level of control via in-line monitoring greatly improves our ability to effect multistep sequences using segmented flow processing.

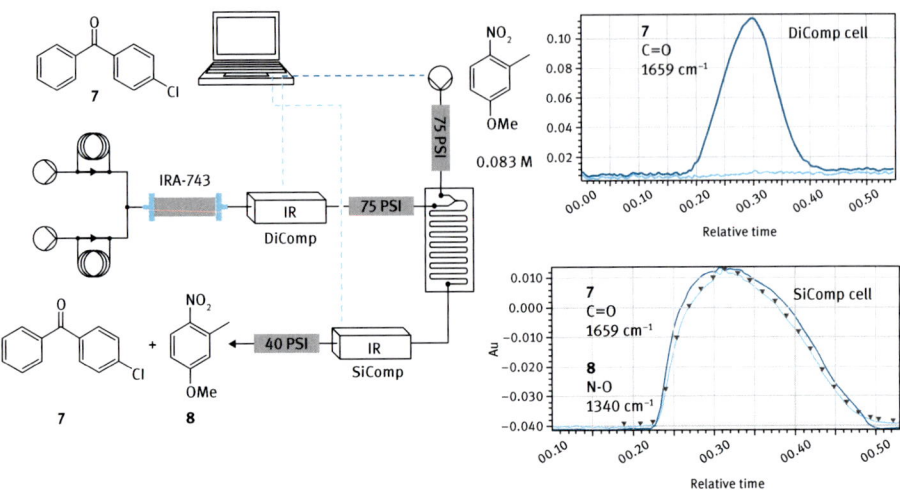

Fig. 8.11: Proof of principle experiment illustrating precise addition of a third stream with 1:1 stoichiometry. Modified after [6].

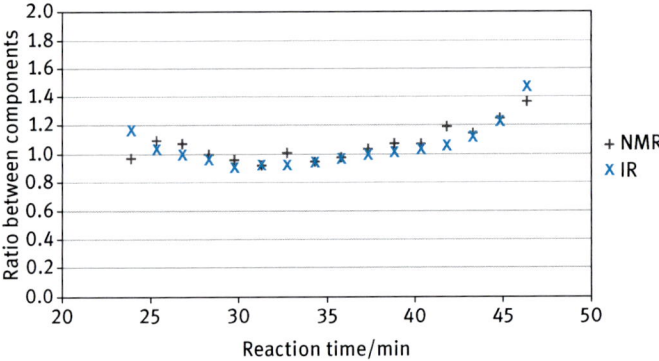

Fig. 8.12: Graph illustrating correlation between ^{1}H NMR and IR analysis of accuracy. Modified after [6].

8.1.3 Nuclear magnetic resonance (NMR)

As of today, NMR spectroscopy has become an indispensable technique in chemistry, biology, and medicine. Although it is the most information-rich analysis technique for molecular structure determination, it suffers from the intrinsically low sensitivity resulting in large-size NMR-instruments. A small-size, but sensitive NMR device applicable with microflow system would be welcomed by the flow chemists. The following developments are addressing this goal.

8.1.3.1 Bench-top NMR: picoSpin™

In 2009, picoSpin™, (Colorado, USA) launched a small (20 × 14 × 29 cm) 45 MHz spectrometer with good resolution (< 1.8 Hz) and mid-to-low-range sensitivity that weighs only 4.76 kg and can acquire a 1D ^{1}H or ^{19}F spectrum (Figure 8.13). In August 2013, a second version was introduced, the picoSpin™ 80, that operates at 82 MHz with a resolution of less than 1 Hz and weighs 19 kg.

Instead of the traditional static 5 mm NMR tubes, both picoSpin spectrometers have a flow-through system with sample injection into a 1/16″ capillary (~ 0.3 mm inner diameter). This injection system makes picoSpin™ particularly interesting for flow chemistry applications [7].

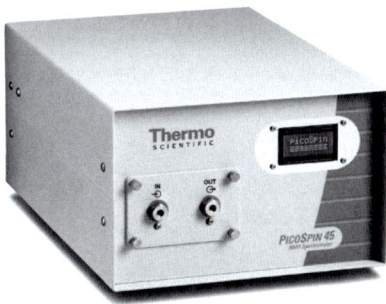

Fig. 8.13: picoSpin™ 45 NMR instrument. Loops can be easily attached via the front-panel fittings.

Liquid samples are simply injected into an internal capillary via front-panel fittings, and only 30 microliters of sample fluid is needed to obtain a spectrum. The unit's fluid capillary is contained within a cartridge, so it can be replaced easily by the user in case it becomes blocked or contaminated.

A highly stable, temperature-controlled permanent magnet ensures easy maintenance-free operation without the need for liquid cryogens. Deuterated solvents are optional due to the presence of a software lock. The electronic circuits are all miniaturized using techniques similar to those used in cell phones. The system is stabilized by a magnet temperature controller and by software techniques.

The picoSpin™ bench-top top NMR needs only a web browser on any external computer or mobile device for control as the spectrometer has a built-in web server board; no installed software on a dedicated PC is required.

If the sample itself behaves in a time-independent way, any number of single scans can be averaged together to improve SNR (Signal-to-Noise Ratio). A single scan is acquired in less than a second; however, for most samples multiple scans are necessary.

How to measure flowing samples?

During the NMR data acquisition, the sample should be stationary. In a moving phase, spectroscopic resolution will suffer unless the flow itself is extremely slow (µl/min range) compared to the time of the data acquisition. This problem can be circumvented with a multiport valve and automated injection system that allows the sample to reach the picoSpin™ capillary.

8.1.3.2 Stripline high resolution probe

NMR probes are the interface between a sample and the spectrometer. The size of the probe is dependent on the size and the arrangement of the sample. Traditionally, NMR on small volumes is performed by means of small solenoids wrapped around a capillary, or planar coils on glass chips containing microfluidic channels. The most important problem of these approaches is that the radiofrequency (rf) coil induces static field distortions that limit the resolution and the Signal-to-Noise (SNR) performance.

Striplines are a promising alternative to solenoid and planar coils for NMR detection of volume restricted samples. The stripline design consists of a flat strip confined between two metal shielding planes, which in microwave electronics is usually called a stripline configuration (Figure 8.14). To obtain a high rf-field strength, which is necessary for high sensitivity, the strip is constricted in the center as is shown in Figure 8.14 (a)–(c). With this type of configuration, minimal static field distortions are induced ensuring high spectral resolution [8, 9].

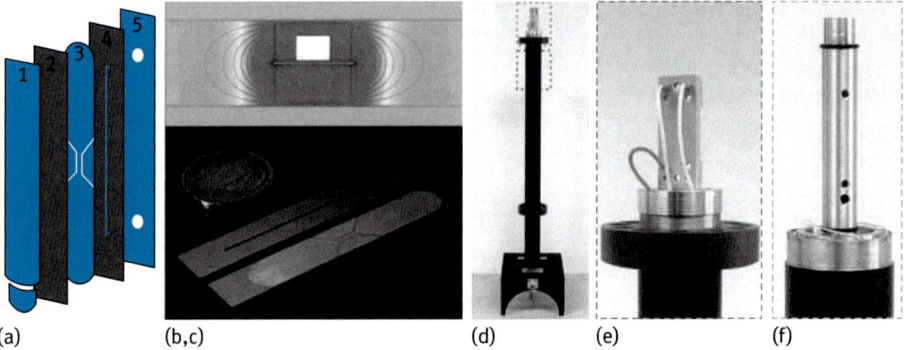

Fig. 8.14: (a) Exploded view of the microfluidic NMR chip. Layers 1, 5: grounded copper layers, layers 2, 4: low-loss silicon platelets; layer 4 contains a microfluidic channel; Layer 3: copper stripline. (b) Cross-section of stripline chip, showing magnetic field lines. White rectangle: microfluidic channel, outlined box: area of uniform magnetic field. (c) Photograph of microfabricated chip parts before bonding. (d) The custom-made microfluidic probe. The dashed line indicates the position of the NMR chip. (e) Close-up view of the microreactor holder mounted on top of the probe. (f) Close-up view of the stripline chip holder. Figure modified after [9].

The 1D ^{1}H-NMR experiments on the stripline probe were carried out on a Chemagnetics CMX-Infinity 600 solid-state NMR spectrometer (600 MHz).

For the reaction monitoring, a Micronit microreactor chip (Y-junction type) was implemented on top of the stripline probe as shown in Figure 8.14 (d), (e). Two syringe pumps are used to control the flow of the reactants. The reaction starts in the 142 nL reaction channel and subsequently, the products flow through the stripline where the detection takes place. The total reaction volume is 4.5 µL. By changing the flow rate, the reaction time can be varied from several seconds up to 10 minutes.

Figure 8.15 shows several spectra obtained during real-time monitoring of the acetylation of benzyl alcohol with acetyl chloride in the presence of N,N-diisopropylethylamine (DIPEA) at different reaction times. The spectra demonstrate that the high NMR resolution can be maintained with the stripline probe. Changes in the reaction mixture can be easily monitored by the peak of the starting materials (2.7 and 4.7 ppm) and the product (5.2 ppm) as well.

These results demonstrate the possibility of tracking intermediates using fast *in situ* analysis. In this particular case, the flowing sample was examined without loosing the high resolution due to the very low flow and the very fast single-scan data acquisition.

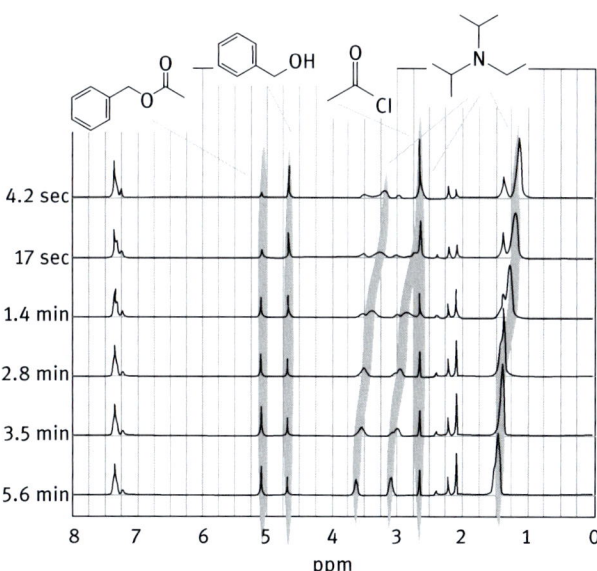

Fig. 8.15: Spectra of the acetylation of benzyl alcohol in the presence of DIPEA, reaction time from 4.2 seconds to 5.6 minutes. Figure modified after [9].

8.2 Automation and self optimization

Several automated microreactor systems have been developed that are capable of sampling a series of predetermined reaction conditions. These systems often employ a univariable approach in which only a single variable is adjusted at a time. Such exhaustive searches are inherently inefficient since they are likely to gather a significant percentage of data points that are far away from the desired maximum.

A more sophisticated automated microreactor system based on Design of Experiment methods has also been developed. Although this system is ideally suited for response surface modeling, the approach is less efficient in terms of the number of reactions required for an optimization.

Integrating feedback into the reaction optimization could significantly increase the speed and efficiency of the overall process by directing the system away from lower yielding reaction conditions [10–14]. In such an approach, a specified set of starting experimental conditions are performed in an automated manner and the results are analyzed. Based upon this reaction data, a new set of experimental conditions are proposed and the procedure is repeated until the optimal conditions are determined.

8.2.1 General description of the self-optimization methods

For the better understanding of the following experimental descriptions, we need to define some common expressions related to optimization algorithms. Some examples will be given on the model reaction benzylalcohol oxidation (Figure 8.16).

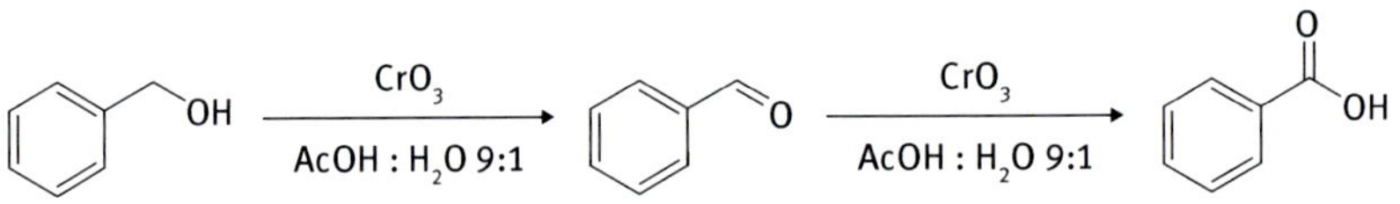

Fig. 8.16: Oxidation of benzylalcohol to benzaldehyde and benzoic acid.

The reaction parameters that we vary during the optimizations are the **variables**. (In our example: temperature, residence time (See Figure 8.17).

The number of variable reaction conditions (variables) will determine the **dimensionality** of the algorithm. Changing only the temperature and the residence time, we have a 2D algorithm. The response surface can be plotted on a 2D diagram (Figure 8.17).

The goal of the optimization process is to find the parameter values that result in a maximum (or minimum) of a function called the **objective function**. In the presented example, we would like to optimize our reaction for the highest benzaldehyde yield. (Other objective functions could be, for example, conversion, or productivity.)

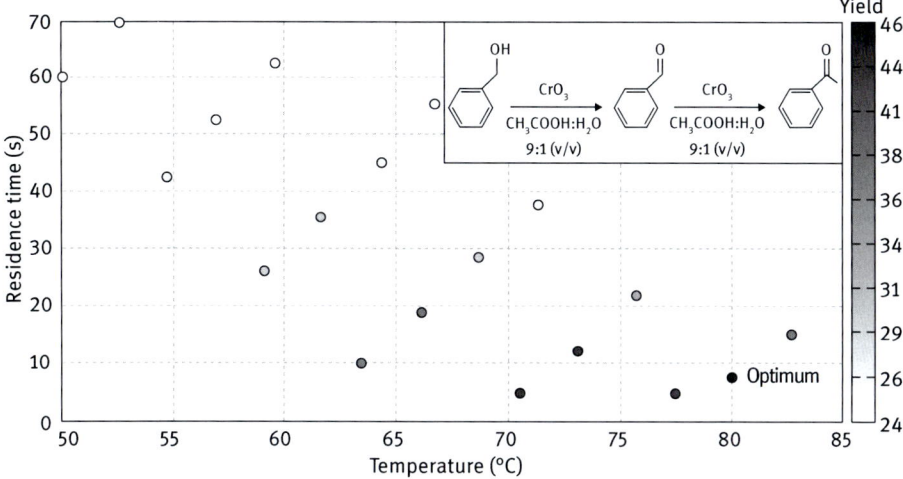

Fig. 8.17: Results from the automated microfluidic platform were used to maximize the yield of the intermediate, benzaldehyde. A 2D optimization procedure varied reaction temperature and residence time to find the optimal benzaldehyde yield of 46% [12, 15]. Figure modified after [15].

Before starting an optimization, one has to determine the objective function and the variables. Some optimization algorithms require defined starting conditions as well.

A **constraint** is a condition defined upon optimization parameters. It defines a range for an optimization parameter. A constraint is a well-formed arithmetic expression describing a relationship between the optimization parameters.

Due to Figure 8.17, the following constraints were given in the benzylalcohol oxidation:

$$50\ °C < T < 85\ °C$$
$$0\ \text{sec} < \tau < 70\ \text{sec.}$$

A **requirement** is an additional restriction imposed on the solution found by the optimization engine. Requirements are checked at the end of each simulation, and if they are not met, the parameters used are rejected. Otherwise the parameters are accepted.
  A feasible solution is one that satisfies all constraints and requirements.

The optimization algorithm may be written to find the global or the local optimum solution. The difference between the two solutions is visualized on Figure 8.18:

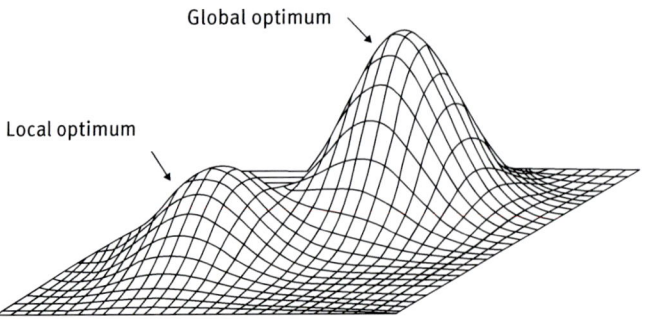

Fig. 8.18: Illustration of local and global optimum.

---

i To ensure that each simulation ends, you must specify a simulation stop condition or **termination criterion**. In general, a simulation stop condition should be specified in such a way that the value of the objective function is significant when simulation stops.

Termination criterion: For example, we can choose to terminate the optimization when the greatest difference between two yields in the simplex was less than the error in the accuracy of the analytical method.

The optimization methods discussed in this book use **black-box** algorithms that appear efficient without a priori reaction or gradient information required.

8.2.2 Automation and feedback control systems

Incorporating an in-line analytical detection method with an appropriate feedback control and logic optimization algorithm significantly improves the reaction optimization procedure [11, 12]. In theory, any type of analytical methods can be used here that give result in a sufficiently short time. However, the most widespread techniques that are automated today are GC, GLC and HPLC methods.

A flowsheet of self-optimization [11] with automated sampling is described on Figure 8.19.

Experimental hardware and data measurements are interfaced with LabVIEW programs. Matlab scripts regulate the temperature controller, to execute the optimization algorithm, and to perform advanced mathematical operations such as nonlinear regression and solving multiple equations simultaneously.

To start the optimization process, all the necessary criteria for the optimization and control methods were inputted into the program. After providing this information, the algorithm was started and the process became completely automated and user intervention was obviated.

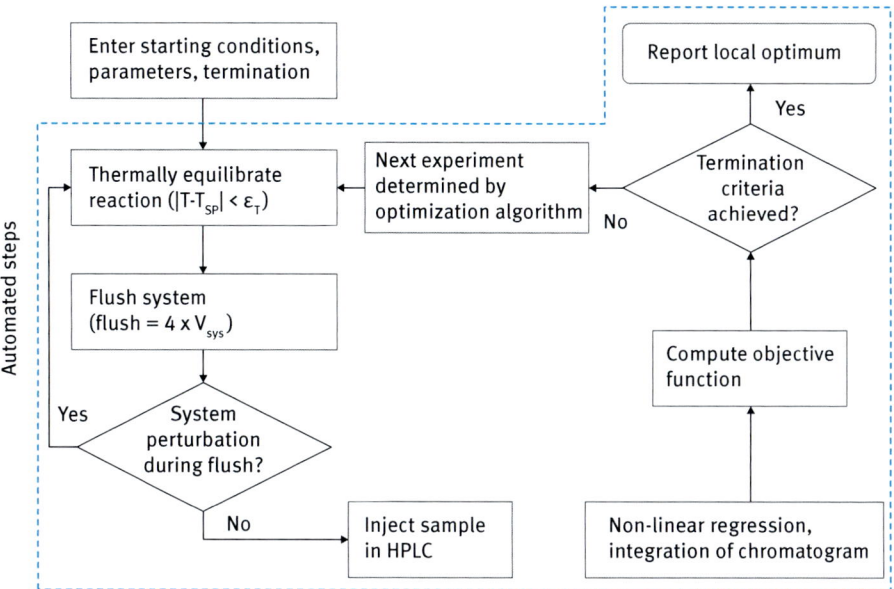

Fig. 8.19: Flowsheet description of operations implemented into automated microfluidic system [11]. Figure reproduced with permission from the American Chemical Society.

Each experiment began by thermally equilibrating the microreactors defined as the absolute difference between the reaction temperature and the set point temperature (T_{SP}) being less than a specified tolerance, nominally 3 °C. The system was then flushed adequately, generally four system volumes (V_{sys}), to ensure steady-state data collection. A sample was then injected into the HPLC system, and the chromatographic data were recorded with the LabVIEW hardware. The area of the chromatogram was computed to determine the concentration of the different components. After using these measurements to calculate the objective function, the optimization algorithm determined the next sequential experiment in the procedure or terminated if the termination criteria was achieved.

8.2.2.1 An example to compare three different black-box algorithms: Knoevenagel synthesis

Three black-box optimization algorithms were implemented to demonstrate the ability of rapidly optimizing a chemical synthesis [11]:
- Two local search methods, the Nelder–Mead Simplex Method and the Steepest Descent Method with response surfacing modeling
- A global search technique, SNOBFIT, was also implemented to illustrate the potential to use the microreactor system for the optimization of highly nonlinear reaction systems.

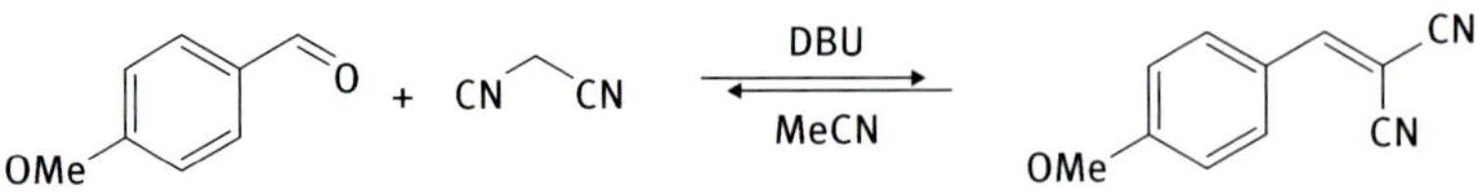

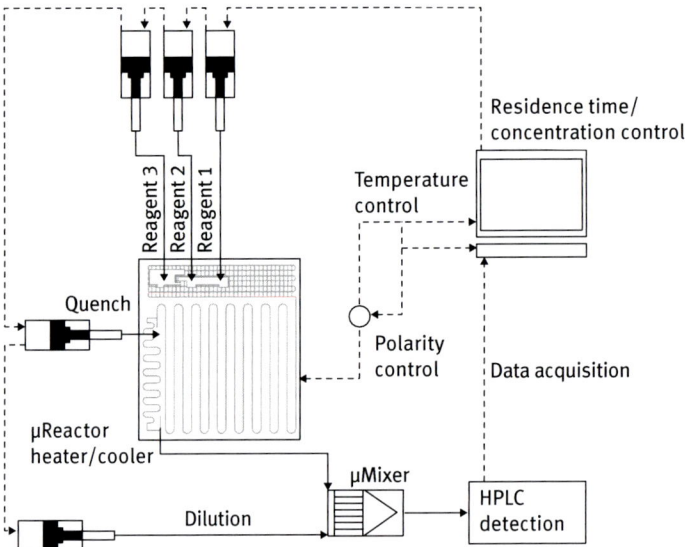

Fig. 8.20: Scheme of Knoevenagel reaction between *p*-anisaldehyde and malonitrile.

Fig. 8.21: Schematic of automated microfluidic system consisting of syringe pumps, microreactor, micromixer, HPLC, and computer with associated LabVIEW interface hardware [11]. Figure reproduced with permission from the American Chemical Society.

The reversible, condensation reaction involving *p*-anisaldehyde **12** and malononitrile **13** catalyzed by 1,8-diazabicyclo-[5.4.0]undec-7-ene (DBU) was selected as the model reaction (Figure 8.20). The weighed objective function, which is a combination of the production throughput and the yield, biases the system to look at shorter residence times without sacrificing yield.

Each optimization procedure varied the temperature (T) and residence time (τ) to maximize the objective function within the feasible space enclosed by box constraints given as:

$$40\,°C < T < 100\,°C$$
$$30\,s < \tau < 300\,s.$$

For the Simplex Method and Steepest Descent Method, the optimization initiated from 70 °C and 180 s, while the SNOBFIT Method required no initial condition information.

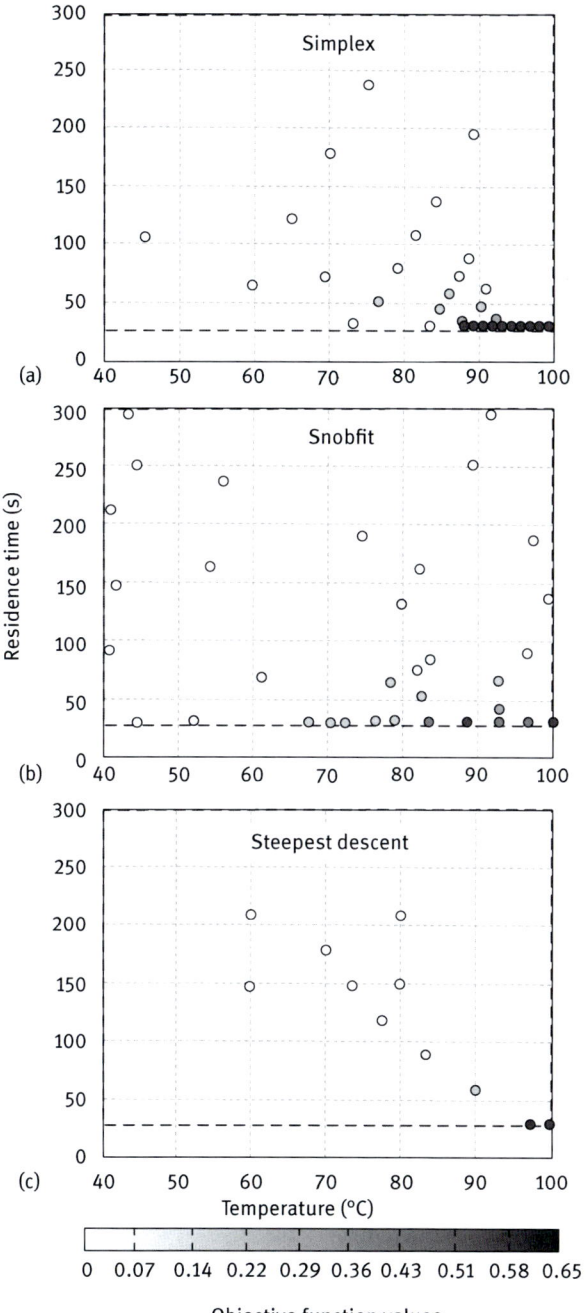

Fig. 8.22: Optimization results for Knoevenagel example using Simplex Method (a), SNOBFIT (b), and Steepest Descent Method (c). Objective function values are denoted by the color bar and range from 0 (poor) to 0.65 (good). Boundaries on the reaction variables are denoted by black dashed lines [11]. Figure reproduced with permission from the American Chemical Society.

The inlet concentration of **12** and **13**, just as their 1:1 equimolar ratio was kept constant during the experiments. The reaction mixture was quenched and diluted before automatically injecting reaction samples into the HPLC for analysis.

Results from each optimization procedure are shown in Figure 8.22. From the specified starting point, the Simplex Method maximized the objective function by methodically selecting experiments at higher temperatures and lower residence times. After attempting to select experiments outside of the feasible space, the simplex contracted in order to hone in on the optimum. As determined by the Simplex Method, the optimum for the objective function was located at a temperature of 99 °C and a residence time of 30 s.

As shown by Figure 8.22 (b), the local fitting feature of the SNOBFIT algorithm preferentially selected experiments at low residence times and higher temperatures. After performing 36 automated experiments, the SNOBFIT method also located the optimum of the objective function, with a temperature 99 °C and a residence time of 30 s. Additionally, because the SNOBFIT method performed experiments in unexplored regions of the parameter space, more confidence that these conditions correspond to the global maximum was gained.

For the specified objective function and inputted parameters, the optimum was located in the fewest number of required experiments using the Steepest Descent Method. The program progressed along a gradient towards experiments at higher temperatures and lower residence times.

As Figure 8.22 shows, all the methods were able to find the optimal parameters, although the number of experiments and the parameter space that they examined are different.

In the next chapter, we attempt to explain the Nelder–Mead Simplex algorithm in details.

8.2.3 Nelder–Mead Simplex method

8.2.3.1 Description of optimization method

This method, illustrated below, creates an initial simplex (points A_S, B_S, C_S) consisting of $k + 1$ vertices, where k is equal to the number of parameters or variables (Figure 8.23). For example, when optimizing the reaction by adjusting two variables, $k = 2$ and the simplex is composed of three points ($k + 1 = 3$). The initial three points of the simplex are defined by the starting point and the step size [10].

The objective function value (for simplication here: yield) of the product is evaluated at each of these conditions and the next experimental conditions are determined by reflecting the point with the worst yield (A_S) about the midpoint of the remaining k points (B_S, C_S). This newly examined point (A_R) will replace the point (A_S) in the simplex if its yield is greater ($A_R > A_S$). In this case, a new simplex is created (A_R, B_S, and C_S). However, if the yield at this newly examined point is lower than the yield

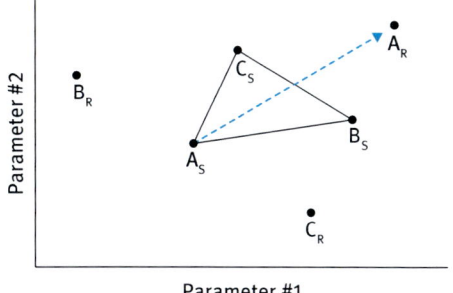

Fig. 8.23: Example of simplex where $A_S < B_S < C_S$ (Modified after [10]).

of the point that was reflected within simplex ($A_S > A_R$), this point is not included in the simplex.

Rather, the reflection procedure is repeated using the point with the next lowest yield (B_S). This sequential process continues until it has been determined that none of the possible reflections in the simplex will result in a greater yield ($A_S > A_R$, $B_S > B_R$, $C_S > C_R$). Under these circumstances, the simplex is said to have surrounded the optimum.

To locate the optimum with greater accuracy, the size of simplex contracts and the procedure is repeated until a specified termination criterion is achieved.

8.2.4 Multidimensional optimization

Obviously, the optimization algorithms are able to handle more then two variables as well [11]. However, having more variables will increase the number of steps that the algorithm need to perform and the process will take longer, using more starting compounds.

In the next described experiment, oxidation of benzylalcohol to benzaldehyde (Figure 8.16) was performed in a 4D optimization using the Simplex Method. In this investigation, the researchers considered four parameters that can influence the reaction: temperature, residence time, concentration of benzylalcohol, and CrO_3 equivalents were varied to maximize the benzaldehyde yield (Table 8.1).

The automated microreactor system performed 46 sequential experiments to determine the local optimum of benzaldehyde yield at 80% (Figure 8.24). The range of values that the algorithm investigated are shown in Table 8.1, with the optimal conditions corresponding to $T = 88\,°C$, $\tau = 48\,s$, benzaldehyde concentration 8.2 mM, and 0.65 equiv. of CrO_3.

The results indicated that the reaction was enhanced at higher temperatures with shorter reaction times. Traditionally, however, this reaction is performed at longer residence times and at lower temperatures to prevent subsequent oxidation of the alde-

Table 8.1: Range of values for each reaction parameter varied during 4D optimization.

Reaction parameter	Min.	Max.
Temperature (°C)	50	94
Residence time (s)	25	79
$[PhCH_2OH] \times 10^{-3}$ (M)	6.7	9.3
CrO_3 equivalence	0.25	1.37

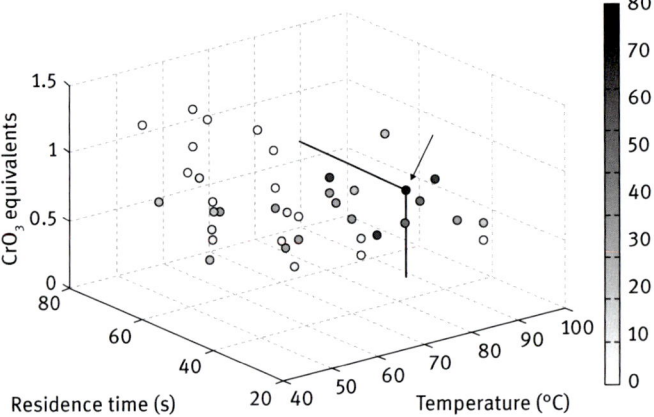

Fig. 8.24: Benzaldehyde yield measured during 4D optimization by Simplex algorithm. Inlet reactor concentrations of benzyl alcohol are not shown in this graph in order to present the benzaldehyde yield data in the clearest possible form [11]. Figure reproduced with permission from the American Chemical Society.

hyde to the carboxylic acid. However, the increased control over reaction conditions in microreactors allowed the reaction to be performed at more aggressive conditions.

8.2.5 Optimization and scale-up

Since reaction optimization results obtained in a flow microreactor are typically not limited by mass or heat transfer effects, moving from optimal laboratory conditions to larger scale flow reactors can be more easily achieved by integrating the observed chemical understanding with established chemical engineering reactor design methods [10].

The Heck reaction of 4-chlorobenzotrifluoride (**15**) and 2,3-dihydrofuran (**16**) was investigated in the next example (Figure 8.25). The optimization process was directed by the Nelder–Mead Simplex Method. Since the desired product **17** readily reacts with a second equivalent of the aryl chloride, the yield of the reaction is highly dependent upon the number of equivalents of **16**.

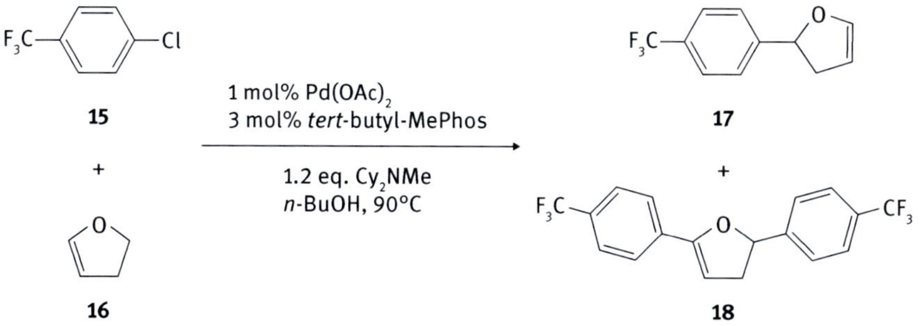

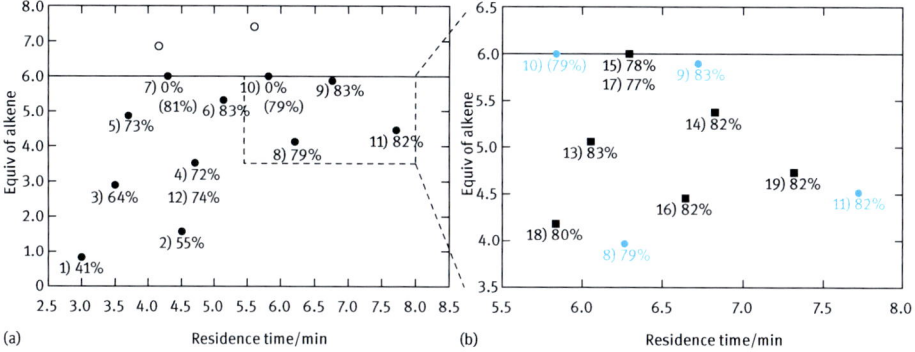

Fig. 8.25: Heck reaction of monoarylated product **17**; Cy = cyclohexyl.

Fig. 8.26: Optimization results for the Heck reaction depicted in Figure 8.25. Results are indexed by experiment number (1–19) and the corresponding yield of **17**. (a) The data points for experiments that lie above the 6.0 equiv of alkene line (empty dots) were projected toward the boundary (6.0 M) and penalized with a 0% yield in the optimization program. (b) Extended view of experiments within the contracted simplex. Reaction yield values that were obtained with the larger simplex (blue dots) included for reference [10]. Reproduced with permission from Wiley-VCH Verlag GmbH & Co.

An upper boundary of 6.0 was placed on the equivalents of **16**. Any potential reaction conditions selected by the optimization algorithm exceeding 6.0 equivalents were projected onto this upper boundary. Although the yields at this boundary line were performed and analyzed, these points were "penalized" by substituting 0% into the optimization algorithm. The use of a penalty function improves the efficiency of the algorithm by preventing the simplex from prematurely contracting or collapsing.

To allow the system to find optima that are located on the boundary, this penalty function was removed after the simplex had contracted. This contracted simplex (Figure 8.26, Exps. 13–19) located the optimal conditions near a residence time of 6 minutes and 5.0 equivalents of **16**; corresponding to yield of 83%.

To evaluate the Heck reaction of **15** and **16** at a 50-fold larger scale, a 7 mL Corning Advanced-Flow Glass Reactor module was employed.

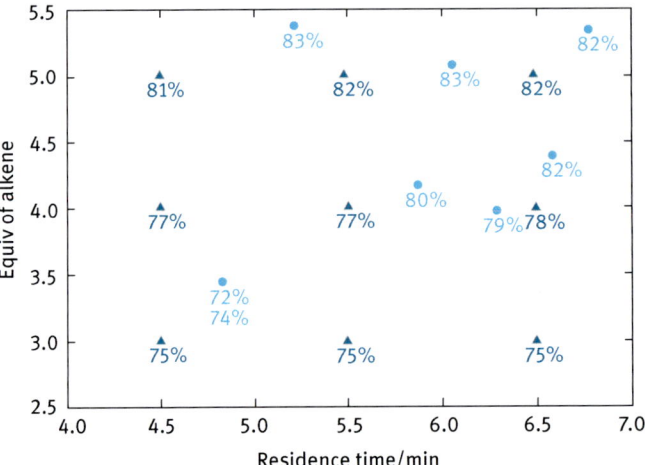

Fig. 8.27: Yields of 50-fold reaction scale up in mesoscale flow reactor (triangles). The reaction yields (dots) obtained in the microreactor are shown for comparison [10]. Reproduced with permission from Wiley-VCH Verlag GmbH & Co.

Nine different reaction conditions were selected that encompassed the region containing the optimum yields as previously determined in the microreactor. These results indicate that the optimal conditions were successfully translated from the microreactor to the mesoscale system (Figure 8.27).

To validate the yields obtained by HPLC analysis, the reaction was run under the optimal conditions of 5.5 minutes residence time and 5.0 equivalents of alkene for over two hours, during which time 168 mL of crude solution was collected. The monoarylated product was isolated by distillation and chromatography to provide 26.9 g of **17** with a yield of 80%. This is in good agreement with the yields determined by HPLC analysis and corresponds to an annual production rate of 114 kg/year.

8.2.6 Flow reactors with built-in optimization

For faster optimization, ThalesNano offers the H-Cube Pro™ flow reactor with built-in optimization algorithm. The software implements the Nelder–Mead Simplex method. The reactor was originally designed for hydrogenation reactions, but due to the different reaction modules compatible with the H-Cube Pro™, the optimization algorithm is available for other chemistries as well.

The built-in software generates a regular simplex into the geometrical center of the user-defined parameter space. The system displays the experiment parameters with an accuracy defined by the user. If the parameters given by the algorithm exceed the boundaries previously defined, the software modifies the exceeding value to the maximum boundary limit value. The objective function values are entered into the system manually, after analysis of the product solution with an external analytical tool.

Further readings on Self Automation and Optimization: references [13–15]

8.3 In-line separation

8.3.1 Liquid-liquid separators

Membrane-based separation relies on the exploitation of surface forces and the use of a membrane wetted by one of the phases; Membrane-based separators are therefore ideal candidates for operating at both microscale and milliscale. There are several products on the market (e.g., Syrris FLLEX, Zaiput) implementing membrane-based separation. However, successful separation requires accurate control of pressures, making the operation and implementation cumbersome.

8.3.1.1 Technical description

An improved separator design was presented recently by Zaiput Flow Technology that integrates a pressure control element to ensure that adequate operating conditions are always maintained [16, 17].

Figure 8.28 shows a schematic of the membrane separator which incorporates a pressure control segment immediately following the membrane. The pressure control is made up of a diaphragm stretched over the retentate stream with the permeate stream flowing on the reverse side. The diaphragm is made of 50 µm PFA (perfluoroalkoxy) film, thus providing high chemical compatibility (Figure 8.29).

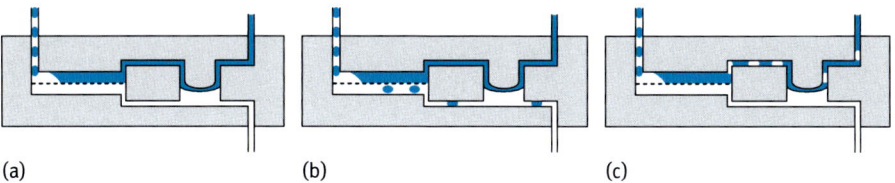

(a)　　　　　　　　　(b)　　　　　　　　　(c)

Fig. 8.28: Integrated pressure control in membrane separator showing deformed diaphragm (heavy curved line) to provide a fixed pressure difference across the membrane (short vertical lines). The aqueous and organic phases are shown in blue and white, respectively. (a) Separator under normal operation. (b) Separator operating with breakthrough of the retained phase. (c) Separator operating with retention of the permeate phase [17]. Figure reproduced with permission from the American Chemical Society.

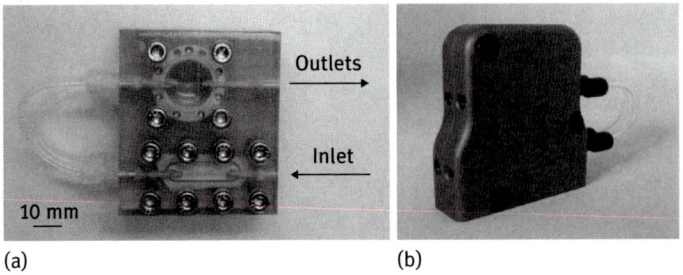

Fig. 8.29: Photograph of polycarbonate membrane separator with integrated pressure control (a). The separator membrane is located on the lower portion, and the pressure control diaphragm is located on the upper portion of the device [17]. Figure reproduced with permission from the American Chemical Society. Photo of the assembled equipment (b).

The role of this flexible diaphragm is to balance out the possible pressure differences on the two sides of the membrane that would result in the breakthrough of one of the phases.

The integrated pressure controller greatly reduces the complexity when implementing separators within chemical reactor systems. It decouples the separator pressures and thus simplifies the on-line control required for operation. The separator allows operation under pressure (300 psi/2 MPa) and in-line operations. The specific design is especially suited to flow rates between 1 and 10 mL/min. The separating membrane is available in different pore sizes to cover a wide range of applications.

8.3.1.2 Example for application

In the following example, a two-stage solvent-swap was tested with the Zaiput continuous flow liquid-liquid separator (Table 8.2).

A stream of 0.34 M benzoic acid in EtOAc was pumped at 1 ml/min into a mixing T-piece to contact a stream of 0.55 M NaOH. The first separator then split the phases with the organic phase passing through the membrane and through an additional 0.6 bar back-pressure controller. (This was added because the organic outlet of the separator generally must be at a higher pressure than the aqueous or else excess pressure on the aqueous outlet (in this case due to the second separator) will cause breakthrough of the aqueous phase by overriding the differential pressure controller.)

The aqueous phase containing benzoic acid then contacted a 1 mL/min stream of toluene. After a short length of tubing, a 1 mL/min stream of 0.6 M HCl was added. The stream was then separated by a second, identical membrane separator. The results of the extraction are summarized in Table 8.2.

The system was operated for 2 h (corresponding to over 60 residence volumes) without any failure in the separation and halted without failure of the separation. The

Table 8.2: Summary of solvent swap results, comparing a batch shake flask and continuous with two membrane separators. Error values are one standard deviation of 4 samples taken over the 2 h run [17]. Figure reproduced with permission from the American Chemical Society.

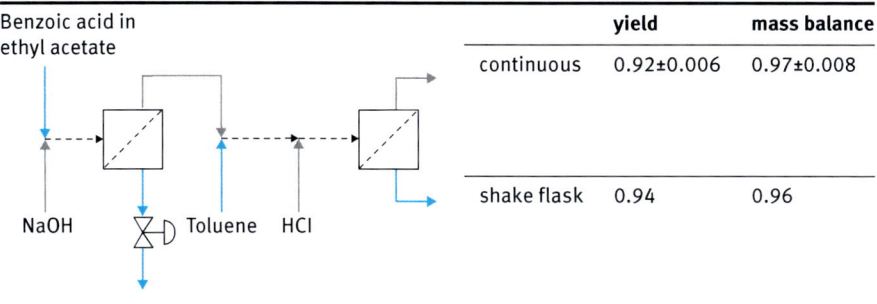

	yield	mass balance
continuous	0.92±0.006	0.97±0.008
shake flask	0.94	0.96

continuous system reproduces the batch performance both in terms of yield and mass balance.

8.3.2 Scavenger and chromatography columns

In recent years, the toolbox of flow processes to effect chemical transformations has become increasingly sophisticated and wide-ranging. For example, an array of immobilized reagents, catch-and-release agents, catalysts, and scavengers has been developed that allows some key reactions to be achieved efficiently and without contamination of the downstream flow.

Here we present some examples for such columns without trying to cover the whole range of possibilities.

8.3.2.1 Scavenger columns for a series of work-up procedure

On Figure 8.30, a key-step of the synthesis of imatinib (the active ingredient in Gleevec, a treatment for chronic myeloid leukaemia developed by Novartis) is presented [18]. In this particular step of the synthesis, a series of scavenger and reactant columns were applied. First, a solution of benzyl chloride **19** and N-methylpiperazine in DMF were reacted upon elution through a column of $CaCO_3$ maintained at 80 °C.

The output of this column was then directed into a column containing a polymer-supported isocyanate to scavenge any unreacted N-methylpiperazine, then into a column of silica-supported sulfonic acid to sequester the product allowing any unreacted **19** to flow to waste.

The product (**20**) was subsequently released in a purified form by treatment with a solution of a base, which also acted as a solvent switch, directly into the next reaction step.

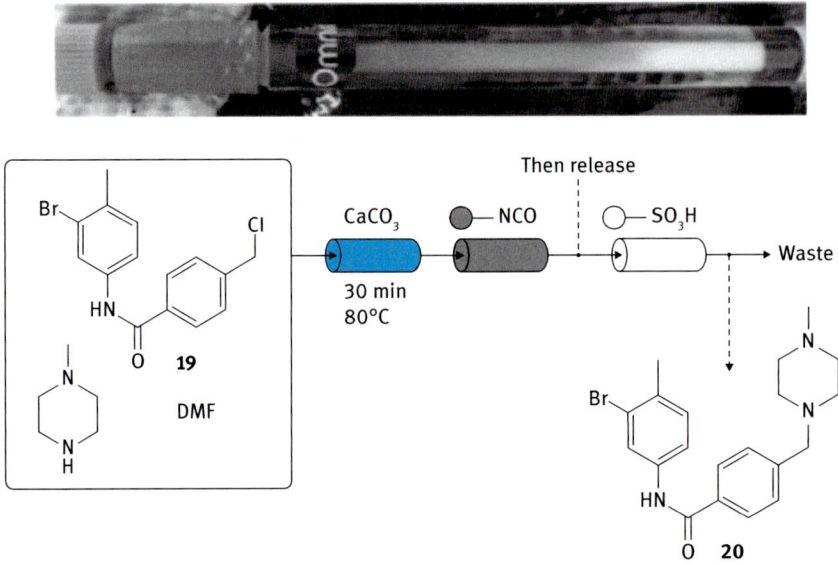

Fig. 8.30: Flow synthesis of an imatinib intermediate with image showing 2 sequestered on silica-supported sulfuric acid. Figure modified after [18].

8.3.2.2 Parallel columns for continuous operation

Griffiths-Jones and coworkers [19] developed a completely automated multistep, continuous-flow system to generate a 48-member sulfonamide reaction library (Figure 8.31).

In this platform, monoalkylation of a Boc-protected primary sulfonamide was performed in a glass reactor column, packed with polystyrene-supported 1,5,7-triazabicyclo[4.4.0]dec-5-ene (PS-TBD), followed by deprotection in a second column packed with the polystyrene-supported sulfonic acid AmberlystTM H-15.

The platform can work fully automated, manual intervention may only be necessary to remove the spent columns.

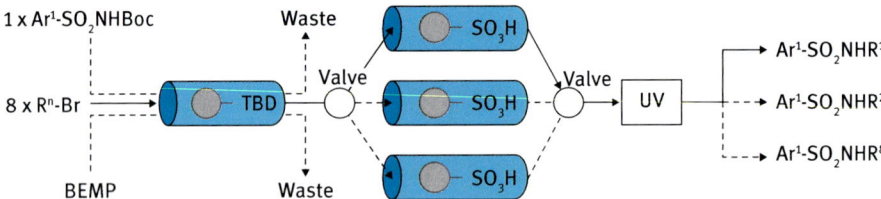

Fig. 8.31: Schematic of an automated platform for reaction screening to create a 48-member sulfonamide reaction library. Abbreviation: BEMP, 2-*tert*-2-diethylamino-1,3-dimethyl-perhydro-1,2,3-diazophosphorine. Figure modified after [15].

The recent example presents a flow synthesis of a reaction library, but the same set-up can be easily applied for the continuous production of one compound as well. One can imagine that the H-15 column with time becomes saturated and needs to be replaced; therefore the continuous system has to be stopped. This problem is easily avoided when several H-15 columns are connected parallel to each other (as shown on Figure 8.31.). When the first column becomes saturated (for example, the pressure drop on the column increases), the connecting multiport valve closes and directs the liquid to the second parallel column. Thereafter, the first column can be regenerated, or manually changed. The system is able to operate on a fully automated manner, thus circumventing the limited lifetime of the scavenger column.

8.3.3 Simulated moving Bed Chromatography

Purification is often the bottleneck in a chemical synthesis and can negate the benefits ascribed to flow reactors, unless side-products can be removed by crystallization or liquid–liquid extraction. Separating complex mixtures of products often necessitates recourse to batch column chromatography.

Simulated Moving-Bed (SMB) Chromatography, a form of continuous countercurrent chromatography widely used industrially, offers a solution [20].

8.3.3.1 True Moving-Bed Chromatography – an imaginary process

In the following discussion, True Moving-Bed chromatography will first be described (a very difficult process to put into practice) and from this, the more practical Simulated Moving-Bed Chromatography (SMB) process can be derived and understood [21].

First imagine a resin filled column formed as a toroid. Then imagine a rapid, continuous flow of eluent (water) in one direction inside the loop. Now imagine that the chromatographic resin contained in the toroid is rapidly circulating in the loop in the opposite direction from the water circulation.

Recall that this process is, for the time being, imaginary so how the two opposing flows of water and resin are enabled need not be explained. Setting up these two opposing internal flows results in a key characteristic of a true moving-bed. If a binary mixture to be separated is added continuously at one point of the loop, the component more readily adsorbed by the resin will tend to move with the resin and the component less readily adsorbed will move with the water. If the two opposing internal flows are balanced carefully, the two components can be continuously separated and recovered.

So here we have a type of continuous chromatography. The benefits observed, in addition to continuous operation, are less eluent and less resin than used for batch chromatography. Eluent must also be added continuously to the system so the overall process looks like the following figure (Figure 8.33).

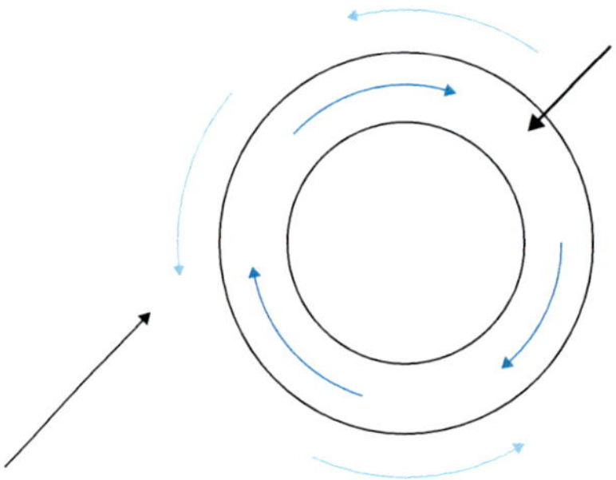

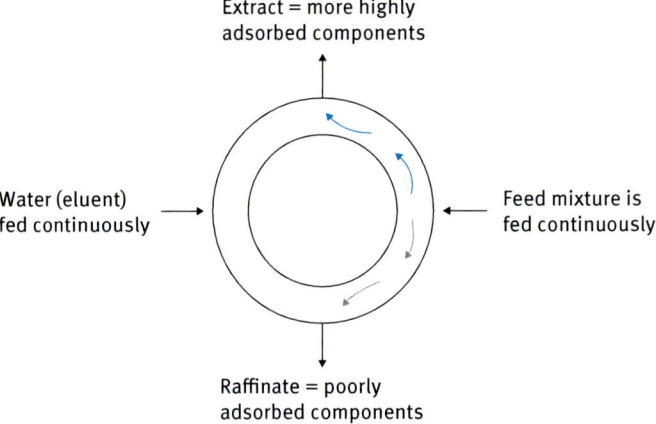

Water circulating rapidly in one direction inside the loop.

Resin inside the loop is circulating rapidly countercurrent to the water.

Fig. 8.32: Theory of True Moving-Bed Chromatography. Eluent and resin circulate in the opposite direction. Figure modified after [21].

Extract = more highly adsorbed components

Water (eluent) fed continuously

Feed mixture is fed continuously

Raffinate = poorly adsorbed components

Fig. 8.33: Theory of True Moving-Bed Chromatography. Component more adsorbed by the resin will move with the resin; component less adsorbed will move with the eluent. Figure modified after [21].

> **i** Note that the highly adsorbed component is commonly called "extract" and the less adsorbed component is called "raffinate".

Unlike batch chromatography which can separate many components from a mixture, the moving-bed method is a binary separation technique – the material which moves with the circulating eluent and the material (or, group of materials) which moves with the circulating resin.

8.3.3.2 Simulated Moving-Bed Chromatography (SMB) – a practical process

Simulated Moving-Bed technology is used to implement a very close approximation of True Moving-Bed Chromatography. SMB exhibits the same valuable characteristics of the true moving bed concept. The following discussion will develop the SMB concept based on the previous discussion.

When encountering SMB in the literature, often a set of cells or columns is illustrated parallel to each other. Note that the exit from each cell enters the top of the next cell and all the cells are linked in this manner into a loop (Figure 8.34).

In order to understand how the imaginary true moving-bed can be simulated, it is useful to draw the common depiction of an SMB as a loop. In the following figure, the concept is still the same, that is, the exit from each cell enters the top of the next cell and all the cells are linked in this manner into a loop. We are just drawing the SMB differently to make a correspondence with our discussion of true moving bed technology.

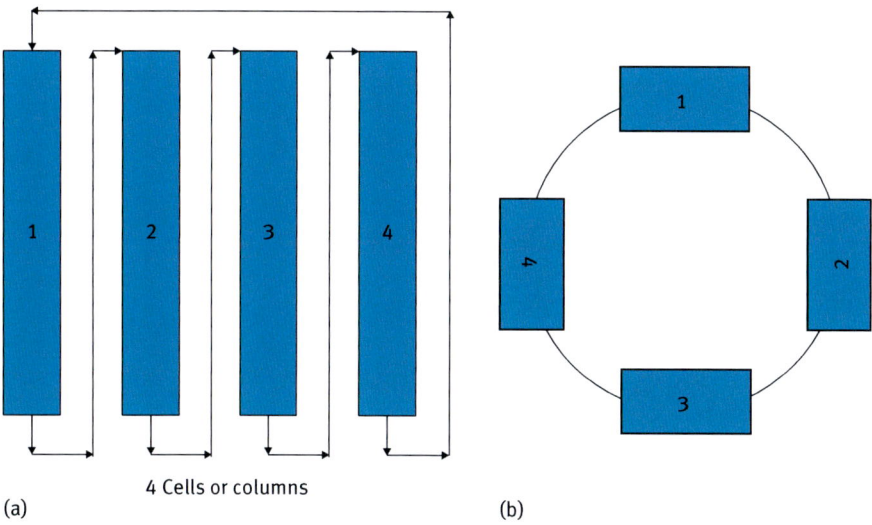

Fig. 8.34: Illustration of 4-cell columns on two different ways. Figure modified after [21].

(However the resin is residing in cells and is therefore stationary. Unlike the true moving-bed, we have no way to move the resin in a direction opposite to the circulating liquid. If a feed mixture is added at a point between cells, everything would simply move in the direction of the liquid flow.)

The secret of Simulated Moving-Bed Chromatography operation is that the resin movement can be simulated in the direction opposite to the internal circulation if the feed valve is occasionally switched in the same direction as the liquid flow. The other 3 valves – eluent, extract and raffinate, must also move in this manner (Figure 8.35).

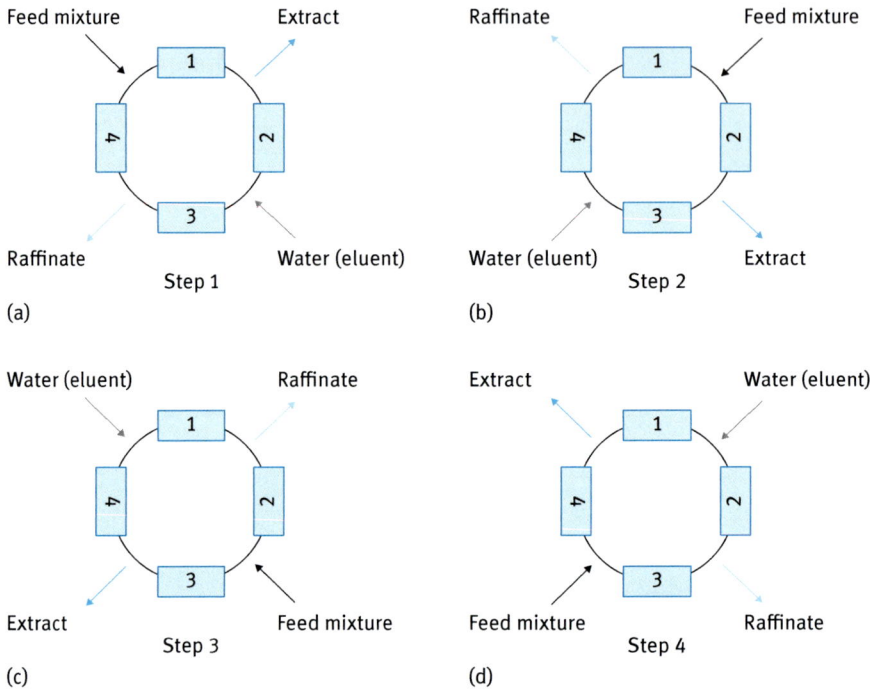

Fig. 8.35: For a person standing in the frame of reference of the feed mixture valve it would appear that the circulation liquid is moving ahead and the resin is moving back. Figure modified after [21].

With respect to the figures below, for a person standing in the frame of reference of the feed mixture valve (not aware that it is moving around the loop but thinking it is stationary and the loop itself is moving), it would appear that the circulation liquid is moving ahead and the resin is moving back. This is the same as a true moving-bed.

Note, therefore, that for SMB operation the switching of valve locations takes the place of the internal resin flow.

Important: You can best understand how a true moving-bed is simulated by placing yourself in the frame of reference of the rotating valves and assume that the presently working valves (feed, water, extract or raffinate) for a given material are always in the same location in space and the rest of the SMB equipment – cells and so on – moves relative to you, or in other words, the resin appears to move counter-current to the internal liquid flow because of your frame of reference.

8.3.3.3 SMB in the practice

Today, some companies have advances applications of SMB Chromatography (Orochem Technologies, ChromWorks). The technology is used for the industrial-scale pu-

Fig. 8.36: Pilot scale SMB unit at Orochem Technologies. Figure modified after [22].

Fig. 8.37: Picture of a multifunctional valve of the KNAUER® CSEP C9812 unit.

rification of a variety of different molecules such as omega-3 fatty acids, sugars, biofuels and APIs.

One of the most important parts of the SMB unit is the multifunctional valve (Figure 8.37, KNAUER, Germany), which consists of a rotor and a stator with 24 ports each. The ports are connected to each other by continuous channels. To realize a proper simulated solid flow rate, the switching of the columns should be precisely performed at every switching time interval [22].

A detailed chemical example for SMB application

Recently, O'Brien *et al.* reported the first successful coupling of flow synthesis and SMB chromatography to continuously produce pure product [20].

The nucleophilic aromatic substitution (S_NAr) reaction of 2,4-difluoronitrobenzene (**21**) with morpholine (Figure 8.38) affords a mixture of products and thus was selected to demonstrate the approach.

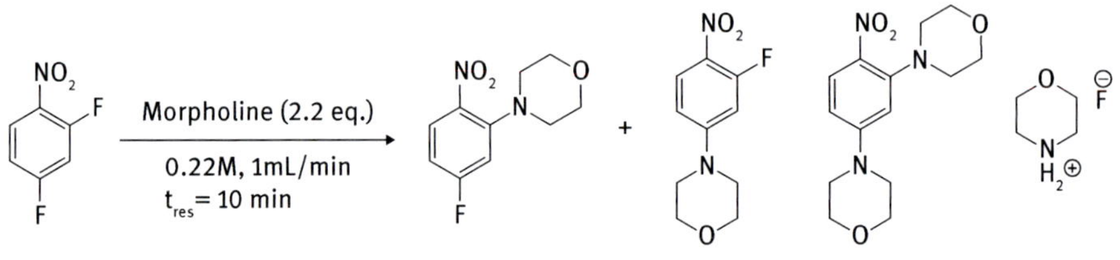

Fig. 8.38: Reaction of 2,4-difluoronitrobenzene with morpholine. *Para*-**22**, **23** and **24** are the side-products.

A commercially available flow reactor (Vapourtec, R2/R4) was used in the experiments. Upon changing the reaction solvent to THF higher selectivity for the product (*ortho*-**22**) was observed (Table 8.3). Increased temperature (165 °C) ensured complete consumption of the starting material. The method provided an already very high quality (92% *ortho*-**22**) crude product that was further purified.

Table 8.3: Reaction conditions and product distribution.

Solvent	T (°C)	*ortho*-22/*para*-22/23	Conversion (%)
EtOH	100	74 : 13 : 13	96
THF	165	92 : 5 : 3	98

Thermodynamic and kinetic parameters for the Simulated Moving-Bed Chromatography design were determined from preliminary batch chromatography experiments. Finally, the SMB was performed on a reversed-phase manner.

The mixture from the flow reactor was then diluted with eluent and fed into the SMB system that was arranged in a 6 column (1-2-2-1) configuration (Figure 8.39). Fractions of the extract and the raffinate were collected after every cycle (six port shifting intervals, 171 s) and the product distributions were determined.

The extract contained only *ortho*-**22** product with over 99% purity. The system reached steady state after several cycles and was able to operate continuously to provide the target compound with 89% yield (Figure 8.40).

8.3 In-line separation — 249

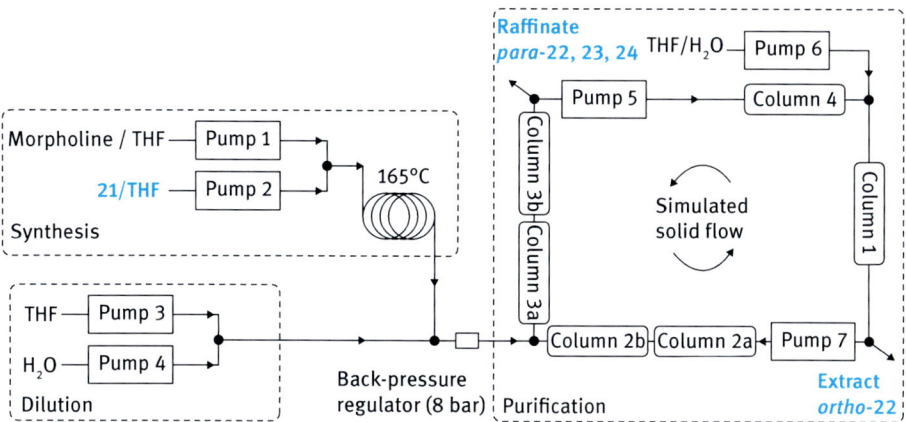

Fig. 8.39: Schematic diagram of the directly coupled system [20]. Reproduced with permission from Wiley-VCH Verlag GmbH & Co.

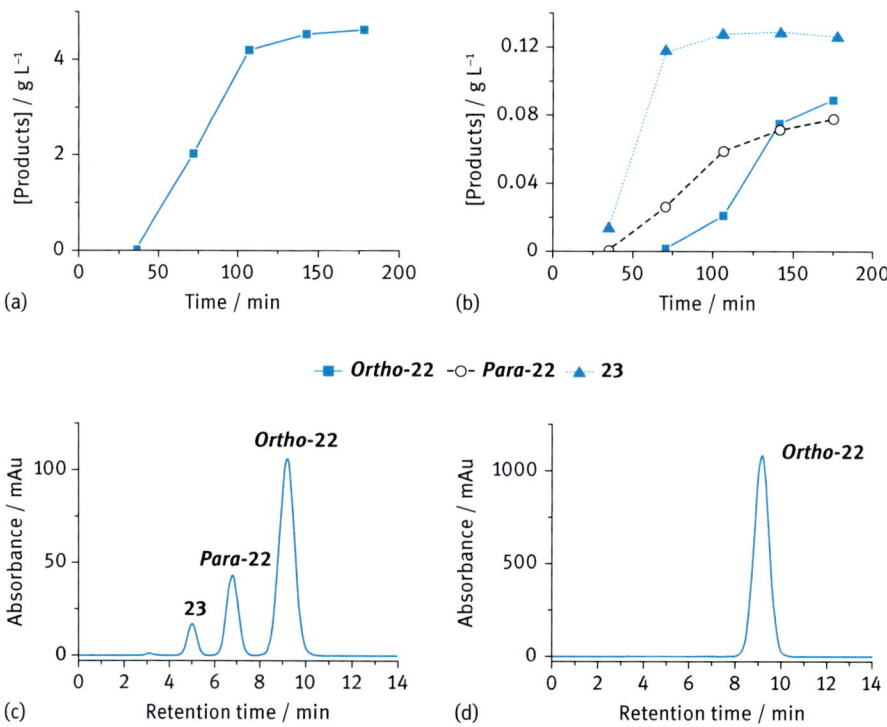

Fig. 8.40: Concentration profiles at the a) extract port in the coupled system; b) raffinate port in the coupled system; c) chromatogram of the feed composition from the flow reactor; d) chromatogram of extract fraction over the final shift in the coupled system [20]. Reproduced with permission from Wiley-VCH Verlag GmbH & Co.

Study questions

8.1. Which of the following analytical instruments could be used in an on-line/in-line manner: IR, MS, FID-GC, UV-VIS, Fluorescent spectroscopy?

8.2. Determine the optimal temperature for the preparation of **6**, using Figure 8.4!

8.3. Imagine that during the hydrogenation of nicotinate, we decide to connect the ReactIR device behind the reactor. (After the back-pressure valve, so no harm will be made on the instrument by high pressure.) Suppose that the reaction performs in the reactor with 100% conversion. Compared to Figure 8.8, what kind of scheme would you be able to draw? Consider the starting material and the product as well.

8.4. Imagine a set of reactions that you carry out in the lab. The reaction needs to be optimized for the yield, so you decide to make a temperature screen between 20–100 °C. On the next day, you decide to change the flow rate of your reaction, but you keep all the other conditions constant. Draw the diagram of your optimization on a 2D graph. What is the problem with the parameter space you used? How should you design the experiment to obtain the global maximum?

8.5. Remember the benzylalcohol oxidation reaction we discussed in Section 8.2.3 (Figure 8.16). You managed to optimize the reaction with high selectivity for benzaldehyde, but unfortunately the conversion of the reaction is not so good. You decide to separate the two compounds by SMB chromatography on silica, using an appropriate organic solvent as eluent (Let's ignore the fact that the reaction solvent was water-acetic acid.) Which component will be in the raffinate and in the extract?

Bibliography

[1] Browne DL, Wright S, Deadman BJ, Dunnage S, Baxendale IR, Turner RM, Ley SV. Continuous flow reaction monitoring using an on-line miniature mass spectrometer. Rapid Commun. Mass Spectrom., 2012, Vol. 26, pp. 1999–2010.

[2] Mills PL, Nicole JF. Multiple automated reactor systems (MARS). 1. A novel reactor system for detailed testing of gas-phase heterogeneous oxidation catalysts. Ind. Eng. Chem. Res., 2005, Vol. 44, pp. 6435–6452.

[3] Welch CJ, Gong X, Cuff J, Dolman S, Nyrop J, Lin F, Rogers H. Online analysis of flowing streams using micro-flow HPLC. Org. Process Res. Dev., 2009, Vol. 13, pp. 1022–1025.

[4] Baxendale IR, Griffiths-Jones CM, Ley SV, Tranmer GK. Preparation of the neolignan natural product grossamide by a continuous flow process. 2006, Synlett, pp. 427–430.

[5] Carter CF, Lange H, Ley SV, et al. ReactIR Flow Cell: A New Analytical Tool for Continuous Flow Chemical Processing. Organic Process Research & Development, 2010, Vol. 14, pp. 393–404.

[6] Lange H, Carter CF, Hopkin MD et al. A breakthrough method for the accurate addition of reagents in multi-step segmented flow processing. Chem. Sci., 2011, Vol. 2, pp. 765–769.

[7] Thermo Scientific™ picoSpin™ NMR 45 Spectrometer. [Online] [Cited: 01 13, 2014.] http://picospin.com/products/picospin-45/.

[8] Bart J, Kolkman AJ, Oosthoek-de Vries AJ, et al. A Microfluidic High-Resolution NMR Flow Probe. J. Am. Chem. Soc., 2009, Vol. 131, pp. 5014–5015.

[9] Bart J, Oosthoek-de Vries AJ, Tijssen K, Janssen JWG, van Bentum PJM, Gardeniers JGE, Kentgens APM. In-Line NMR Analysis Using Stripline Based Detectors. 14th International Conference on Miniaturized Systems for Chemistry and Life Sciences, 2010, pp. 2086–2088.

[10] McMullen JP, Stone MT, Buchwald SL, Jensen KF. An Integrated Microreactor System for Self-Optimization of a Heck Reaction: From Micro- to Mesoscale Flow Systems. Angew. Chem. Int. Ed., 2010, Vol. 49, pp. 7076–7080.
[11] McMullen JP and Jensen KF. An Automated Microfluidic System for Online Optimization in Chemical Synthesis. Org. Process Res. Dev., 2010, Vol. 14, pp. 1169–1176.
[12] Rasheed M and Wirth T. Intelligent Microflow: Development of Self-Optimizing Reaction systems. Angew. Chem. Int. Ed., 2011, Vol. 50, pp. 357–358.
[13] Parrott AJ, Bourne RA, Akien GR, Irvine DJ, Poliakoff M. Self-Optimizing Continuous Reactions in Supercritical Carbon Dioxyde. Angew. Chem. Int. Ed., 2011, Vol. 50, pp. 3788–3792.
[14] Darvas F. Application of the Sequential Simplex Method in Designing Drug Analogs. J. Med. Chem., 1974, Vol. 17, pp. 799–804.
[15] McMullen JP and Jensen KF. Integrated microreactors for Reaction Automation: New Approaches to Reaction Development. Annual Review of Analytical Chemistry, 2010, Vol. 3, pp. 19–42.
[16] Liquid-Liquid Separators. [Online] [Cited: 01 13, 2014.] http://www.zaiput.com/liquid-liquid-separators.
[17] Adamo A, Heider PL, Weeranoppanant N, and Jensen KF. Membrane-Based, Liquid-Liquid Separator with Integrated Pressure Control. Ind. Eng. Chem. Res., 2013, Vol. 52, pp. 10802–10808.
[18] Hopkin MD, Baxendale IR, Ley SV. The Lab of the Future. The importance of remote monitoring and control. chimica oggi/Chemistry Today, 2011, Vol. 29, pp. 28–32.
[19] Griffiths-Jones CM, Hopkin MD, Jonsson D, Ley SV, Tapolczay DJ, et al. Fully automated flow-through synthesis of secondary sulfonamides in a binary reactor system. J. Comb. Chem., 2007, Vol. 9, pp. 422–30.
[20] O'Brien AG, Horváth Z, Lévesque F, Lee JW, Seidel-Morgenstern A, and Seeberger PH.Continuous Synthesis and Purification by Direct Coupling of a Flow Reactor with Simulated Moving-Bed Chromatography. Angew. Chem. Int. Ed., 2012, Vol. 51, pp. 7028–7030.
[21] What is Simulated Moving Bed Chromatography? [Online] [Cited: 01 13, 1014.] http://www.arifractal.com/technical-library/chromatography.
[22] Simulated moving bed chromatography overview. [Online] [Cited: 01 13, 2014.] http://orochem.com/index.php?route=information/information{&}path=3{_}17{&}information{_}id=61.

Jean-Christophe Monbaliu, Ana Cukalovic, and Christian V. Stevens
9 Safety aspects related to microreactor technology

9.1 Introduction

9.1.1 Chemical processes

Chemical processes convert raw or starting materials into more elaborated compounds of increased utility or economic value using a wide range of methods and means. Two extreme scenarios are often encountered: (a) chemical processes using highly reactive starting materials and (b) chemical processes using rather stable materials. Although a high intrinsic reactivity allows materials to undergo reactions under moderate operating conditions (temperature and pressure), a hazard exists if the reaction is strongly exothermic or entropically favored. On the other hand, a moderate intrinsic reactivity often requires drastic operating conditions, that is, elevated temperatures and pressures, to force the reaction within a practical timescale, which, again, raises operational hazards.

Process industry is a major economical player with tremendous applications in every single action of the human being [1]. Large amounts of raw chemicals such as oil and minerals are being processed for centuries into a tremendous variety of commodity and specialty chemicals with applications ranging from drugs to fertilizers. Rapidly rising needs continuously and increasingly stimulate industrial production: new facilities with larger capacity sprout up all over the globe. Although modern society couldn't survive without industry, their overly vicinity creates a background threat for the population and the environment, also leading to the NIMBY (Not In My Backyard) syndrome. Major industrial disasters with tragic impacts on population and environment have been reported in the past (e.g, Flixborough, England; Bhopal, India; Seveso, Italy; Toulouse, France). However, most of the chemical incidents directly affect operators and workers present on-site rather than the surrounding population. Loss of containments of toxic substances, direct exposure, hazards of explosions and fires are amongst the most threatening risks of accidents to operators and on-site personnel. In the US only, the Chemical Safety and Hazard Investigation Board (CSB) [2] examined 85 serious chemical incidents involving hundreds of casualties and hundreds of millions of dollars in property damage from August 2000 to August 2013. In Europe, 111 major incidents involving chemicals were reported [3] within the same period of time. Although goals and scales are diametrically opposed, the hazards related to chemical processes remain and have led to numerous tragic incidents in academic labs as well: in the US only, the CSB has collected data on more than 120 incidents. The death of a lab research assistant in a flash fire at the University of California, Los Angeles, in 2008 and a 2010 explosion at Texas Tech University that severely injured a graduate student are amongst the most notorious [4].

9.1.2 Safety in chemical processes

> **Safety** or **loss prevention:** the prevention of accidents through the use of appropriate technologies to identify the hazards of a chemical plant and eliminate them before an accident occurs.
>
> **Hazard:** a chemical or physical condition that has the potential to cause damage to people, property, or the environment.
>
> **Risk:** a quantitative or qualitative measure of human injury, environmental damage, or economic loss in terms of both the incident likelihood and the magnitude of the loss or injury. The risk is usually defined as $R = P \times S$ (P = probability and S = Severity).

Nowadays, most industrial and academic labs have developed a strong *safety culture* as an answer to a long series of tragic chemical incidents[5]. Safety and training programs are implemented to provide laboratory personnel with safety consciousness and appropriate lab practices. Industrial facilities are designed to be inherently safe(r) [6, 7]: *hazards* are thus tracked down in order to eliminate them rather than to patch up existing facilities/processes with "add-ons" safety features in order to render the risk acceptable. This obviously comes with a non-negligible cost, especially for established processes. Hazard assessments are being performed in the early stages of R&D in order to shrink the *risk* associated with such an activity as much as possible. A wide variety of qualitative and quantitative hazard assessment methods are now available to assess reactive hazards during storage, transport and processing: Risk Matrix, HAZOP, Inherent Safety Index, Worst Case and Consequence Analysis, the Dow Fire and Explosion Index, and so on [5]. These assessment methods are applied at different stages of the process development and usually take a variety of parameters into account (chemical compatibility, operating conditions, process equipment, and so on). Not only has the reactor to be thoroughly reviewed, but also auxiliaries, stockpiles, transportation routes and a variety of unpredictable parameters, such as natural disasters, and terrorism. These assessments are slow, resource intensive and yet can mislead the conclusions [8]. Since MRT is a relatively new area of expertise, some of these hazards assessment methods use indexes that are not appropriately defined as a consequence of a lack of practical knowledge.

9.2 Inherently safer processes using microreaction technology

9.2.1 Advantages of microreaction technology to safety

Micro Reaction Technology (MRT) offers significant assets for designing inherently safer processes: efficient heat management and small internal volumes are among the most important [9]. The high surface-to-volume ratio ensures improved control over the operating conditions and reduces the risk of thermal runaway during the process itself or downstream processing steps. The ability to efficiently remove heat allows for exothermic reactions to be carried out at higher temperatures and concentrations, which has a beneficial impact on the global efficiency of the process (reaction kinetics and waste reduction). Unstable materials are immediately processed rather than stored or shipped. Small volumes significantly reduce the hazards associated with the handling of toxic/reactive materials or gases under high temperatures and pressures. In the following sections, we will be focusing on these aspects. Other parameters, such as efficient and fast mixing, have been treated in other sections.

9.2.1.1 Heat exchange

Arrhenius law: the temperature dependence of a reaction with a specific rate k is defined by Equation (9.1). A and E_A are the pre-exponential factor and the activation energy, respectively.

$$k(T) = A e^{(-\frac{E_A}{RT})}. \tag{9.1}$$

Heat release rate: for a zero order reaction, the heat release rate varies as a function of the temperature (Equation (9.2)). Q is the heat of reaction.

$$Q_r = k_0 e^{(-\frac{E_A}{RT})} Q. \tag{9.2}$$

Newtonian cooling: states that the rate of heat loss of an object is proportional to the difference between temperature of the object (T_1) and its surrounding environment (T_2) (Equation (9.3)). U and A are the heat transfer coefficient and the heat transfer area, respectively.

$$Q_{ex} = UA(T_2 - T_1). \tag{9.3}$$

Exothermic reactions are of particular concern upon the scaling up process. While lab-scale exothermic reactions are efficiently managed at low temperatures and dilution, larger batches are inevitably linked with a surface-to-volume dilemma [10]. The reaction heat released by a reaction is proportional to the amount of chemicals being processed, and this heat has to be removed from the system. Strong dilution has an impact on heat exchange – the solvent conveying the heat – but significantly reduces the industrial applicability with tremendous waste generation and low global efficiency. The removal of the reaction heat proceeds mainly through the surface of the reactor, that is, the heat exchanger, where thermofluids are usually circulated to keep the reactor isothermal. The ratio between the surface of the reactor and its volume decreases with an increase in reactor size, and strongly affects the efficiency of heat removal. Thus, in practice, smaller volumes and larger surfaces make it easier to manage heat removal.

Heat balance in a reactor is a complex process involving several parameters. In a simplified approach, the heat balance ΔQ (Equation (9.4)) is defined as a difference between the heat loss Q_{ex} (via the heat exchanger, Equation (9.3)) and the heat release

Q_r (from the reaction, Equation (9.2))

$$\Delta Q = Q_{ex} - Q_r. \tag{9.4}$$

These concepts are conveniently summarized in the Semenov diagram (Figure 9.1) that superimposes the plots of Q_{ex} and Q_r. This diagram illustrates the effect of a change in the heat transfer for a given reaction. Equilibrium for the heat balance is reached for $Q_{ex} = Q_r$, that is, at the intersection of both plots (Figure 9.1). The lower intersection (S) is a stable equilibrium point, while the upper intersection (R) is critical: a smaller deviation to a higher temperature will develop a runaway regimen. A runaway regimen is characterized by an accumulation of heat in the system, a subsequent acceleration of the reaction that releases more heat to the system; overpressure (mostly coming from the heated solvent) can lead to a disastrous rupture of the reaction vessel. Several operational factors affect the positioning of Q_{ex} vs Q_r (see Equation (9.3) and Figure 9.1), therefore impacting on their intersections: (a) a lower temperature of the thermofluid (T_2 in Equation (9.3)) will shift Q_{ex} (green dash-dot lines in Figure 9.1 (b)); (b) the slope of Q_{ex} will vary (blue dashed lines in Figure 9.1 (b)) if the heat transfer coefficient (U) or the heat transfer area (S) are affected (Equation (9.3)). A large heat transfer area increases the slope of Q_{ex} and therefore pushes away the runaway area.

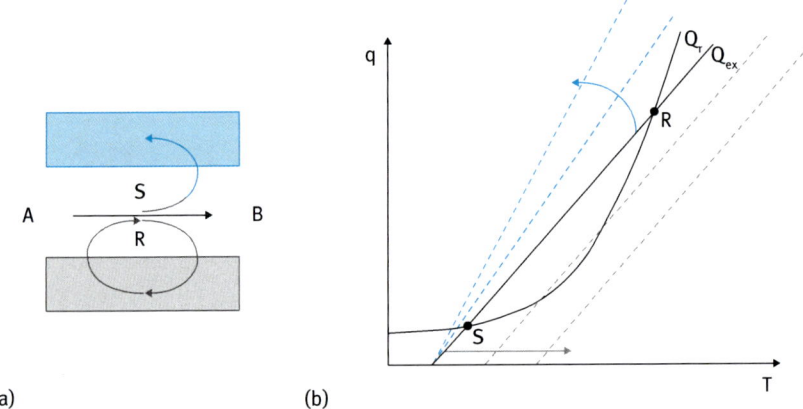

(a)   (b)

Fig. 9.1: Illustration of (a) an exothermic process carried out with an efficient (S) or a defective (R) heat exchanger and (b) the corresponding Semenov diagram. S: Safe reactor operation – the reaction heat escapes the reactor via the heat exchanger; R: Typical runaway conditions – the heat exchanger fails to remove reaction heat leading to an exponential increase in reaction rate.

A large heat transfer area is, for instance, one of the main features of MRT devices: specific surface areas of microreactors are traditionally very high (Table 9.1) while specific surface areas for batch reactors are much lower. Consequently, efficient and accurate heat control precludes the formation of hotspots in MRT devices, reducing undesirable side-reactions and keeping the reactor in a safe operating regimen.

Table 9.1: Surface-to-volume ratio for a MRT device (~ 100 µm channel size) and batch reactors.

Reactor Type	S/V (cm^{3}cm^{-3})
µ-reactor	200
batch – 100 mL	1
batch – 1 m^3	0.06

Thermal runaways in batch reactors have led to a variety of industrial disasters. In 2007, a thermal runaway during the production of a fuel additive (methylcyclopentadienyl manganese tricarbonyl) destroyed T2 Laboratories in Jacksonville, Florida, killing four people and seriously injuring 32 other [11]. A wide variety of strongly exothermic reactions has been safely and successfully developed using MRT devices, such as processing organolithium or organomagnesium derivatives, strong bases, or nitrations. Relevant examples from the literature are detailed in Section 9.2.2.

9.2.1.2 Small volumes, containment and multistep on-site capability

With small internal reactor volume, usually ranging from 1 µL to a few mL, MRT guarantees that only limited amounts of reactive material are being processed at a given time. Even in the case of reactor failure, the hazards remain low as material release and area exposure are much smaller – while keeping the same production capability. Pyrophoric materials are safely handled in MRT devices due to their sealed nature. The lack of reactor head-space provides an immediate advantage over batch conditions: much higher temperatures are possible, without increased risk of explosion or exposure. For instance, the continuous flow production of solketal tert-butyl ether (STBE) processing 0.01 mL s^{-1} of isobutene has the same daily throughput as a 1000 L batch reactor under pressure (17 bar), with significantly reduced operational hazard [12]. The combination of small reaction volumes and large surface area impact on the termination of radical chains by surface quenching: explosive mixtures can be handled in MRT devices since the nonexplosive region is significantly broader than in conventional batch reactors (see for example the oxidation of ethylene) [13].

MRT offers another safety asset: in-line quenching or processing of hazardous materials, that is, hazardous compounds, are immediately reacted and transformed into more advanced nonhazardous materials, therefore eliminating the need for stockpiling and transportation (Figure 9.2) [8]. Various examples of continuous multistep chemical transformations have illustrated the benefits of in-line processing (extraction/separation) of hazardous and potentially explosive chemicals [14] [15]. Storage of reactive material has led to various industrial disasters: Bhopal (India, 1984), AZF (France, 2001), Bayer (USA, 2008) [16] and West Fertilizer (USA, 2013) are amongst the most notorious. A selection of recent literature examples is provided in Section 9.2.2.

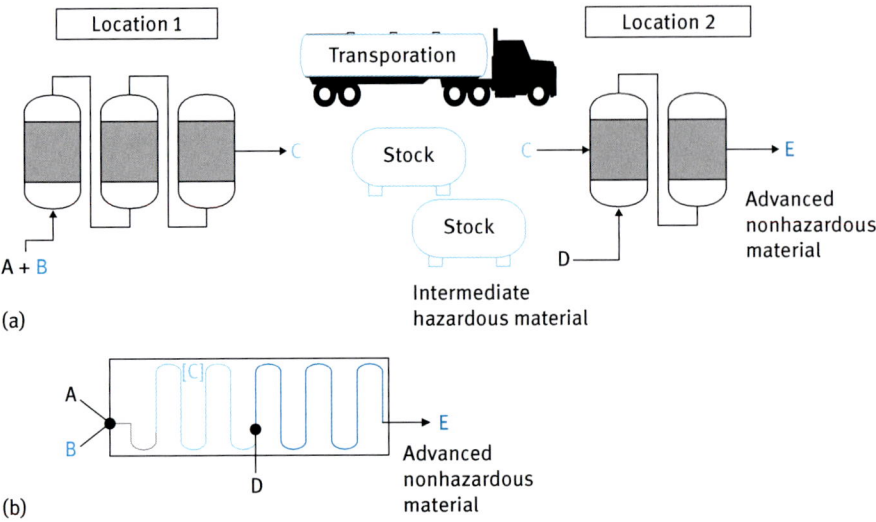

Fig. 9.2: Illustration of a hypothetical process using a hazardous material **C**. Transportation and/or stockpiling issues of hazardous material in the batch process (a) are eliminated in the MRT process (b) with the *in situ* transformation of **C**.

9.2.1.3 Embedded controls, automation and expansion of reaction space

MRT devices usually come with a variety of embedded controls allowing a close monitoring of process parameters (temperature, pressure, flow rates), and a variety of analytical tools are now available on the market for real-time in-line quantitative and qualitative reaction monitoring. MRT devices with embedded feedback control have been reported [17]. Although these safety features are also widely available on batch reactors, their presence on MRT devices adds an ultimate layer of safety by process control and management.

The intrinsic safety and transport properties of MRT devices have enabled their use to explore unconventional and/or harsh reaction conditions (so-called "expansion of reaction space") including the use of high temperatures, high pressures, high concentration, handling of transient, explosive or hazardous species. A variety of recent examples is presented in Section 9.2.3.

9.2.2 Recent examples of processes involving dangerous reagents/reactions under MRT conditions

Hazards associated with a specific process can be associated with the reaction's intrinsic kinetics (very fast exothermic reaction), with the nature of any intermediate (explosive, toxic) or the reaction conditions themselves (high pressure or temperature). In the next subsections, recent examples of MRT processes enhancing process safety are presented. The following classification is used: (a) exothermic reactions and highly reactive materials (Section 9.2.2.1); (b) toxic reagents and intermediates (Section 9.2.2.2) and (c) unusual or harsh reaction conditions (Section 9.2.3).

9.2.2.1 MRT processes involving exothermic reactions and highly reactive, potentially explosive materials

Organometallic transformations

Organometallic transformations are usually carried out under cryoscopic conditions in batch to avoid undesirable side-reactions such as isomerization, premature quenching and decomposition. The scaling up of these transformations is usually fairly difficult to foresee. Besides, many organometallic compounds are pyrophoric and need to be handled under highly controlled conditions. For many of these fast and exothermic transformations, high mixing efficiency and heat transfer are essential to maintain the exotherm below the decomposition area. These reactions are usually performed at higher temperature in MRT devices than in batch, therefore lowering the energy demand.

Reactions under continuous-flow conditions including metal-halogen exchange, addition, addition-elimination, conjugate addition and partial reduction have been reported with impressive multigram productivities. A variety of dangerous organometallics such as Grignard reagents [18–22], alkyllithiums [18] [23–25], and DIBAL-H [26, 27] have been safely used in MRT devices. For example, the scalable preparation of ketones through the nucleophilic addition of Grignard reagents to aromatic and/or aliphatic nitriles and subsequent hydrolysis of the intermediate imine was reported by Mateo in plug flow reactors [19]. The strongly exothermic hydrolysis of the intermediate imines could be performed continuously and the entire process was safely scaled up to 144 g day^{-1}. Carbonyl compounds were reacted with Grignard reagents at room temperature under flow conditions for the preparation of a small collection of secondary and tertiary alcohols. Excellent yields and general applicability were reported by Rencurosi and coworkers [21]. The procedure was also applied for the preparation of Tramadol, an analgesic drug belonging to the opioid group.

Recently, researchers at the Fudan University studied Li/Br exchanges for a library of substituted heteroaromatics. The lithium halogen exchange was performed at higher temperatures (up to 20 °C) than usual (−78 °C in batch) in a MRT device with high mixing, yet producing multigram quantities of pharmaceutical intermediates [23]. A telescoped route was designed for the continuous production of (E/Z)-Tamoxifen, an important breast cancer drug by Ley and coworkers [18]. The sequence involves a lithium-halogen exchange, the addition of an aryllithium intermediate on a ketone and an elimination step. Only 80 minutes of reactor operation produced sufficient material (12.43 g) for one patient's treatment for over 900 days, equivalent to one daily dose every 5 seconds (Figure 9.3)!

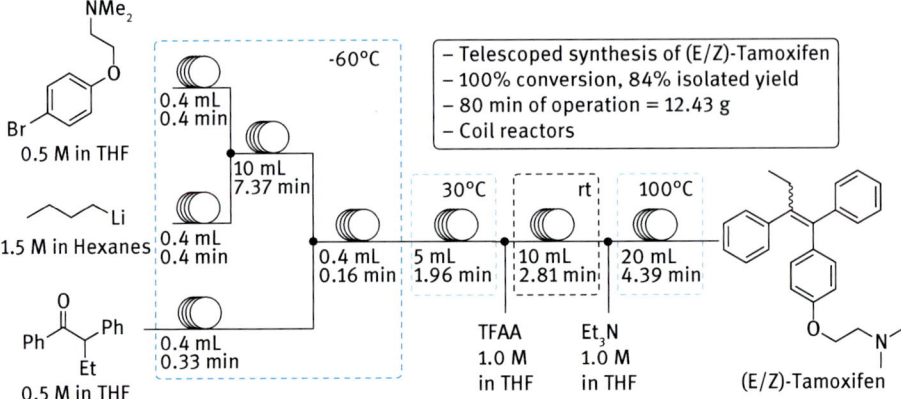

Fig. 9.3: Telescoped route towards the continuous production of (E/Z)-Tamoxifen (adapted from [18]).

Nitrations, related reactions and sulfonations

Nitration reactions belong to basic reactions in chemical synthesis, used for the synthesis of pharmaceuticals, agricultural chemicals, pigments, precursors for polymers or explosives. However, these reactions are considered highly hazardous due to their highly exothermic nature, the corrosive nature of the nitration reagent and the autocatalytic nature of the reaction [28]. Another problem is the reaction selectivity and product degradation: high temperatures can easily lead to decomposition and formation of other side-products. Diverse MRT devices have been reported for the safe and selective nitration of a variety of organic molecules, hence allowing for the development of industrial processes at rather large scales.

Various examples of continuous flow nitrations illustrated the nitration of a variety of aromatics [29–32]. Scientists at AstraZeneca reported the nitration of 3-alkylpyrazoles on a 100-gram scale with a productivity of $0.82\,\mathrm{g\,h^{-1}}$ [33]. The Novartis Preparation Laboratories have demonstrated the scale-up of three nitration reactions using a commercial coil MRT device at high production rates. Hundred grams of advanced nitro intermediates were obtained by homogeneous nitration using an acetic acid or acetic anhydride with fuming nitric acid mixture. This continuous procedure affords high comparable yields to that of batch, but significantly improves the purity profile and the safety of the process [34]. More recently, the herbicide pendimethalin was prepared according to a continuous and highly selective dinitration in a 0.2 mL stainless steel MRT device within short residence times (> 10 s). A productivity of up to $4.32\,\mathrm{t\,year^{-1}}$ could be achieved [35].

An impressive example for a nitration at production scale was reported by scientists at the Corning European Technology Center and DSM [36]. A key intermediate for the synthesis of naproxcinod was scaled to 25 tons by DSM and subsequently performed under cGMP using a millistructured glass MRT device. The hazardous step in this reaction was not the nitration itself, but the subsequent highly exothermic quenching. The nitration utilized 65% nitric acid in excess and involved subsequent quenching by a stepwise addition of water and sodium hydroxide streams (Figure 9.4). Similarly, alkyl nitrites were obtained at an impressive 10 t y^{-1} scale by mixing acidic streams of various alcohols and aqueous sodium nitrite [37].

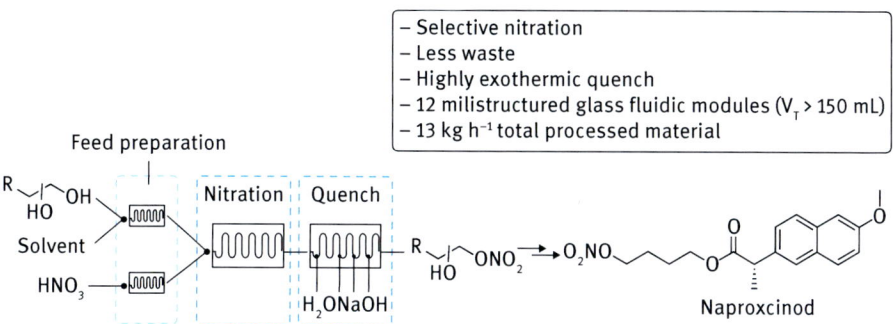

Fig. 9.4: Simplified flow chart for the selective nitration of an intermediate in the synthesis of naproxcinod (adapted from [36]).

Sulfonations are quite similar to nitrations: the reaction conditions require aggressive starting materials and are usually exothermic. Again, the low heat and mass transfer capabilities of conventional reactors may give rise to the insufficient mixing and local overheating, and thus lead to the formation of many impurities or even lead to runaway conditions. Examples of sulfonation exploiting the benefits or MRT devices appeared recently in the literature [38, 39].

Direct halogenations

Among direct halogenations [40], direct fluorination is the most challenging to implement conventionally on a macroscopic scale due to the high exothermicity of the reaction, the acute toxicity of F_2 and the corrosion issues associated with the use of fluorine gas. Low selectivity is often reported in batch and runaway conditions can develop. Feasibility studies for the direct fluorination of various organic substrates using the assets of MRT devices were demonstrated in the groups of Jähnisch (falling film and stainless steel microbubble reactor) [41], Chambers and Standford (micromachined nickel reactor) [42], and Jensen (nickel-coated silicon wafer microreactor) [43] in the early 2000s. These studies have shown the feasibility of running direct fluorination reactions in a microreactor and have demonstrated that the size reduction of the reactor significantly improved reaction control and selectivity. Mainly substituted aromatic compounds [41–45] or molecules containing carbonyl groups were used [46–48]. For instance, Lang and coworkers recently reported a Ni-coated chemical minireactor (channel size: 1 mm × 1 mm) for the direct fluorination of ethylene carbonate. [47] The reactor allowed for gas-liquid flow in a slug-flow regime. Continuous and direct fluorination of ethylene carbonate was performed at 22 °C with up to 88% fluorine concentration. The chemically resistant nickel coating showed no corrosion after 100 h of operation.

Hydrogen and organic peroxides

Peroxides have been intensively used in the chemical industry, especially as initiators in polymer production, as bleaching agents or as cheap and clean oxidants for a variety of organic transformations. Despite their intensive industrial use, their instability is far from negligible: organic peroxides may decompose violently or explode under heat stress or in the presence of infinitesimal amounts of metallic impurities.

The challenging direct synthesis of hydrogen peroxide from hydrogen and oxygen has been explored for decades. The reaction is linked to an inherent risk associated with the handling of highly explosive gas mixtures. A variety of dedicated MRT devices were explored to render the process safe. Inoue and coworkers managed to obtain H_2O_2 in up to 10 wt.-% concentrations, by the reaction of hydrogen and oxygen over a Pd/Al_2O_3 catalyst packed in a glass-fabricated microreactor [49]. The achieved productivity is of industrial interest and represented a significant breakthrough, since usual productivities range are below 5 wt.-%. Other organic peroxides such as *tert*-butyl peroxypivalate [50] and peracetic acid [8] were also safely produced in MRT devices.

Recent examples illustrated the safe use of hydrogen peroxide and peracetic acid as cheap and clean oxidants in MRT devices. Park and coworkers reported a catalytic diacetoxylation using hydrogen peroxide and peracetic acid using MRT devices made of polytetrafluoroethylene (PTFE) coils [51]. Excellent yield and selectivity were achieved in significantly shortened reaction times without the decomposition of explosive oxidants and further transformation of unstable products, offering a safe and efficient alternative to traditional methods for alkene diacetoxylation. Kappe [52] and Hessel [53] independently studied the continuous production of adipic acid from cyclohexene and hydrogen peroxide under high temperature and pressure explosive regimes. Tungstic acid was used as homogeneous catalyst [52]. Due to the exceptional mass and transport capability of the small diameter channels, no phase transfer catalyst was required. The unusual conditions allowed by the specificity of the device bring significant benefits in terms of process efficiency and safety (Figure 9.5). Kappe and his coworkers also studied the use of organic peroxides for copper-catalyzed C-H activations in flow [54].

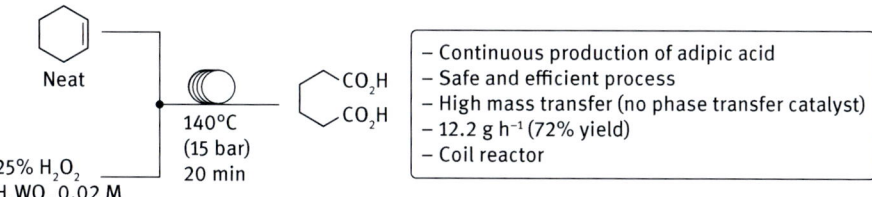

Fig. 9.5: Continuous-flow production of adipic acid using a MRT device (adapted from [52]).

Synthesizing and handling azides

Inorganic azides and hydrazoic acid are acutely toxic. The main safety concerns regarding the generation and use of azide promoters is, however, usually associated with their high explosive potential. Organic azides raise similar safety issues, in particular for compounds with a C/N ratio < 3. Detonation can occur under shock or heat stress or in the presence of metal impurities.

In a series of publications, researchers at the University of Graz and at Lonza AG studied hydrazoic acid in a commercial coil reactor [55, 56]. Hydrazoic acid was preformed *in situ* by mixing of independent flows of NaN_3/Brownsted acid or $TMSN_3$/MeOH. Hydrazoic acid was then reacted with organic nitriles or 2-oxazolines at elevated temperature and pressure regimes. The scalability of both protocols was demonstrated for selected examples. The excess of hydrazoic acid was quenched by reaction with aqueous sodium nitrite or ceric ammonium nitrate. Jamison avoided the direct generation of hydrazoic acid by reacting organic nitriles and inorganic azides in protic solvents (H_2O/NMP or IPA/NMP) under rather extreme conditions (high temperature, see also Section 9.2.3) [57]. The process was studied at the multigram scale and safely produced up to $116\,g\,day^{-1}$ using a commercially available MRT (Figure 9.6).

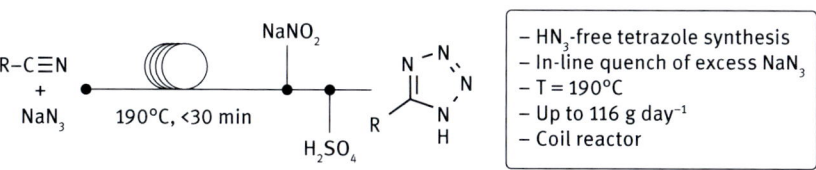

Fig. 9.6: HN_3-free continuous production of tetrazoles [57].

Organic azides have found a wide variety of applications in organic chemistry. Rutjes and coworkers reported their synthesis from amines and the explosive diazo-transfer reagent imidazole-1-sulfonyl azide hydrochloride under MRT conditions [58]. Among transformations processing organic azides in MRT devices, the Curtius rearrangement is widely represented in recent reports. The usual concerns linked to the generation and use of potentially explosive and toxic intermediates are alleviated by the use of a variety of MRT devices (coil reactors, [59] monolithic reactors [60], or silicon-based chips [15]). High yields are usually reported within short reaction times at high temperature, and additional layers of safety are enabled by a variety of embedded cartridges with scavengers and/or liquid-liquid separators [15] [59]. Recently, Hayashi and coworkers used a coil MRT device to perform a Curtius rearrangement to a key intermediate (10 g scale) for the production of (−)-oseltamivir (Figure 9.7) [61]. The reaction was safely and efficiently scaled up using parallel experiments.

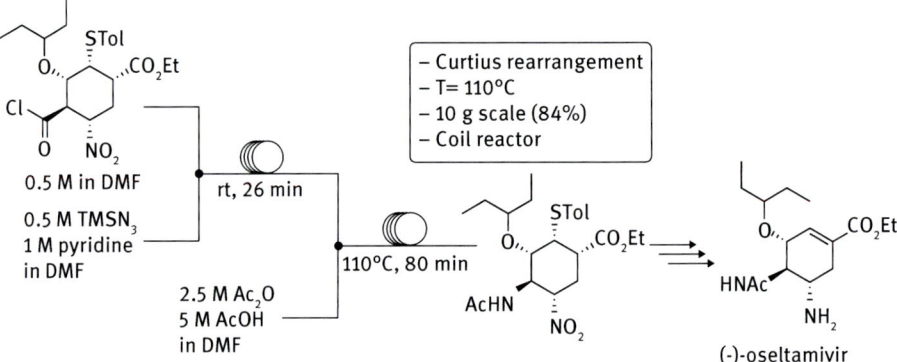

Fig. 9.7: MRT-mediated synthesis of a key intermediate for the total synthesis of (−)-Oseltamivir (adapted from [61]).

Another transformation processing organic azides that benefits from MRT device is the dipolar addition of acetylenes with azides to give substituted 1,2,3-triazoles. Homogeneous [62] and heterogeneous [63] catalysts were successfully reported for flow applications. High yields were reported within short reaction times and high temperatures [62]. Safety was improved by scavenging the excess of organic azides with a phosphine resin located at the outlet of the reactor [63].

Preparation of diazomethane and other diazonium derivatives

Diazonium derivatives, and in particular diazomethane, are considered quite problematic owing to the safety concerns associated with their preparation and use on laboratory or pilot plant-scale.

Different research groups have considered MRT devices to produce diazomethane. Stark and coworkers employed a MRT device to produce diazomethane from N-methyl-N-nitroso-p-toluenesulfonamide (Diazald, a commercially available precursor). Diazomethane was then reacted *in situ* with benzoic acid. The Stark research group went a step further and employed the technology to continuously produce

Diazald itself from p-toluenesulfonyl chloride (> 90% and up to 9 kg L^{-1} h^{-1}) [64, 65]. Similarly, researchers at the Corning European Technology Center and the University of Padova have developed and scaled up a MRT process for the continuous production of diazomethane from N-methyl-N-nitrosourea [66]. The reaction was first optimized in a lab scale MRT mesostructured glass MRT device. Diazomethane was quenched *in situ* with benzoic acid. A ten-fold scale production increase was subsequently achieved by translating the conditions in a production-scale mesostructured glass MRT device. Process scale-up was efficiently and safely achieved allowing the production and use of diazomethane up to 19 mol day^{-1} at a total flow rate of 53 mL min^{-1}, while keeping low *in situ* levels of reacting diazomethane. More recently, Kim and coworkers developed a MRT device featuring two microchannels separated by a diazomethane-porous polydimethylsiloxane (PDMS) membrane. The PDMS membrane allowed diazomethane, generated *in situ* from the reaction of Diazald and potassium hydroxide, to diffuse to the second microchannel where it reacted with various substrates. Inorganic salts and side material remained in the primary channel and were conveyed to a waste tank (Figure 9.8) [67].

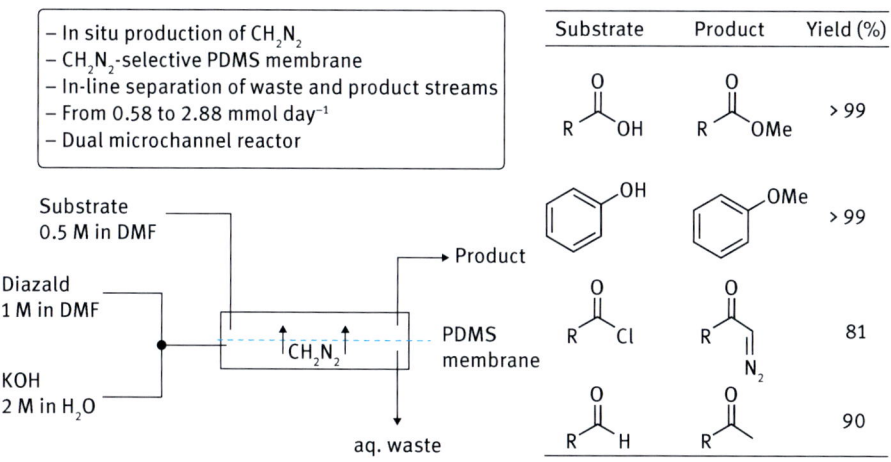

Fig. 9.8: *In situ* production and reaction of diazomethane (adapted from [67]).

Ley and Martin studied the diazotation of a variety of organic substrates. The diazotation of a variety of aminoacids in flow allowed the multigram production of chiral hydroxyacids. The use of a flow apparatus allowed to safely conduct the diazotization reactions at 60 °C, allowing the reactions to complete between 10 and 60 min in the reactor coil before in-line multistage extraction [68]. A method for the continuous synthesis of aliphatic and aromatic diazoketones from acyl chloride precursors and trimethylsilyldiazomethane using a tubular commercial reactor was developed and used to prepare quinoxalines in a multistep sequence without isolation of the potentially explosive diazoketone. The reaction of the acyl chloride precursors with trimethylsilyldiazomethane was completed at room temperature within 2 h (up to 90 h at 0 °C in batch) using a polystyrene (PS)-tetraalkylammonium fluoride resin. The protocol showcased an efficient in-line purification using supported scavengers with time-saving and safety benefits [69]. Aryldiazonium intermediates were used in the continuous synthesis of various aromatic compounds [70–72].

Ozonolysis

Ozone is a common clean, convenient, efficient and low cost oxidant that introduces oxygen-containing groups (alcohols, aldehydes, ketones, or acids) by oxidative cleavage of either double or triple carbon-carbon bonds. The high exotherms associated with the initial reaction of substrates with ozone, and the inherent instability of intermediate ozonides raise significant hazards at larger (production) scales. Several research groups have demonstrated that the benefits of continuous flow (efficient heat exchange and in-line quenching of explosive ozonides) could alleviate these inherent hazards (see Figure 9.9 for a general flow ozonolysis setup).

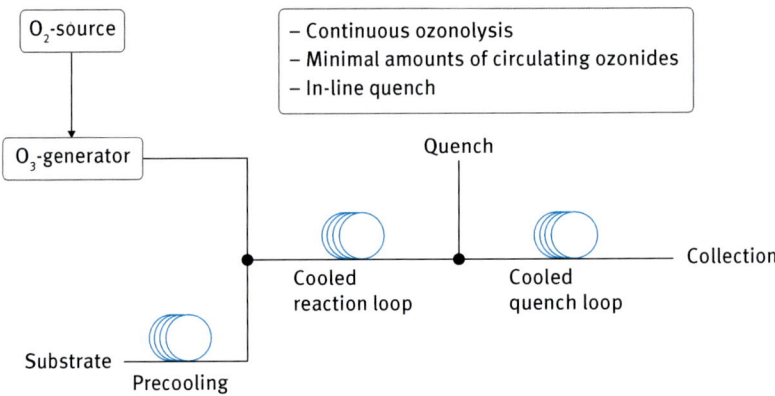

Fig. 9.9: General flow ozonolysis setup.

Jensen and coworkers at Massachusetts Institute of Technology pioneered ozonolysis reactions in MRT devices [73]. Various organic substrates were oxidized in multichannel microreactors (silicon and Pyrex wafers). Contact times as low as 1 s and conversions and selectivities up to 97% were reported. Other types of reactors have been reported, such as falling film microreactors [74, 75] and glass reactors with high mixing efficiency [76]. Commercial MRT devices fully dedicated to ozonolysis are available [77].

The Ley group has used a semipermeable membrane capillary (Teflon AF-2400) to effect ozonolysis [78]. The semipermeable tube allows gas but not liquid to cross the membrane. Full conversion of various alkenes was achieved with residence times within 1 h. Researchers at University College London used a commercial coil reactor to oxidize n-decene with ozone [79]. Intermediate ozonides were quenched in-line with triphenylphosphine at −10 °C. The configuration of the system allowed the production of chemically significant amounts of nonanal (1.8 g h^{-1} at 1.3 ozone equivalents), with minimal amounts of ozonides present at any time, significantly reducing the explosion hazards. Besides, the reaction was carried out at −10 °C, instead of −78 °C in batch, which represents considerable energy savings. Ozonolysis reactions at much larger scales were also reported: scientists at Lonza [80] developed a large scale ozonolysis for the production of an insecticide key intermediate. A 450 L loop reactor was used for the conversion of chrysanthemic acid with ozone. The process was optimized to produce 0.5 t day^{-1} in continuous mode. Earlier, scientists at Bayer Schering Pharma studied the feasibility of the continuous synthesis of intermediates of vitamin D analogues (0.76 kg day^{-1} isolated product) [76].

Hydrogenations

Macroscale hydrogenation in MRT devices offers obvious safety advantages since hydrogen is an extremely flammable gas and hydrogenations are exothermic in nature. Besides, heterogeneous catalytic hydrogenation reactions benefit from the additional advantages of MRT resulting from the high specific interfacial area. A variety of reactors were developed to effect hydrogenation reactions at various scales, ranging from classical chip-, tubular or plate microfluidic reactors to coil-, tubular and plate mesoreactors. Commercial dedicated reactors are available. These reactors and their uses have been extensively reviewed by Kappe and Hessel [81, 82].

Heterogeneous or homogeneous catalysts have been utilized in MRT devices. Heterogeneous catalysts have usually been incorporated in the reactor wall or structure (e.g., monolithic reactors [83]). As an example, Galarneau and coworkers developed a silica monolithic reactor with embedded Pd nanoparticles for hydrogenations. This reactor was operated under smooth conditions (25 °C, 2–3 bars H_2) under continuous flow and offered excellent conversions and selectivities over a test period of up to 70 h on selected model substrates [83].

There are several approaches to induce mass transfer between the gas phase and the substrate in the liquid phase. Among these, mechanical mixing of the two phases (including segmented flow regimes) is the most represented, while some reported the use of semi-permeable membranes to introduce gas into the reagent stream [84, 85]. For instance, a research group at the University of Connecticut reported a modified "tube-in-tube" MRT device where the inner gas-porous tube (Teflon AF-2400) contains the gaseous reagent, while the liquid reaction flows around it contained in an outer stainless steel reactor coil. The device was used for studying the continuous-flow ho-

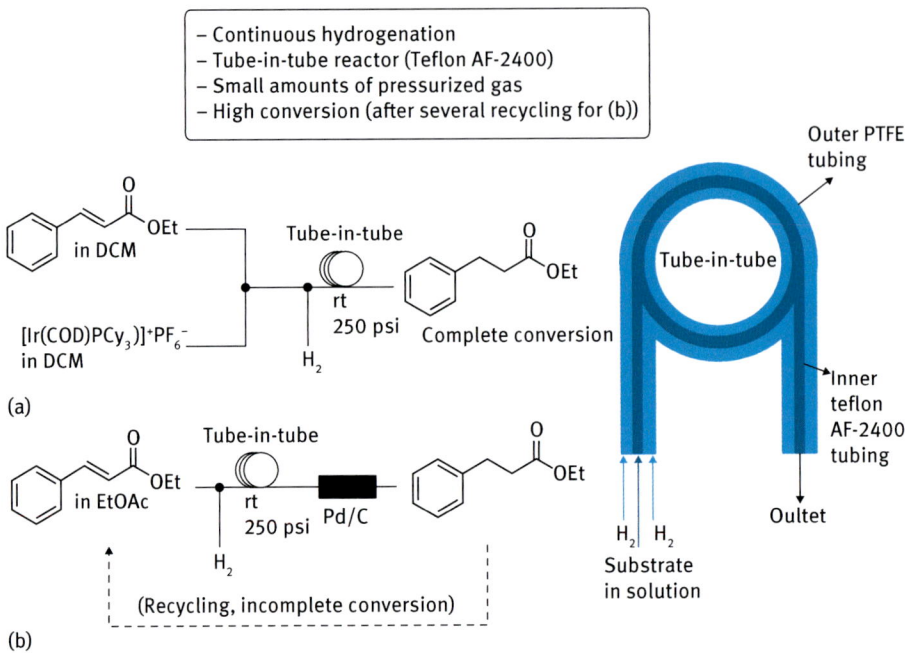

Fig. 9.10: Simplified view of a tube-in-tube prototype for the continuous (a) homogeneous or (b) heterogeneous hydrogenation of organic substrates (adapted from [85]).

mogeneous hydrogenation of a variety of alkenes utilizing Wilkinson's catalyst. By contrast to Ley's "tube-in-tube" prototype, [85] the outer stainless steel tube allows for heat transfer and temperature control. The hydrogenation of various substrates has been performed successfully and safely at 125 °C on small scale and could be substantially scaled up (up to 45 mmol h^{-1}, 99% yield), while keeping only a minimal gas volume in the system and therefore improving the safety profile of the system. In Ley's report [85], the setup was compatible with both heterogeneous and homogeneous hydrogenation reactions under pressure (250 psi), but could only be used at room temperature (Figure 9.10).

9.2.2.2 MRT processes involving toxic materials

Hydrogen cyanide and other cyanides

Stevens and collaborators at Ghent University studied MRT devices to perform reactions involving the use of highly toxic HCN [86]. Commercial stainless steel MRT devices were used to generate small amounts of HCN *in situ* in a safe and controlled manner by mixing an inorganic source of cyanide and an organic acid, or acetone cyanohydrin and an organic base. The applicability and safety of the technique was

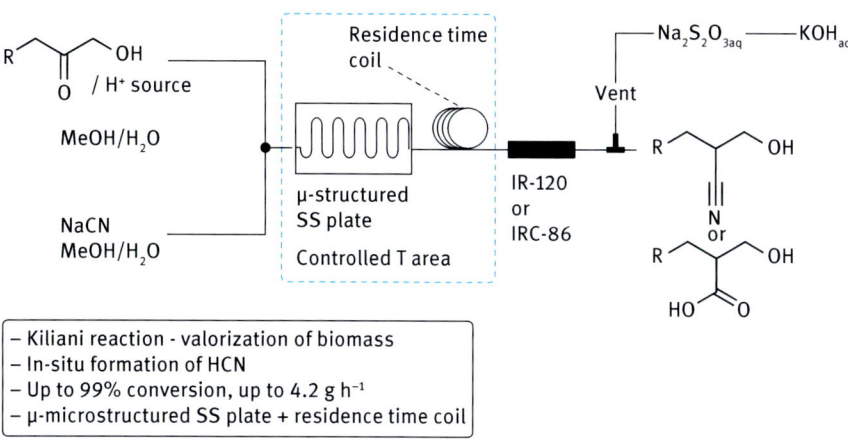

Fig. 9.11: *In situ* generation of HCN and valorization of biomass (adapted from [88]).

demonstrated for the flow synthesis of isochromen-1-ones (Strecker reaction) [87] and for the flow synthesis of α-hydroxycarboxylic acids (Kiliani reaction) (Figure 9.11) [88]. An integrated borosilicate glass microreactor setup with an embedded catalyst was developed by Watts to perform multicomponent Strecker reaction using TMSCN towards a variety of α-aminonitriles [89].

Carbonylations with carbon monoxide

Carbon monoxide is colorless, odorless and tasteless, but highly toxic. Large volumes are used in industry for the production of phosgene (see below) and various carbonylation reactions employing carbon monoxide have been studied.

Ley and coworkers used their tube-in-tube prototype MRT device (see Section 9.2.2.1 and Figure 9.10) to effect palladium-catalyzed methoxycarbonylation of aryl, heteroaromatic and vinyl halides [90]. A Teflon AF-2400-based tube-in-tube reactor was used to mediate the selective permeation of carbon monoxide into solution at elevated pressures. The low volume of pressurized gas within the reactor enhances the safety profile compared to batch processes. High conversion and yield was usually observed. More recently, the Yoshida group developed a photochemical MRT device to study a Pauson–Khand reaction. The reaction was carried out at room temperature and afforded much higher yield than the batch method. Collaboration between researchers at the Aarhus University and at the Osaka Prefecture University led to the development of a promising carbonylation method using a dual chamber microstructured MRT device [91]. Carbon monoxide was produced *in situ* by dehydration of formic acid by sulfuric acid in a Teflon AF-2400 inner tube that allows coaxial permeation of CO (Figure 9.12). Various carbonylation reactions, such as Heck, Sonogashira, and radical carbonylations, were successfully carried out.

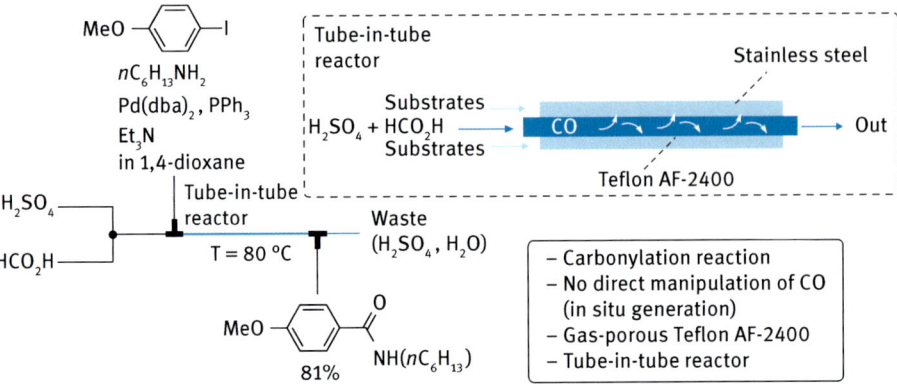

Fig. 9.12: Safe carbonylation in a tube-in-tube reactor with *in situ* generation of carbon monoxide (adapted from [91]).

Ammonia

Anhydrous ammonia is a dangerous substance: the gas is flammable and can form explosive mixtures with air at low concentration. Its toxicity and danger for the environment is often cited as a hazard for large scale productions. The Ley group at the University of Cambridge studied the applicability of their tube-in-tube prototype reactor based on an inner Teflon AF-2400 tube for gas diffusion. They reported an efficient MRT device for the continuous synthesis of a series of pyrroles through the Paal–Knorr reaction of 1,4-diketones with gaseous ammonia [92]. The latter technique was further used in the multigram scale production (~ 240 g day^{-1}) of the anti-inflammatory fanetizole (Figure 9.13) [93].

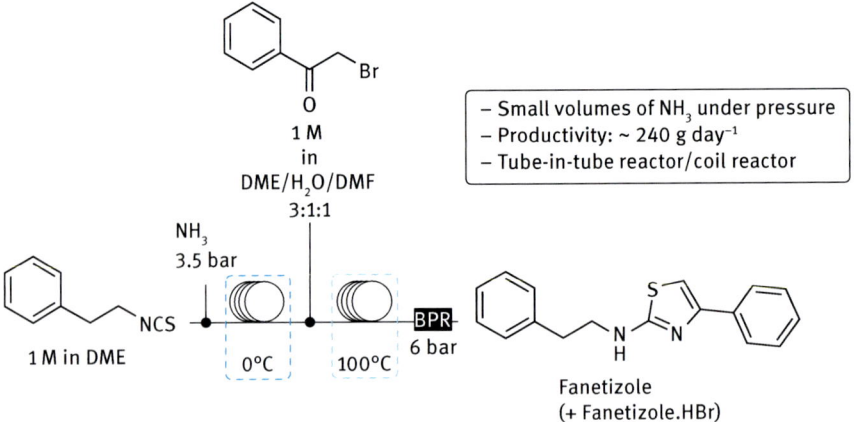

Fig. 9.13: Continuous production of fanetizole (adapted from [93]).

Osmium tetroxide

Osmium tetroxide is an expensive volatile and toxic compound, yet very useful as a catalyst for the dihydroxylation and oxidative cleavage of olefins. The usual strategies to overcome these issues include its immobilization or encapsulation on polymers and inorganic substrates. Kim and coworkers at the Pohang University of Sciences and Technology developed a microreactor featuring polysilane-based (PVSZ/PDMS) microchannels [94]. The wall surface of the microreactor is covered with a nanobrush of polymer coating (poly(4-vinylpyridine) – P4VP) that immobilizes OsO_4 (Figure 9.14). N-methylmorpholine-N-oxide (NMO) was used as a co-oxidant for dihydroxylation reactions, while sodium periodate was used as a co-oxidant for oxidative cleavage. High conversion and yields were reported for a variety of alkenes with this safe, cost-effective and scalable device. The results also revealed the durability and reusability of the microreactor without catalyst leaching.

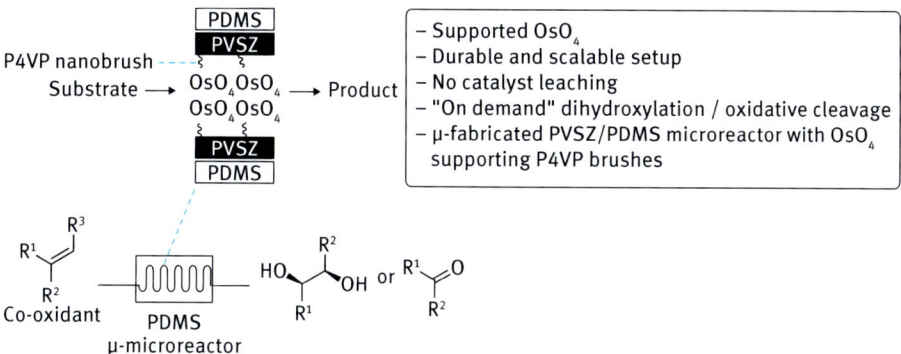

Fig. 9.14: Continuous dihydroxidation or oxidative cleavage using a polysilane-based (PVSZ/PDMS) microreactor (adapted from [94]).

Preparation and use of phosgene

Phosgene is an important chemical intermediate for the production of polyurethane foams, the synthesis of pharmaceuticals and pesticides. Its insidious toxicity raises significant safety hazards for macroscale utilization: most industrial phosgene is consumed at the point of production to avoid stockpiling and/or transportation. The Jensen group at Massachusetts Institute of Technology pioneered the continuous synthesis of phosgene from carbon monoxide and chlorine with a silicon-based micropacked-bed reactor [95]. This example illustrated the potential of MRT devices for safe on-site/on-demand production of a highly hazardous, yet widely used compound.

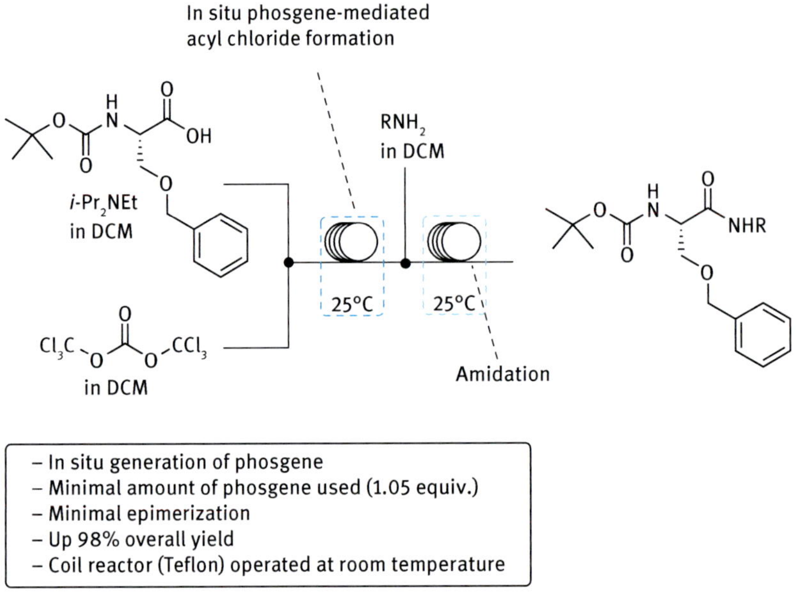

Fig. 9.15: Safe and scalable use of phosgene in a tubular MRT device for amidation reactions (adapted from [96]).

More recently, scientists at Tokyo Institute of Technology designed and studied a MRT device for the continuous *in situ* generation of phosgene and its use for amidation reactions (Figure 9.15) [96].

Phosgene was generated *in situ* from triphosgene/ and N,N–diisopropylethylamine (i–Pr$_2$NEt) and used to convert amino acid derivatives into their corresponding acyl chlorides that were subsequently coupled with amines. The entire sequence was telescoped in a coil reactor (perfluoro polymer) at 20 °C. The outstanding control of reaction conditions (residence time, local stoichiometry) significantly reduced substrate epimerization. Amides were obtained within short residence time (< 10 s) in high yields (up to 98% overall).

Radiolabeling using MRT Devices

The field of positron emission tomography (PET) radiochemistry has also benefited from the distinctive assets of MRT. Radiochemistry handles highly hazardous short-lived species such as ^{11}C or ^{18}F ($t_{1/2}$ = 20.4 and 109.7 min for ^{11}C and ^{18}F, respectively) for the preparation of radiolabeled molecules using automated batch sequences at the nano- or microgram scale. The inherent ability of MRT devices to manipulate and process accurately nano- or micro-volumes of highly hazardous short-lived species under a wide variety of controlled conditions within defined environments makes them an obvious tool for the radiochemist. Microchips can easily be shielded and since higher and faster conversions can be achieved within MRT devices, less [11C]- or [18F]- precursors are consumed. Radiolabeling of various organic substrates has been convincingly demonstrated in MRT devices [97, 98].

A variety of materials have been used to construct MRT devices for radiochemistry [99] and dedicated commercial MRT devices exist [98] [100, 101]. For instance, the [18F]-radiopharmaceuticals fluorodeoxyglucose (FDG) and Annexin-V were synthesized in a glass-plates microreactor using ^{18}F and ^{124}I, respectively [102]. Polymer reactors have also been utilized: PDMS [103] and polyether ether ketone (PEEK) [104].

As an example for [18F]-labeling, Pike and coworkers at the Molecular Imaging Branch of the National Institute of Mental Health used a commercial MRT device for the micro and macroscale production of PET imaging probes (Figure 9.16) [98]. Optimal radiochemical yields were achieved for the microflow production of radiofluo *meta*-substituted [18F]-fluoroarenes from diaryliodonium derivatives. The same group reported other recent MRT applications for the synthesis of [18F]-fallypride, [18F]-BPR and [18F]-SL702 [99]. These molecules were obtained consistently, rapidly and with high specific radioactivities.

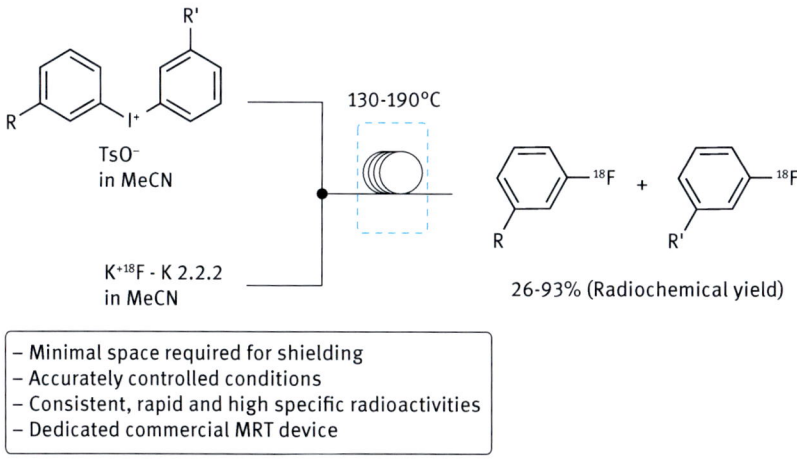

Fig. 9.16: [18F]-labeling of diaryliodonium derivatives (adapted from [98]).

A MRT device was utilized by Long and coworkers to effect a rapid multiphase carbonylation with [11C]-carbon monoxide exploiting the distinct benefits of MRT devices in terms of high mass transfer, very large interfacial areas and inherent safety (see also Section 9.2.2.2). A tubular microreactor was packed with a silica-supported Pd catalyst and provided an effective MRT device for the continuous flow [11C]-carbonylation of a variety of aryl halides. The microtubular device was reusable for a number of runs, and much higher yields were achieved than in batch [105].

9.2.3 MRT processes involving harsh conditions (elevated temperatures and pressures)

MRT devices allow for an easier, safer and faster access to unconventional process conditions, that is, high temperatures and pressures [106]. Organic transformations normally performed in a high-boiling solvent at reflux temperature or under sealed-vessel microwave conditions can be readily converted to a flow regime by using lower boiling solvents in or near their supercritical state, thus offering a safe and scalable alternative to hazardous batch processes [107]. Working under supercritical conditions at microscale (supercritical microfluidics) combines the advantages of size reduction provided by microsystems to the unique properties of supercritical fluids (SCFs), that is, extremely fast coaxial diffusion and tunable flow regimes [108]. The main challenge of implementing such extreme temperature and pressure conditions in microfluidic systems is to design microreactors capable of sustaining high heat and mechanic stress in addition to a high chemical compatibility. Stainless steel capillary reactors have been used up to 450 °C and 950 bar [109].

Researchers under the guidance of Kappe at the Karl-Franzens-University Graz have studied the applicability of high temperatures and pressures under flow conditions to a variety of important pharmaceutical and bulk chemical industry intermediates [110–113]. For instance, the industrially important building block 7-ethyltryptophol was obtained by direct reaction between ethylphenylhydrazine hydrochloride, dihydrofuran and a basic catalyst under temperatures of 150–170 °C (under 30–50 bars) in a 9.8 mL Hastelloy coil reactor [110]. Reactions were complete within very short residence times, allowing for high throughput even with rather compact reactors. The same group developed a catalyst- and solvent-free protocol for the synthesis of n-alkyl chlorides [111]. Direct chlorodehydroxylation of the corresponding n-alcohols was achieved using high-temperature/high-pressure regime with 30% aqueous hydrochloric acid. Optimum conditions for the preparation of n-butyl and n-hexyl chloride involve the use of a glass microreactor chip, a reaction temperature of 160–180 °C (20 bar backpressure) and a residence time of 15 min.

Other groups have studied organic transformations under flow and extreme temperature/pressure conditions. Among many others: Dormán and coworkers performed thermal cyclization at temperatures in the range 300–360 °C at 100–150 bar [114]; Larhed developed a high throughput borosilicate tubular reactor able to sustain up to 270 °C for specific organic transformations [115]; Sanderson used superheated conditions for synthesizing the key aminoalcohol in the synthesis of fluoxetine (Prozac) [116] (Figure 9.17) and Seeberger reported the thermolysis of azidoacrylates for the synthesis of indoles under high temperature flow conditions [117].

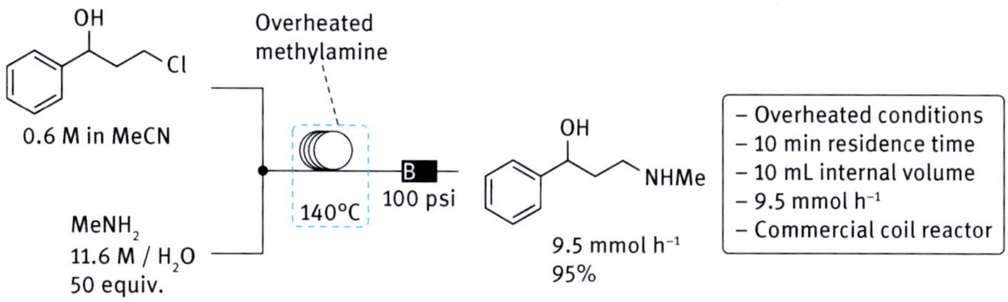

Fig. 9.17: Continuous flow process under overheated conditions for a key intermediate in the synthesis of fluoxetine (adapted from [116]).

Supercritical microfluidics [108] are a blooming research area. For instance, Pucheault, Aymonier and coworkers at the University of Bordeaux designed a microfluidic process to access a library of tunable functionalized palladium-based nanocrystals stabilized by various ligands. Stabilized Pd-nanocrystals were synthesized at 100 °C and 25 MPa in a co-flow capillary microsystem using supercritical CO_2 [118].

9.3 Conclusions

MRT offers significant assets for the development of inherently safer chemical processes: hazardous chemicals, intermediates and processes are kept under highly controlled conditions in small volumes with no head-space. Besides, unconventional or extreme conditions are easily accessible and offer a redoubtable yet safe access to process intensification. A variety of chip, tubular and plate micro- and mesoreactors can be designed to combine the required chemical and mechanical resistances for a given process. The implementation of in-line extractions, separations or multistep reaction telescoping alleviates the exposition of workers and surrounding populations to hazardous chemicals by eliminating transportation and stockpiling issues.

Study questions
9.1. What are the definitions of Safety, Hazard, and Risk?
9.2. What are the advantages of Microreaction Technology regarding process safety?
9.3. Which makes the exothermic reactions safe if performed in MRT?
9.4. What are the 3 major hazard categories in organic chemistry practice where MRT has significant advantages?
9.5. Give examples for the safe realization of nitration and azide formation!

Further readings

- Ref #9: Kockmann N, Roberge DM. Harsh reaction conditions in continuous-flow microreactors for pharmaceutical production. Chem Eng Technol 2009, 32, 1682–94.
- Ref #10a: Fogler HS. Essentials of chemical reaction engineering. Pearson Education, 2011.
- Ref #10b: Stoessel F. Thermal safety of chemical processes: risk assessment and process design. Wiley-VCH Verlag GmbH & Co. KGaA, Weinheim, 2008.
- Ref #14: Poechlauer P, Braune S, Dielemans B, Kaptein B, Obermüller R, Thathagar M. On-site-on demand production of hazardous chemicals by continuous flow processes. Chim Oggi/Chem Today 2012, 30, 51–4.
- Ref #15: Sahoo HR, Kralj JG, Jensen KF. Multistep continuous-flow microchemical synthesis involving multiple reactions and separations. Angew Chem Int Edn 2007, 46, 5704–8.
- Ref #17: McMullen JP, Jensen KF. Integrated microreactors for reaction automation: new approaches to reaction development. Annu Rev Anal Chem 2010, 3, 19–42.

Bibliography

[1] The European Chemical Industry Council. (Accessed October 5, 2013 at http://www.cefic.org/)

[2] The US Chemical Safety and Hazard Investigation Board. (Accessed October 5, 2013 at http://www.csb.gov/)

[3] Major Accident Reporting System of the European Commission. (Accessed October 5, 2013 at https://emars.jrc.ec.europa.eu)

[4] US Chemical Safety and Hazard Investigation Board videos: Safety in academic labs. (Accessed October 5, 2013 at http://www.youtube.com/watch?v=ALBWxGik64A)

[5] Further reading: (a) Arendt JS, Lorenzo DK. Evaluating process safety in the chemical industry: a user's guide to quantitative risk analysis. American Institute of Chemical Engineers, 3 Park Avenue, New York, New York 10016–5991, 2000. (b) Sanders RE. Chemical process safety: learning from case histories, Elsevier Butterworth– Heinemann, 2005. (c) Kletz T. What went wrong? Case histories of process plant disasters and how they could have been avoided, 5th edn. Elsevier Trevor, 2009. (d) Crowl DA, Louvar, JF. Chemical Process Safety: Fundamentals with Applications, 3rd edn. Prentice Hall, 2011.

[6] Srinivasan R, Natarajan S. Developments in inherent safety: A review of the progress during 2001–2011 and opportunities ahead. Process Saf Environ Prot 2012, 90, 389– 403.

[7] US Chemical Safety and Hazard Investigation Board videos: Inherently safer processes. (Accessed October 5, 2013 at http://www.youtube.com/watch?v=h4ZgvD4FjJ8{#}at=244)

[8] Ebrahimi F, Kolehmainen E, Turunen I. Safety advantages of on-site microprocesses. Org Process Res Dev 2009, 13, 965–969.

[9] Kockmann N, Roberge DM. Harsh reaction conditions in continuous-flow microreactors for pharmaceutical production. Chem Eng Technol 2009, 32, 1682–1694.

[10] Further reading: (a) Fogler HS. Essentials of chemical reaction engineering. Pearson Education, 2011. (b) Stoessel F. Thermal safety of chemical processes: risk assessment and process design. Wiley-VCH Verlag GmbH & Co. KGaA, Weinheim, 2008.

[11] US Chemical Safety and Hazard Investigation Board videos: Runaway: explosion at T2 laboratories. (Accessed October 5, 2013 at http://www.youtube.com/watch?v=C561PCq5E1g)

[12] Monbaliu JCM, Winter M, Chevalier B, et al. Effective production of the biodiesel additive STBE by a continuous flow process. Bioresour Technol 2011, 102, 9304–9307.

[13] Heinrich S, Edeling F, Liebner C, Hieronymus H, Lange T, Klemm E. Catalyst as ignition source of an explosion inside a microreactor. Chem Eng Sci 2012, 84, 540–543.

[14] Poechlauer P, Braune S, Dielemans B, Kaptein B, Obermüller R, Thathagar M. On-site- on demand production of hazardous chemicals by continuous flow processes. Chim Oggi/Chem Today 2012, 30, 51–54.
[15] Sahoo HR, Kralj JG, Jensen KF. Multistep continuous-flow microchemical synthesis involving multiple reactions and separations. Angew Chem Int Edn 2007, 46, 5704–5708.
[16] US Chemical Safety and Hazard Investigation Board videos: Rupture of MIC tank at Bayer facility in West Virginia. (Accessed October 5, 2013 at http://www.youtube.com/watch?v=bcfvzGtuamM)
[17] McMullen JP, Jensen KF. Integrated microreactors for reaction automation: new approaches to reaction development. Annu Rev Anal Chem 2010, 3, 19–42.
[18] Murray PRD, Browne DL, Pastre JC, Butters C, Guthrie D, Ley SV. Continuous flow- processing of organometallic reagents using an advanced peristaltic pumping system and the telescoped flow synthesis of (E/Z)-tamoxifen. Org Process Res Dev 2013, 17, 1192–1208.
[19] Mateos C, Rincón J a., Villanueva J. Efficient and scalable synthesis of ketones via nucleophilic Grignard addition to nitriles using continuous flow chemistry. Tetrahedron Lett 2013, 54, 2226–2230.
[20] Brodmann T, Koos P, Metzger A, Knochel P, Ley SV. Continuous preparation of arylmagnesium reagents in flow with inline IR monitoring. Org Process Res Dev 2012, 16, 1102–1113.
[21] Riva E, Gagliardi S, Martinelli M, Passarella D, Vigo D, Rencurosi A. Reaction of Grignard reagents with carbonyl compounds under continuous flow conditions. Tetrahedron 2010, 66, 3242–3247.
[22] Ducry L, Roberge DM. Dibal-H reduction of methyl butyrate into butyraldehyde using microreactors. Org Process Res Dev 2008, 12, 163–167.
[23] Liu B, Fan Y, Lv X, Liu X, Yang Y, Jia Y. Generation and reactions of heteroaromatic lithium compounds by using in-line mixer in a continuous flow microreactor system at mild conditions. Org Process Res Dev 2013, 17, 133–137.
[24] Nagaki A, Kim H, Yoshida JI. Aryllithium compounds bearing alkoxycarbonyl groups: generation and reactions using a microflow system. Angew Chem Int Edn 2008, 47, 7833–7836.
[25] Usutani H, Tomida Y, Nagaki A, Okamoto H, Nokami T, Yoshida J. Generation and reactions of o-bromophenyllithium without benzyne formation using a microreactor. J Am Chem Soc 2007, 129, 3046–3047.
[26] Webb D, Jamison TF. A Continuous homologation of esters: an efficient telescoped reduction-olefination sequence. Org Lett 2012, 14, 2465–2467.
[27] Webb D, Jamison TF. Diisobutylaluminum hydride reductions revitalized: a fast, robust, and selective continuous flow system for aldehyde synthesis. Org Lett 2012, 14, 568–571.
[28] Ducry L, Roberge DM. Controlled autocatalytic nitration of phenol in a microreactor. Angew Chem Int Edn 2005, 44, 7972–7975.
[29] Yu Z, Lv Y, Yu C, Su W. A high-output, continuous selective and heterogeneous nitration of pdifluorobenzene. Org Process Res Dev 2013, 17, 438–442.
[30] Gage JR, Guo X, Tao J, Zheng C. High output continuous nitration. Org Process Res Dev 2012, 36, 930–933.
[31] Knapkiewicz P, Skowerski K, Jaskólska DE, Barbasiewicz M, Olszewski TK. Nitration under continuous flow conditions: convenient synthesis of 2-isopropoxy-5- nitrobenzaldehyde, an important building block in the preparation of nitro- substituted Hoveyda–Grubbs metathesis catalyst. Org Process Res Dev 2012, 16, 1430–1435.
[32] Kulkarni AA, Nivangune NT, Kalyani VS, Joshi RA, Joshi RR. Continuous Flow Nitration of Salicylic Acid. Org Process Res Dev 2008, 12, 995–1000.
[33] Pelleter J, Renaud F. Facile, fast and safe process development of nitration and bromination reactions using continuous flow reactors. Org Process Res Dev 2009, 13, 698–705.

[34] Brocklehurst CE, Lehmann H, La Vecchia L. Nitration chemistry in continuous flow using fuming nitric acid in a commercially available flow reactor. Org Process Res Dev 2011, 15, 1447–1453.

[35] Chen Y, Zhao Y, Han M, Ye C, Dang M, Chen G. Safe, efficient and selective synthesis of dinitro herbicides via a multifunctional continuous-flow microreactor: one-step dinitration with nitric acid as agent. Green Chem 2013, 15, 91–94.

[36] Braune S, Pöchlauer P, Reintjens R, et al. Selective nitration in a microreactor for pharmaceutical production under cGMP conditions. Chem Oggi/Chem Today 2009, 27, 26–29.

[37] Monbaliu JCM, Jorda J, Chevalier B, Stevens CV, Morvan B. Continuous-flow production of alkyl nitrites. Chem Oggi/Chem Today 2011, 29, 80–82.

[38] Chen Y, Su Y, Jiao F, Chen G. A simple and efficient synthesis protocol for sulfonation of nitrobenzene under solvent-free conditions via a microreactor. RSC Adv 2012, 2, 5637–5644.

[39] Muller A, Cominos V, Hessel V, et al. Fluidic bus system for chemical process engineering in the laboratory and for small-scale production. Chem Eng J 2005, 107, 205–214.

[40] Löb P, Löwe H, Hessel V. Fluorinations, chlorinations and brominations of organic compounds in micro reactors. J Fluor Chem 2004, 125, 1677–1694.

[41] Baerns M, Hessel V, Ehrfeld W, Haverkamp V. Direct fluorination of toluene using elemental fluorine in gas / liquid microreactors. J Fluor Chem 2000, 105, 117–128.

[42] Chambers RD, Holling D, Spink RCH, Sandford G. Gas-liquid film microreactors for selective direct fluorination. Lab Chip 2001, 1, 132–137.

[43] de Mas N, Günther A, Schmidt MA, Jensen KF. Microfabricated multiphase reactors for the selective direct fluorination of aromatics. Ind Eng Chem Res 2003, 42, 698–710.

[44] de Mas N, Günther A, Schmidt MA, Jensen KF. Increasing productivity of microreactors for fast gas – liquid reactions?: the case of direct fluorination of toluene. Ind Eng Chem Res 2009, 48, 1428–1434.

[45] Chambers RD, Fox MA, Sandford G, Trmcic J, Goeta A. Elemental fluorine. J Fluor Chem 2007, 128, 29–33.

[46] Navarrini W, Venturini F, Tortelli V, et al. Direct fluorination of carbon monoxide in microreactors. J Fluor Chem 2012, 142, 19–23.

[47] Lang P, Hill M, Krossing I, Woias P. Multiphase minireactor system for direct fluorination of ethylene carbonate. Chem Eng J 2012, 179, 330–337.

[48] Chambers RD, Fox MA, Holling D, Nakano T, Okazoe T, Sandford G. Elemental fluorine. Part 16. Versatile thin-film gas-liquid multi-channel microreactors for effective scale-out. Lab Chip 2005, 5, 191–198.

[49] Inoue T, Ohtaki K, Murakami S, Matsumoto S. Direct synthesis of hydrogen peroxide based on microreactor technology. Fuel Process Technol 2013, 108, 8–11.

[50] Illg T, Hessel V, Löb P, Schouten JC. Continuous synthesis of tert-butyl peroxypivalate using a single-channel microreactor equipped with orifices as emulsification units. ChemSusChem 2011, 4, 392–398.

[51] Park JH, Park CY, Song HS, Huh YS, Kim GH, Park CP. Green diacetoxylation of alkenes in a microchemical system. Org Lett 2013, 15, 752–755.

[52] Damm M, Gutmann B, Kappe CO. Continuous-flow synthesis of adipic acid from cyclohexene using hydrogen peroxide in high-temperature explosive regimes. ChemSusChem 2013, 6, 978–982.

[53] Shang M, Noël T, Wang Q, Hessel V. Packed-bed microreactor for continuous-flow adipic acid synthesis from cyclohexene and hydrogen peroxide. Chem Eng Technol 2013, 36, 1001–1009.

[54] Kumar GS, Pieber B, Reddy KR, Kappe CO. Copper-catalyzed formation of C-O bonds by direct α-C-H bond activation of ethers using stoichiometric amounts of peroxide in batch and continuous-flow formats. Chemistry 2012, 18, 6124–6128.

[55] Gutmann B, Roduit J-P, Roberge D, Kappe CO. Synthesis of 5-substituted 1H- tetrazoles from nitriles and hydrazoic acid by using a safe and scalable high- temperature microreactor approach. Angew Chem Int Edn 2010, 49, 7101–7105.
[56] Gutmann B, Obermayer D, Roduit JP, Roberge DM, Kappe CO. Safe generation and synthetic utilization of hydrazoic acid in a continuous flow reactor. J Flow Chem 2012, 2, 8–19.
[57] Palde PB, Jamison TF. Safe and efficient tetrazole synthesis in a continuous-flow microreactor. Angew Chem Int Edn 2011, 50, 3525–3528.
[58] Delville MME, Nieuwland PJ, Janssen P, Koch K, van Hest JCM, Rutjes FPJT. Continuous flow azide formation: optimization and scale-up. Chem Eng J 2011, 167, 556–559.
[59] Baumann M, Baxendale IR, Ley SV, Nikbin N, Smith CD, Tierney JP. A modular flow reactor for performing Curtius rearrangements as a continuous flow process. Org Biomol Chem 2008, 6, 1577–1586.
[60] Baumann M, Baxendale IR, Ley SV, Nikbin N, Smith CD. Azide monoliths as convenient flow reactors for efficient Curtius rearrangement reactions. Org Biomol Chem 2008, 6, 1587–1593.
[61] Ishikawa H, Bondzic BP, Hayashi Y. Synthesis of (-)-oseltamivir by using a microreactor in the Curtius rearrangement. Eur J Org Chem 2011, 6020–6031.
[62] Varas AC, Noël T, Wang Q, Hessel V. Copper(I)-catalyzed azide-alkyne cycloadditions in microflow: catalyst activity, high-T operation, and an integrated continuous copper scavenging unit. ChemSusChem 2012, 5, 1703–1707.
[63] Smith CD, Baxendale IR, Lanners S, Hayward JJ, Smith SC, Ley SV. [3 + 2] Cycloaddition of acetylenes with azides to give 1,4-disubstituted 1,2,3-triazoles in a modular flow reactor. Org Biomol Chem 2007, 5, 1559–1561.
[64] Struempel M, Ondruschka B, Stark A. Continuous Production of the diazomethane precursor N-methyl-N-nitroso-p-toluenesulfonamide: batch optimization and transfer into a microreactor setup. Org Process Res Dev 2009, 13, 1014–1021.
[65] Struempel M, Ondruschka B, Daute R, Stark A. Making diazomethane accessible for R&D and industry: generation and direct conversion in a continuous micro-reactor set-up. Green Chem 2008, 10, 41–43.
[66] Rossi E, Woehl P, Maggini M. Scalable in situ diazomethane generation in continuous-flow reactors. Org Process Res Dev 2012, 16, 1146–1149.
[67] Maurya RA, Park CP, Lee JH, Kim D-P. Continuous in situ generation, separation, and reaction of diazomethane in a dual-channel microreactor. Angew Chem Int Edn 2011, 50, 5952–5955.
[68] Hu DX, Brien MO, Ley SV. Continuous multiple liquid-liquid separation: diazotization of amino acids in flow. Org Lett 2012, 14, 4246–4249.
[69] Martin LJ, Marzinzik AL, Ley SV, Baxendale IR. Safe and reliable synthesis of diazoketones and quinoxalines in a continuous flow reactor. Org Lett 2011, 13, 320– 3.
[70] Yu Z, Lv Y, Yu C, Su W. Continuous flow reactor for Balz–Schiemann reaction: a new procedure for the preparation of aromatic fluorides. Tetrahedron Lett 2013, 54, 1261–1263.
[71] Weber M, Yilmaz G, Wille G. Azide synthesis in microstructured flow systems. Chem Oggi/Chem Today 2011, 29, 8–10.
[72] Malet-Sanz L, Madrzak J, Ley SV, Baxendale IR. Preparation of arylsulfonyl chlorides by chlorosulfonylation of in situ generated diazonium salts using a continuous flow reactor. Org Biomol Chem 2010 8, 5324–5332.
[73] Wada Y, Schmidt MA, Jensen KF. Flow distribution and ozonolysis in gas – liquid multichannel microreactors. Ind Eng Chem Res 2006, 45, 8036–8042.
[74] Steinfeldt N, Abdallah R, Dingerdissen U, Jähnisch K. Ozonolysis of acetic acid 1-vinyl- hexyl ester in a falling film microreactor. Org Process Res Dev 2007, 11, 1025–1031.

[75] Steinfeldt N, Bentrup U, Ja K. Reaction mechanism and in situ ATR spectroscopic studies of the 1-decene ozonolysis in micro- and semibatch reactors. Ind Eng Chem Res 2010, 49, 72–80.

[76] Sandra H, Bentrup U, Budde U, et al. An ozonolysis – reduction sequence for the synthesis of pharmaceutical intermediates in microstructured devices. Org Process Res Dev 2009, 13, 952–960.

[77] Irfan M, Glasnov TN, Kappe CO. Continuous flow ozonolysis in a laboratory scale reactor. Org Lett 2011, 13, 984–987.

[78] O'Brien M, Baxendale IR, Ley SV. Flow ozonolysis using a semipermeable Teflon AF-2400 membrane to effect gas-liquid contact. Org Lett 2010, 12, 1596–1598.

[79] Roydhouse MD, Ghaini A, Constantinou A, Cantu-Perez A, Motherwell WB, Gavriilidis A. Ozonolysis in flow using capillary reactors. Org Process Res Dev 2011, 15, 989–996.

[80] Nobis M, Roberge DR, Nobis M, Roberge DM. Mastering ozonolysis: production from laboratory to ton scale in continuous flow. Chim Oggi/Chem Today 2011, 29, 56–58.

[81] Irfan M, Glasnov TN, Kappe CO. Heterogeneous catalytic hydrogenation reactions in continuous-flow reactors. ChemSusChem 2011, 4, 300–316.

[82] Dencic I, Hessel V, de Croon MHJM, Meuldijk J, van der Doelen CWJ, Koch K. Recent changes in patenting behavior in microprocess technology and its possible use for gas-liquid reactions and the oxidation of glucose. ChemSusChem 2012, 5, 232–245.

[83] Sachse A, Linares N, Barbaro P, Fajula F, Galarneau A. Selective hydrogenation over Pd nanoparticles supported on a pore-flow-through silica monolith microreactor with hierarchical porosity. Dalton Trans 2013, 42, 1378–1384.

[84] Mercadante MA, Kelly CB, Lee C, Leadbeater NE. Continuous flow hydrogenation using an on-demand gas delivery reactor. Org Process Res Dev 2012, 16, 1064–1068.

[85] O'Brien M, Taylor N, Polyzos A, Baxendale IR, Ley SV. Hydrogenation in flow: Homogeneous and heterogeneous catalysis using Teflon AF-2400 to effect gas–liquid contact at elevated pressure. Chem Sci 2011, 2, 1250–1257.

[86] Heugebaert TSA, Roman BI, De Blieck A, Stevens CV. A safe production method for acetone cyanohydrins. Tetrahedron Lett 2010, 51, 4189–4191.

[87] Acke DRJ, Stevens CV. A HCN-based reaction under microreactor conditions: industrially feasible and continuous synthesis of 3,4-diamino-1H-isochromen-1-ones. Green Chem 2007, 9, 386–390.

[88] Cukalovic A, Monbaliu JCM, Heynderickx GJ, Stevens CV. User friendly and flexible Kiliani reaction on ketoses using microreaction technology. J Flow Chem 2012, 2, 43–46.

[89] Wiles C, Watts P. An integrated microreactor for the multicomponent synthesis of alpha-aminonitriles. Org Process Res Dev 2008, 12, 1001–1006.

[90] Koos P, Gross U, Polyzos A, O'Brien M, Baxendale I, Ley SV. Teflon AF-2400 mediated gas-liquid contact in continuous flow methoxycarbonylations and in-line FTIR measurement of CO concentration. Org Biomol Chem 2011, 9, 6903–6908.

[91] Brancour C, Fukuyama T, Mukai Y, Skrydstrup T, Ryu I. Modernized low pressure carbonylation methods in batch and flow employing common acids as a CO source. Org Lett 2013, 15, 2794–2797.

[92] Cranwell PB, O'Brien M, Browne DL, et al. Flow synthesis using gaseous ammonia in a Teflon AF-2400 tube-in-tube reactor: Paal-Knorr pyrrole formation and gas concentration measurement by inline flow titration. Org Biomol Chem 2012, 10, 5774–5779.

[93] Pastre JC, Browne DL, O'Brien M, Ley SV. Scaling up of continuous flow processes with gases using a tube-in-tube reactor: inline titrations and fanetizole synthesis with ammonia. Org Process Res Dev 2013, 17, 1183–1191.

[94] Basavaraju KC, Sharma S, Maurya RA, Kim D-P. Safe use of a toxic compound: heterogeneous OsO_4 catalysis in a nanobrush polymer microreactor. Angew Chem Int Edn 2013, 125, 6867–6870.
[95] Ajmera SK, Losey MW, Jensen KF, Schmidt M a. Microfabricated packed-bed reactor for phosgene synthesis. AIChE J 2001, 47, 1639–1647.
[96] Fuse S, Tanabe N, Takahashi T. Continuous in situ generation and reaction of phosgene in a microflow system. Chem Commun 2011, 47, 12661–12663.
[97] Selivanova SV, Mu L, Ungersboeck J, et al. Single-step radiofluorination of peptides using continuous flow microreactor. Org Biomol Chem [Internet] 2012, 10, 3871–3874.
[98] Chun J-H, Lu S, Pike VW. Rapid and efficient radiosyntheses of meta-substituted [F]fluoroarenes from [F]fluoride ion and diaryliodonium tosylates within a microreactor. Eur J Org Chem 2011, 4439–4447.
[99] Lu S, Chun J, Pike V. Fluorine-18 chemistry in micro-reactors. J Label Comp Radiopharm 2010, 53, 234–238.
[100] Chun J-H, Lu S, Lee Y-S, Pike VW. Fast and high-yield microreactor syntheses of ortho- substituted [(18)F]fluoroarenes from reactions of [(18)F]fluoride ion with diaryliodonium salts. J Org Chem 2010, 75, 3332–3338.
[101] Pascali G, Mazzone G, Saccomanni G, Manera C, Salvadori PA. Microfluidic approach for fast labeling optimization and dose-on-demand implementation. Nucl Med Biol 2010, 37, 547–555.
[102] Gillies JM, Prenant C, Chimon GN, Smethurst GJ, Dekker BA, Zweit J. Microfluidic technology for PET radiochemistry. Appl Radiat Isot 2006, 64, 333–336.
[103] Lee C-C, Sui G, Elizarov A, et al. Multistep synthesis of a radiolabeled imaging probe using integrated microfluidics. Science 2005, 310, 1793–1796.
[104] Lebedev A, Miraghaie R, Kotta K, et al. Batch-reactor microfluidic device: first human use of a microfluidically produced PET radiotracer. Lab Chip 2013, 13, 136–145.
[105] Miller PW, Long NJ, de Mello AJ, et al. Rapid multiphase carbonylation reactions by using a microtube reactor: applications in positron emission tomography 11C- radiolabeling. Angew Chem Int Edn Engl 2007, 46, 2875–2878.
[106] Hessel V, Kralisch D, Kockmann N, Noël T, Wang Q. Novel process windows for enabling, accelerating, and uplifting flow chemistry. ChemSusChem 2013, 6, 746–789.
[107] Glasnov TN, Kappe CO. The microwave-to-flow paradigm: translating high- temperature batch microwave chemistry to scalable continuous-flow processes. Chemistry 2011, 17, 11956–11968.
[108] Marre S, Roig Y, Aymonier C. Supercritical microfluidics: opportunities in flow- through chemistry and materials science. J Supercrit Fluids 2012, 66, 251–264.
[109] Tidona B, Urakawa A, von Rohr RP. High pressure plant for heterogeneous catalytic CO_2 hydrogenation reactions in a continuous flow microreactor. Chem Eng Process 2013, 65, 53–57.
[110] Gutmann B, Gottsponer M, Elsner P, Cantillo D, Roberge DM, Kappe CO. On the Fischer indole synthesis of 7-ethyltryptophol – mechanistic and process intensification studies under continuous flow conditions. Org Process Res Dev 2013, 17, 294–302.
[111] Reichart B, Tekautz G, Kappe CO. Continuous flow synthesis of n-alkyl chlorides in a high-temperature microreactor environment. Org Process Res Dev 2013, 17, 152–157.
[112] Viviano M, Glasnov TN, Reichart B, Tekautz G, Kappe CO. A scalable two-step continuous flow synthesis of nabumetone and related 4-aryl-2-butanones. Org Process Res Dev 2011, 15, 858–870.
[113] Obermayer D, Glasnov TN, Kappe CO. Microwave-assisted and continuous flow multistep synthesis of 4-(pyrazol-1-yl)carboxanilides. J Org Chem 2011, 76, 6657–6669.

[114] Lengyel L, Nagy TZ, Sipos G, et al. Highly efficient thermal cyclization reactions of alkylidene esters in continuous flow to give aromatic/heteroaromatic derivatives. Tetrahedron Lett 2012, 53, 738–743.

[115] Öhrngren P, Fardost A, Russo F, Schanche J-S, Fagrell M, Larhed M. Evaluation of a nonresonant microwave applicator for continuous-flow chemistry applications. Org Process Res Dev 2012, 16, 1053–1063.

[116] Ahmed-Omer B, Sanderson AJ. Preparation of fluoxetine by multiple flow processing steps. Org Biomol Chem 2011, 9, 3854–3862.

[117] O'Brien AG, Lévesque F, Seeberger PH. Continuous flow thermolysis of azidoacrylates for the synthesis of heterocycles and pharmaceutical intermediates. Chem Commun 2011, 47, 2688–2690.

[118] Gendrineau T, Marre S, Vaultier M, Pucheault M, Aymonier C. Microfluidic synthesis of palladium nanocrystals assisted by supercritical CO_2: tailored surface properties for applications in boron chemistry. Angew Chem Int Ed 2012, 51, 8525–8528.

Volker Hessel, Qi Wang, and Dana Kralisch

10 From green chemistry principles in flow chemistry towards green flow process design in the holistic viewpoint

In this subchapter, major Green Chemistry metrics are described (atom economy, reaction mass efficiency, environmental (E)-factor, etc.). Instructive flow chemistry applications which claim greenness of their development are spotlighted and, where adequate, viewed under the umbrella of the green chemistry metrics or more holistic, life-cycle based evaluation methods. This will reveal major trends in developing sustainable green synthetic routes in flow. With relation to the greenness, the specific characteristics of the flow chemistry approach will be outworked. From there, the view will be drawn on the level of systems engineering and process design. The much increased complexity here demands for an adequate metering counterpart. The Green Engineering principles, as for example given by the ACS Pharmaceutical Roundtable and derived (industrial) definitions, will be introduced. It will be shown how these govern current instructive flow process design developments and that greenness demands process integration beyond the single green efforts on the reaction level.

The choice of examples aim to interconnect with the other chapters of the graduate textbook. After each subsection, some questions and answers and homework will be given. For quick and efficient information, advice on further reading in terms of reviews, books, websites will be given.

10.1 Introduction of Green Chemistry principles

10.1.1 Green principles

Green chemistry is the design of chemical products and processes that reduce or eliminate the use and generation of hazardous substances (Anastas and Warner, 1998). Green Engineering is the development and commercialization of industrial processes that are economically feasible and reduce the risk to human health and the environment (Anastas and Zimmerman, 2003). The 12 principles of green chemistry and 12 principles of green engineering are listed in Table 10.1. They can be seen as qualitative measures of the greenness of chemical syntheses and processes. Typically, more than one of these principles has to be met for the development of a new alternative, which is substantially more environmentally benign than established benchmarks.

Table 10.1: 12 Principles of Green Chemistry and 12 Principles of Green Engineering

12 Principles of Green Chemistry	12 Principles of Green Engineering
1. Prevention: prevent waste but not treat or clean up waste afterwards.	1. Designers need to strive to ensure that all material and energy inputs and outputs are as inherently nonhazardous as possible.
2. Atom Economy: Design synthetic methods to maximize the incorporation of all materials used in the process into the final product.	2. It is better to prevent waste than to treat or clean up waste after it is formed.
3. Less Hazardous Chemical Syntheses: Design synthetic methods to use and generate substances that minimize toxicity to human health and the environment.	3. Separation and purification operations should be designed to minimize energy consumption and materials use.
4. Designing Safer Chemicals: Design chemical products to affect their desired function while minimizing their toxicity.	4. Products, processes, and systems should be designed to maximize mass, energy, space, and time efficiency.
5. Safer Solvents and Auxiliaries: Minimize the use of auxiliary substances wherever possible, make them innocuous when used.	5. Products, processes, and systems should be "output pulled" rather than "input pushed" through the use of energy and materials.
6. Design for Energy Efficiency: Minimize the energy requirements of chemical processes and conduct synthetic methods at ambient temperature and pressure if possible.	6. Embedded entropy and complexity must be viewed as an investment when making design choices on recycle, reuse, or beneficial disposition.
7. Use of Renewable Feedstocks: Use renewable raw material or feedstock rather whenever practicable.	7. Targeted durability, not immortality, should be a design goal.
8. Reduce Derivatives: Minimize or avoid unnecessary derivatization if possible, which requires additional reagents and generate waste.	8. Design for unnecessary capacity or capability (e.g., "one size fits all") solutions should be considered a design flaw.
9. Catalysis: Catalytic reagents are superior to stoichiometric reagents.	9. Material diversity in multicomponent products should be minimized to promote disassembly and value retention.
10. Design for Degradation: Design chemical products so they break down into innocuous products that do not persist in the environment.	10. Design of products, processes, and systems must include integration and interconnectivity with available energy and materials flows.
11. Real-time analysis for Pollution Prevention: Develop analytical methodologies needed to allow for real-time, in-process monitoring and control prior to the formation of hazardous substances.	11. Products, processes, and systems should be designed for performance in a commercial "afterlife".
12. Inherently Safer Chemistry for Accident Prevention: Choose substances and the form of a substance used in a chemical process to minimize the potential for chemical accidents, including releases, explosions, and fires.	12. Material and energy inputs should be renewable rather than depleting.

10.1.2 Green flow chemistry

Traditional chemical manufacturing is normally known for the usage of large reactors and associated large operation units as well as corresponding large scale transport and storage of raw materials and products. All these large scale features can lead in case of accidents to major health and safety disasters and also high risk to operators and the neighbor community. Flow chemistry in microreactors shows great promise especially to pharmaceutical and fine chemicals industry as a novel method on which to build new chemical technology and processes. The reaction rate in microreactor is sped up and the yield and purity of desired product is improved. Solvent free mixing, *in situ* reagent generation and integrated separation techniques can make the reactions much easier to control minimizing risk and side-reactions (Haswell and Watts, 2003).

Besides the features mentioned above, green and sustainable chemical synthesis in continuous flow is shown also from several other ways, including improving product selectivity to give less waste, avoiding energy-consuming cryogenic cooling set-ups, and protecting group-free synthesis to improve atom and step economy (Yoshida, Kim and Nagaki, 2011). Notably, flow microreactors enable on-demand and on-site synthesis, which leads to less energy for transportation and easy recycling of substances (Yoshida, Kim and Nagaki, 2011). Therefore, flow microreactors may play a major role in the environmentally benign synthesis and production of useful chemical substances.

Further readings
- P. T. Anastas, J. B. Zimmerman, "Design through the Twelve Principles of Green Engineering", Env. Sci. Tech. 2003, 37(5), 94A-101A.
- S. J. Haswell, P. Watts, Green Chemistry: synthesis in microreactors. Green Chemistry, 2003, 5, 240–249.

10.2 Flow process design and relation to green chemistry/engineering

10.2.1 Flow processing – major means in process intensification

Green flow process design enables process intensification [4] in the sense Górak and Stankiewicz have characterized, namely to match the intrinsic needs of chemistry towards an ideal kind of processing state defined by [5]:
- to give each molecule the same processing experience
- to optimize the driving forces and maximize specific interfacial areas
- to maximize the effectiveness of intra- and intermolecular events
- to maximize the synergistic effects of partial processes

The first two moments refer to transport intensification and the third to chemical intensification, which flow reactors provide [6, 7]. The last moment refers to process-design intensification, which flow process design provides [6, 7].

As one of many evidences for the primary role of flow process design in process intensification, the ACS GCI Pharmaceutical Roundtable has ranked continuous processing on top of ten key green engineering research areas [8]. This multi-industrial platform involves companies such as Boehringer, Pfizer, Eli Lilly, GlaxoSmithKline, DSM, Johnson & Johnson, AstraZeneca, and Merck (US). The key areas 1–5 refer partly to chemical intensification and process-design intensification issues were ranked as 6–10, marking future development milestones [8].

10.2.2 Transport intensification – the flow-scale

Transport intensification arises from the following microfluidic characteristics of flow reactors [6, 9].
- Unique flow – laminar flow for single flows; defined multiphase flow patterns;
- Short diffusion paths for heat and mass transfer;
- High surface-to-volume ratios for phase contact area
- High share of solid wall material – short path to catalyst

Those can be translated into beneficial reaction engineering features of flow processing [6, 9].
- Short residence times – narrow residence time distribution;
- Faster mixing of miscible phases; higher dispersion of immiscible phases;
- Improved heat transport;
- Improved safety.

Such superior processing leads to the following beneficial process engineering features [6, 9].
- Increased selectivity
- Increased reactivity & selectivity
- Increased enantioselectivity – changed isomer ratios

10.2.3 Chemical intensification – the reactor scale

Transport intensification makes chemical intensification possible, which means to use harsh conditions in terms of temperature, pressure, and concentration (novel process windows) to boost chemical reactivity and the other mentioned process engineering features [7, 10]. This as well enables the easy use of tailored processing media such as supercritical fluids. Finally, such superb process control makes the exploration of new reaction paths possible.

10.2.4 Process-design intensification – the full-process scale

Moreover, origin features of flow process design effects were identified and proclaimed as follows, that is, process-design related impacts which flow processing (with microreactors) has and which is not given for conventional processing [7, 11].
- No consistent, gap-free availability of standard modules (currently)
- Reaction dominant process design, whereas today's plants are utility- and purification dominated
- Interactive, synergetic design rather than additive (plug & play) design from components into a chained plant
- Integrated design – the system as interactive platform in between the component and plant level.

The first is negative for flow process design, the other three points define chances for greenness on a holistic level.

10.2.5 Elemental green criteria with proven impact of flow process design

Obviously, all this has utmost consequence on all what is considered to be green in Section 10.1 and in the coming Sections 10.3–10.5. To get an idea how deep the 'green impact' of flow process design is, those Green Chemistry and Green Engineering principles and their evaluation metrics are condensed here to origin, elemental green measures. The aforementioned green principles are a mixture of those elemental measures and advices how to use them in practice, which makes the listing repetitive and narrative. Meanwhile, undoubtedly due to massive evidence both at the laboratory and pilot stage, flow chemistry and flow process design can provide substantial impact to the following elemental green measures [12–14].
1. make maximal use of starting materials (reactants) and minimize waste
2. minimize solvent load
3. have maximal process safety
4. minimize toxicity (to human)
5. maximize process integration

These are the essence of most Green Chemistry and several Green Engineering principles as given before. So far most evidence comes from the level of flow reactors and this is concerned mostly with the measures 1–4. Still, more elaboration on the true process design level is needed to look if there are true, origin effects or if the flow process design just makes the reactor level effects 'happen'. Yet, there is first evidence for process integration effects (measure 5), as given by thermal coupling of reactors [15], reaction network coupling [16], downstream (purification) effects [11], and full-scale heat integration [17]. A full list of all flow process design effects is given in a compilation on this subject published by some of the authors [7].

10.2.6 Elemental green criteria with suspected impact of flow process design

For some elemental green issues on a process design level, the impact of flow processing is not entirely clear at this point of time [12–14]. Yet, the first positive indications are that a use can improve process greenness. These issues are to
6. make maximal use of energy
7. minimize environmental burden (with focus on global warming)
8. make use of smart catalysis
9. enable real-time analysis
10. maximize capacity matching (to market need)

Issue 6 is first approached in a recent pinch analysis study to optimize a heat exchanger network centering around a flow plant. Issue 7 is quite positive for flow process design on a laboratory level; yet studies on a full process design level are still largely missing. Catalysts (issue 8) can be effectively used in flow process design and have green consequences; yet the question here is if that is better than for conventional processing. There is a lack of corresponding investigations and thus evidence at this point of time. Issue 9 is the center of the idea of flow process automation for which activities have just started and are done by very few groups. Issue 10 has just been touched upon recently in the increasing number of studies over modular compact plants ('containers').

10.2.7 Elemental green criteria with uncertainty over impact of flow process design

For the remaining elemental green issues, the impact of flow process design simply cannot be stated, either due to lack of demonstration evidence or because the issue is not coupled to processing, but to product design (like issues 11 and 12) [12–14].

11. maximize use of renewable feedstocks
12. maximize design for degradation
13. minimize embedded entropy and complexity.
14. match targeted durability, not immortality

However, current activities in the field of process intensification of renewable feedstock utilization using flow chemistry provide a first, positive outlook ([11], Kralisch *et al.* 2013). More scientific activities on this field can surely be expected for the next years.

Further readings
- V. Hessel, I. Vural Gursel, Q. Wang, T. Noel, J. Lang, Chem. Eng. Tech., 2012, 35, 1184–1204.
- V. Hessel, B. Cortese, M. H. J. M. de Croon, Chem. Eng. Sci. 2011, 66, 1426–1448.
- V. Hessel, T. Noël, Micro Process Technology, 2. Processing, Ullmann's Encyclopedia of Industrial Chemistry, 2013, DOI: 10.1002/14356007.b16_b37.pub2.
- N. Kockmann (ed.), Micro Process Engineering, Fundamentals, Devices, Fabrication, and Applications; in: Advanced Micro and Nanosystems (issue 5), eds: O. Brand, G. K. Fedder, C. Hierold, J. G. Korvink, O. Tabata, 2006, Wiley-VCH
- V. Hessel, D. Kralisch, N. Kockmann, T. Noël, Q. Wang. ChemSusChem 2013, 6, 746–789.
- C. Wiles, P. Watts, Eur. J. Org. Chem. 2008, 1655–1671.
- J. Wegner, S. Ceylan, A. Kirschning, Chem. Commun. 2011, 47, 4583–4592
- V. Hessel, I. Vural Gursel, Q. Wang, T. Noel, J. Lang, Chem. Eng. Tech., 2012, 35, 1184–1204.
- V. Hessel, B. Cortese, M. H. J. M. de Croon, Chem. Eng. Sci. 2011, 66, 1426–1448.

10.3 Holistic methodology introduction for systematic green flow process design

Taking in mind the information given before about the "green" potentials of flow chemistry, we will now have a closer look at the measurement of this greenness.

It is apparent from the 12 + 12 principles of green chemistry [19] and green engineering [20] that the development of a green flow synthesis process depends on more aspects than the decision for flow processing and the choice of the reaction device. Nevertheless, important parameters such as solvent, reactant and auxiliary demand, process conditions, effort for work-up, recycling and waste disposal, and so on, are directly influenced by this choice. These in turn have an impact on the resulting material and energy flows starting from their extraction from the environment to emissions, waste and energy that are transferred back into the environment at the end of a product life cycle. These flows are small, when laboratory scale experiments are performed for process design. However, decisions concerning specific process configurations taken during this early stage will influence the future environmental impacts of

Table 10.2: Overview of typical green chemistry metrics.

Metric	Calculation	Source
AE	$AE = \dfrac{\text{molecular weight of desired product}}{\text{molecular weight of all products}} \times 100\%$	[23]
E-Factor	$\text{E-Factor} = \dfrac{\text{total mass of waste}}{\text{mass of product}}$	[22]
MI	$MI = \dfrac{\text{mass of all materials excluding water}}{\text{mass of product}}$	[24]
RME	$RME = \dfrac{\text{mass of product}}{\text{mass of all reactants}} \times 100\%$	[24]
	$\text{Energy efficiency} = \dfrac{\text{energy demand}}{\text{chemical amount of product}}$	[26]
PMI	$PMI = \dfrac{\text{total mass in a process or process step}}{\text{mass of product}}$	[25]

the process, once it has been scaled-up, to a great extent [21]. Thus, there is a raising trend in academia and industry to evaluate the greenness of novel approaches starting in early research and development. In order to keep this simple and time efficient, often green chemistry metrics characterizing the mass of waste produced per unit mass of product or the masses of all substances, used in a process, related to the product mass, respectively, are used. Well-known metrics for green chemistry are, for example, the E-Factor by Sheldon [22], atom economy (AE) by Trost [23], and mass-related concepts including several synthetic steps, for example, mass intensity (MI) and reaction mass efficiency (RME) proposed by Curzons and colleagues [24] as well as process mass intensity (PMI) [25]. Furthermore, metrics dealing with the energy demand of a synthesis have been suggested, for example, energy efficiency [26].

These metrics have the further advantage that all required data is easily available and can be calculated without special expertise, which is needed for more sophisticated evaluation methods. The numbers or ratios calculated by means of them point out, for example, high solvent requirements, low selectivities for the final product or a large proportion of waste compared to the product in the total mass balance.

The disadvantage of metrics, however, is that they do not allow any conclusions about potential environmental impacts or any comparison between different environmental effects. In addition, the impact on the environment during the supply of energy, solvents, reactants, and so on, or during their disposal is not included. Thus, the evaluation of green chemistry and engineering by means of single metrics is not holistic. This can easily result in an optimized flow process, for example, using a non-green reactant or solvent, or where yield or selectivity improvements are paid dearly by very high energy consumptions. Often, the effort and environmental impacts for work-up, recycling or disposal are also underestimated during process design. Ultimately, this may require a re-correction of the process design, before the process can be scaled-up and used for industrial production.

That is why Kralisch and colleagues suggested some years ago to accompany green flow process design by means of a holistic Life Cycle Assessment (LCA) [27, 28]. The LCA methodology is standardized by the International Standard Organisation (ISO) [29, 30] and is defined as the "compilation and evaluation of the inputs, outputs and potential environmental impacts of a product system throughout its life cycle". The environmental aspects and potential impacts associated with a process, product or service system are assessed by compiling an inventory of relevant inputs and outputs. This evaluation step is followed by a characterization and quantification of environmental impacts associated with those inputs and outputs. For this, defined environmental impact categories are used covering a wide range of worldwide relevant impacts on the environment, including the depletion of exhaustible raw materials, the emission of hazardous substances as well as land use. At this, the magnitude as well as the potency of the impacts is taken into account using equivalence factors in relation to a reference compound. Common methods for the assessment of these potential environmental impacts are the Eco-Indicator 99 [31], the method proposed by the *Centrum voor Milieukunde Leiden* (CML) [32], or ReCiPe [33]. As an example, environmental impact categories according to CML are, for example, potentials of Abiotic Resource Depletion (ADP), Global Warming (GWP), Ozone Depletion (ODP), Photochemical Ozone Creation (POCP), Acidification (AP), Eutrophication (EP), Human Toxicity (HTP) and Eco-toxicity (ETP) as well as Land Use (LU). The group of Dewulf alternatively uses the category Cumulative Exergy Extraction from the Natural Environment (CEENE) for the impact assessment [34]. They also applied this measure in the context of life-cycle based evaluation of pharmaceutical processes [35, 36].

The holistic cradle-to-grave approach of LCA studies avoids the problems otherwise connected with the optimization of single process steps, which disregard the impact of the respective up-stream or down-stream processes. The use of a common functional unit allows for the comparison of the environmental impacts of process design alternatives. Due to its broad applicability and its validity, the LCA methodology has gained worldwide acceptance as a useful tool for strategic planning, process development as well as policy-making. It is also widely used to assess chemical syntheses on laboratory (e.g., [37–39]) as well as production scale (e.g., [40–42]).

However, the method is not specially designed for the evaluation of chemical processes or its application for decision-making purposes during process design (for more detailed information see [43]). From the plethora of suggested LCA approaches (see e.g., *dynamic LCA* [44], *spatially differentiated LCA* [45], *risk-based LCA* [46], *environmental input-output based LCA* (EIO-LCA) [47] or *hybrid LCA* [48]) one has to select the best-fitting strategies without losing the holistic, comprehensive evaluation idea behind the LCA approach.

In practice, the timeframe to perform a detailed LCA study mostly exceeds the demands of process development, especially in industry. A detailed LCA study further requires an extensive database, which is often not available during the development of a new process, for example, concerning best suited disposal strategies. A Simplified LCA

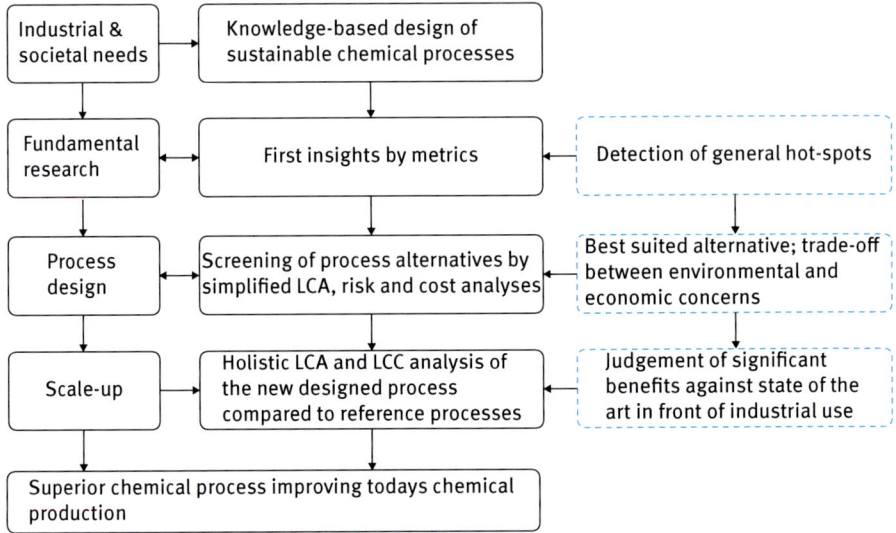

Fig. 10.1: Evaluation approach for sustainable chemical process design [50].

(SLCA) (also called "streamline" LCA) is a suitable alternative at this stage [21]. SLCA needs less time and data input to run the assessment, since it allows for the exclusion of certain life cycle stages, system inputs/outputs or impact categories as well as the use of generic data modules to fill data gaps [49]. However, the initial identification of hot spots within the whole life cycle should result in a comprehensive LCA at the end of the process design step in order to allow for a profound analysis. The overall evaluation approach suggested by Kralisch and colleagues for sustainable chemical process design also including cost and environmental risk criteria is shown in [50]. It has been developed in the context of the evaluation of several flow chemistry case studies [50, 51]. The evaluation starts with the search for a case study with a real need for process design due to existing drawbacks and limitations of the chemical (or pharmaceutical) process. In the next two steps, potential alternatives are evaluated mainly for decision support on experimental scale by means of metrics, SLCA and costs including a broad range of processing and process parameter variations. The outcome is typically a set of parameter configurations with lowest environmental impact potentials and costs. Further development work and scale-up activities can focus on these selected alternatives. Finally, the newly designed flow process is compared with state of the art reference (batch) process by means of a holistic LCA and also by Life Cycle Costing (LCC) in order to quantify savings in environmental impacts and costs as a result of this development.

In the next section, practical examples for green flow process design using metrics as well as sophisticated LCA approaches will be given.

Further readings
- J. Andraos (2009). Application of Green Metrics Analysis to Chemical Reactions and Synthesis Plans. Green Chemistry Metrics, John Wiley & Sons, Ltd., 69–199.
- D. J. C. Constable, C. Jimenez-Gonzalez, et al. (2009). Process Metrics. Green Chemistry Metrics, John Wiley & Sons, Ltd., 228–247.
- M. Eissen, G. Geisler, et al. (2009). Mass Balances and Life Cycle Assessment. Green Chemistry Metrics, John Wiley & Sons, Ltd., 200–227.
- D. Kralisch (2009). Application of LCA in Process Development. Green Chemistry Metrics, John Wiley & Sons, Ltd., 248–271.

10.4 Green flow process design for fine chemicals/pharmaceuticals

Already in 2004, Jenck and colleagues published a first overview about the implementation of sustainability concepts in chemical production processes [52]. Seven years later, a prioritization of industrially relevant topics in the areas of green pharmaceutical and fine chemistry process design were published by the GCI Pharmaceutical Roundtable [53] stressing the importance of a continuous processing and process intensification, but also life-cycle based process evaluation as the key research areas for further development. The latest trends in green chemistry and consideration of sustainability criteria in the pharmaceutical and fine chemicals industry were evaluated by Watson [54]. The analysis shows that especially big companies have already adjusted their business strategies accordingly and integrated green concepts such as continuous processing, alternative solvents or energy sources in their production processes.

But, how to be sure that a fine chemical or pharmaceutical process is a green one? This question can be answered using the green metrics and evaluation methods introduced before. In the following sections, the combination of flow process design or improvement in fine chemistry and pharmacy with evaluation methods will be demonstrated by means of three case examples, a technology comparison, the design of a green fine chemical synthesis procedure and the improvement of an existing pharmaceutical production process.

10.4.1 Technology comparison for green pharmaceutical process design

More than 10 years ago, Jiménez-González and colleagues suggested the concept of a "Clean/Green Technology Guide" (see Table 10.3) as an expert system providing comparative environmental and safety performance information on available technologies for commonly performed unit operations in the pharmaceutical industry [55]. A set of

Table 10.3: Green Technologies' metrics for the scenario considered by [55].

Technological alternative	Environment	Efficiency	Energy	
Microreactors	Green	Gren	Yellow	Green
Minireactors	Green	Green	Green	Green
6000 l batch reactor	Yellow	Yellow	Red	Yellow

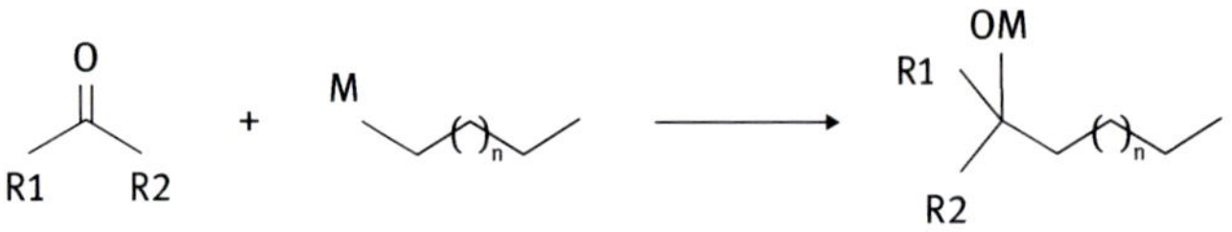

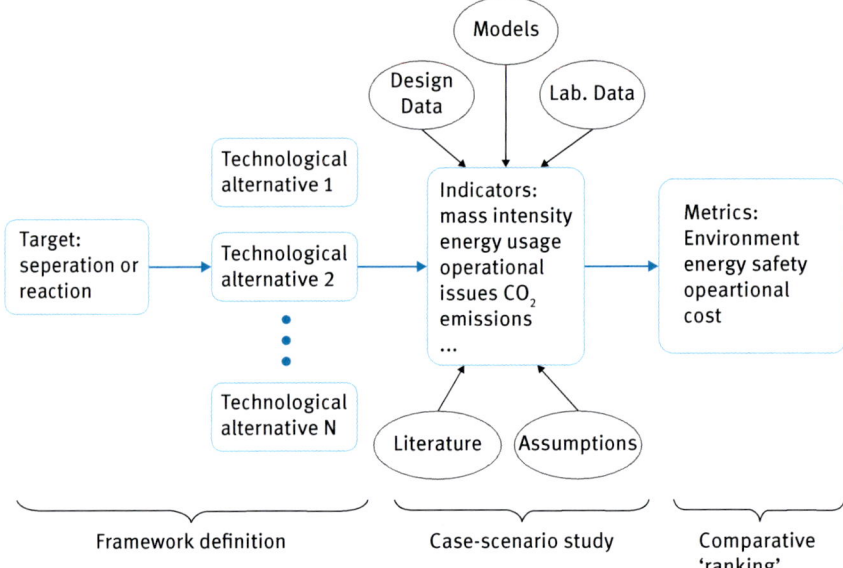

Fig. 10.2: Metal organic synthesis evaluated by metrics [55].

Fig. 10.3: Methodology for the development of the Clean Technology Guide according to [55].

metrics was used to evaluate the technology alternatives batch, mini-, and microreactors for application in a metal organic synthesis.

Based on the defined set of mass, energy, efficiency and safety metrics, they came out with a qualitative ranking of those technology alternatives referred to the model synthesis (see Figure 10.3)

10.4.2 Flow process design of a green biphasic fine chemical synthesis

Huebschmann *et al.* investigated the biphasic esterification of phenol and benzoyl chloride resulting in phenyl benzoate using a Simplified LCA and cost analysis [56]. Phenyl benzoate is used in industry, for example, for polymer modification, as an antioxidant and an intermediate in the production of liquid crystals. The main target of the study was to provide decision support for a sustainable process design for phenyl benzoate synthesis already during the research and design (R&D) stage.

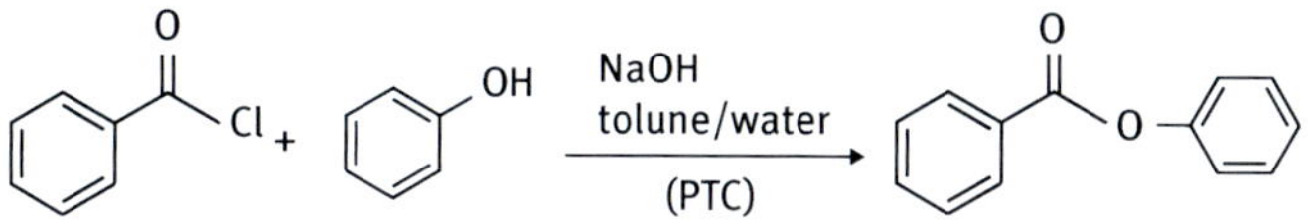

Fig. 10.4: Phase transfer catalysis of benzoyl chloride and phenol evaluated by Simplified LCA [56].

In order to intensify the internal mixing of the biphasic reaction mixture, the synthesis was transferred from batch to flow processing using different types of micromixers. Further, ionic liquids were tested as phase transfer catalysts.

At the beginning of the process design accompanying evaluation, the authors raised the following questions:

"I. Are the continuously running syntheses more environmentally benign than the batch syntheses?
II. What is the influence of the periphery in the energy consumption of the continuously running syntheses compared to the batch reactions?
III. Is the higher yield obtained by the implementation of ionic liquids as phase transfer catalysts compensated by higher environmental burdens caused by the supply of the ionic liquids?
IV. Which mixing structure is the most efficient?
V. Which ionic liquid is more beneficial from an ecological point of view?
VI. How does the consideration of work-up procedures affect the environmental burdens of the process?"

They were able to answer all questions by means of a simplified LCA based on experimental results obtained under laboratory conditions.

First of all, the authors found that the performance of the micromixing structures highly depend on the process parameters chosen, especially on the utilization of ionic liquids such as $[C_{18}MIM]Br$, $[MIM]BuSO_3$] or $[BMIM]Br$ as phase transfer catalysts. These compounds showed strongly positive results on the yield of the esteri-

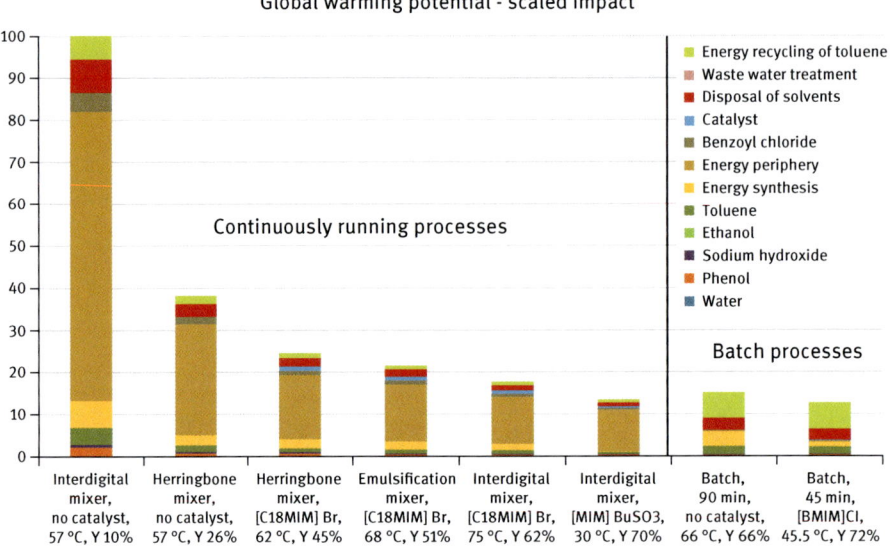

Fig. 10.5: GWP of alternative synthesis concepts of the biphasic reaction of benzoyl chloride and phenol [56].

fication reaction compared to noncatalyzed syntheses. In earlier studies [57, 58], the status of ionic liquids beeing "green"solvents were discussed very critically. Used in small amounts as a phase transfer catalyst instead, the application of ionic liquids turned out to improve the overall environmental balance. This result was found despite the high environmental burden of these compounds resulting from their material and energy demanding synthesis.

The work-up step increased the environmental burden of the process considerably – independently from the decision for batch or flow processing. Thus, it was recommended that future work should also focus on an optimization of this process step.

From a holistic, life-cycle based point of view, the switch from batch mode to a continuously running esterification in an interdigital mixer (best mixing structure found in the experimental tests in combination with a phase transfer catalyst) alone did not result in significant environmental improvements. (see Figure 10.3)

The reason was the higher energy demand of the flow processing plant including pumps and process control systems. As a result, significant savings up to 70% could only be forecasted for the microreaction process in case that the high electricity demand of the peripheral equipment can be reduced in an optimized production process.

10.4.3 Exergetic LCA for improvement of an existing pharmaceutical production process

Van der Vorst *et al.* [35, 36] performed an exergetic LCA of a Galantamine·HBr synthesis for anti-Alzheimer medication. They searched for environmental improvements within the overall synthesis pathway of the active pharmaceutical ingredient (API) in three processing alternatives (first–third generation). In this, thermodynamics and a systematic data inventory methodology for the quantification of the resource efficiency were emerged into single impact values such as exergy loss/mol API or CEENE/mol API for fast benchmarking and evaluation. The systematic data inventory methodology was based on the principle that each chemical production process is a sequence of unit operations in which basic operations at individual equipments take place. The concept has been transferred from simulation software used to split up specific production processes into recurring units. In this case, the basic operations are the building blocks for all production processes helping to simplify the overall evaluation and to harmonize the inventory data [59].

The first generation pathway included nine synthesis steps. In the second generation process, the fourth and fifth synthesis steps were optimized by the replacement of a solvent and by improving the efficiency of both steps (see Figure 10.6). The solvent switch especially had an effect on the resource requirements. In the third generation process pathway, the sixth step, performed under batch conditions, was replaced by a continuous process using a flow reactor (see Table 10.2). The increased resource efficiency by changing from first till third generation was quantitatively evaluated at the process, gate-to-gate and cradle-to-gate level using the CEENE method developed by the group [34].

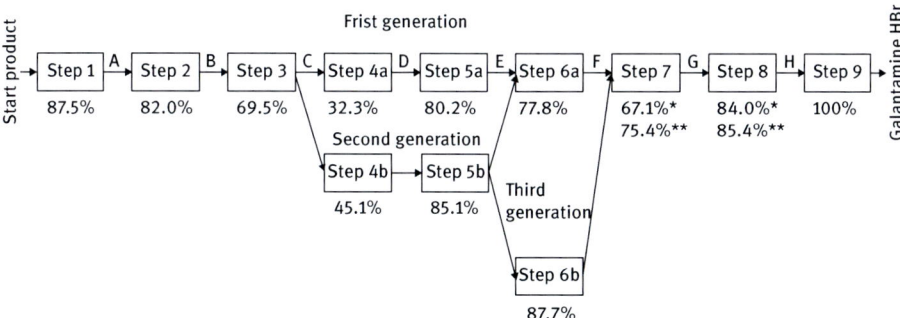

Fig. 10.6: Synthesis steps and their yields for the production of Galantamine·HBr evaluated by the LCA indicator CEENE [60].

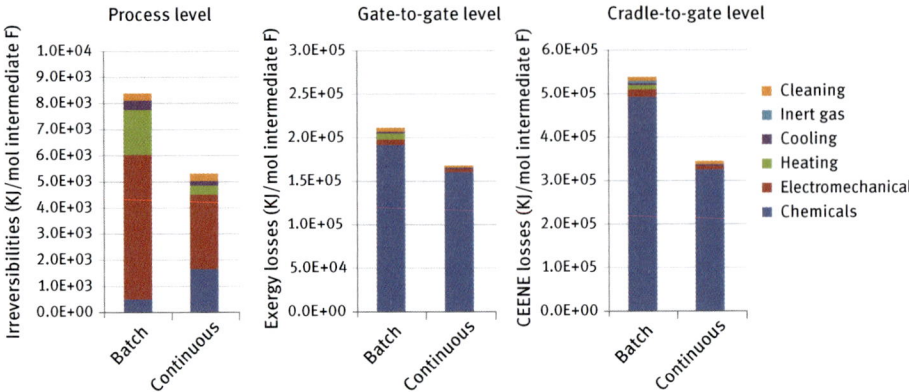

Fig. 10.7: Comparison of the batch and continuous production of 1 mol active intermediate F (process step 6 of Galantamine·HBr production) at three different levels and excluding the previous production steps [60].

Van der Vorst *et al.* [60] calculated a reduction in the overall resource consumption up to 41% due to the optimization of the first generation pathway by new chemistry in combination with flow processing.

Further readings
- C. Jiménez-González, A. D. Curzons, D. J. C. Constable, M. R. Overcash and V. L. Cunningham, *Clean Products and Processes*, 2001, **3**, 35–41.
- S. Huebschmann, D. Kralisch, H. Loewe, D. Breuch, J. H. Petersen, T. Dietrich and R. Scholz, *Green Chem*, 2011, **13**, 1694–1707.
- G. Van der Vorst, W. Aelterman, W. B. De, B. Heirman, L. H. Van and J. Dewulf, *Green Chem.*, 2013, **15**, 744–748.

10.5 Green flow process design for bulk chemicals and benchmark to conventional process

Bulk chemicals, fine chemicals, and pharmaceuticals have very different production volumes, product prices, added value, plant technologies and more, as summarized in Table 10.4 (Cybulski, *et al.* 2001).

Continuous flow processing is considered to be profitable for pharmaceuticals and fine chemicals (Roberge, *et al.* 2005). Such industrial documentation is missing for bulk chemicals and thus continuous flow processes are rarely designed and researched here (with the exception of energy applications such as fuel processing). Yet, this may be a matter of system complexity, which prevents direct insight in cost sav-

Table 10.4: Difference in characteristics of bulk chemicals versus fine chemicals.

Characteristics	Bulk chemical	Fine chemical	Pharmaceutical
Volume (tons/yr)	10^4-10^6	10^2-10^4	$10-10^3$
Price ($/kg)	< 10	> 10	> 100
Added value	Low	High	Very high
Processing	Continuous	Batch-wise	Batch-wise
Plants	Dedicated	Multipurpose	Multipurpose
Flexibility	Low	High	High
Safety and environmental efforts	Relatively low	High	Relatively higher

ings as possible through simple calculation of operational benefits, for example, due to higher selectivity as given in (Roberge, et al. 2005).

To give the first systemic answer on the profitability of bulk-chemical organic processing, the example of the direct synthesis of adipic acid (ADA) in a continuous flow process will be taken in this section beside the analysis of its environmental impacts. A virtual full-scale flow process was designed and benchmarked to the conventional process.

ADA is an important precursor for the production of Nylon 6,6 (Herzog, et al. 2013). The conventional ways to produce adipic acid comprise the two-step catalyzed air/nitric acid oxidation of cyclohexane (see Figure 10.8, left) (Steeman, et al. 1961) and the carboxylation/carbo-methoxylation of butadiene (Castellan, et al. 1991). A direct route was realized through the oxidation of cyclohexene by hydrogen peroxide in batch or membrane reactors (Sato, et al. 1998; Buonomenna, et al. 2010; Deng, et al. 1999) (see Figure 10.8, right). This direct route is process simplified (one-step) with regard to and has higher yield as compared to the conventional route.

10.5.1 Process simulation

Based on the laboratory flow experiments using a milli-packed bed reactor with microsized fluid interstices, a full-chain process simulation was made for the direct synthesis of adipic acid by using Aspen software (see Figure 10.9).

Based on the relevant patents and literatures, the process simulation for the two-step conventional synthesis was built as well (see Figure 10.10).

In view of development time from chemical synthesis to production, the employment of the direct route using microreaction technology is advantageous over developing flow processes or new batch processes with two-step processes. Engineering, procurement and construction phases can be performed significantly faster with much less equipment required. Safety also points to the use of small, yet highly intensified

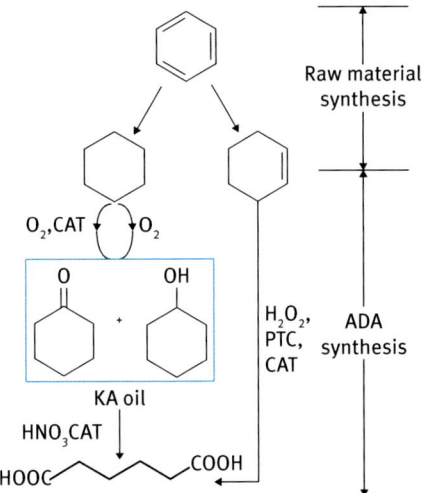

Fig. 10.8: Two different production routes of adipic acid, two- and one-step (with permission from Elsevier (Wang, *et al.* 2013)).

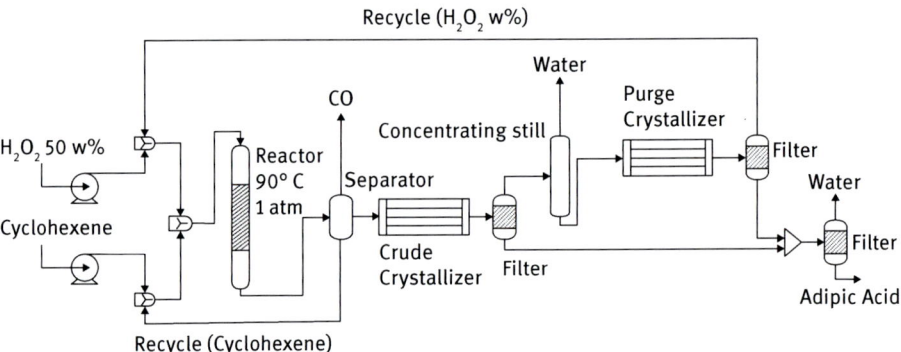

Fig. 10.9: Flow sheet for the direct micro-flow synthesis of ADA (with permission from Elsevier (Wang, *et al.* 2013)).

reactor units. Total capital investment is significantly reduced as well. Green processes have to give proof of their cost competitiveness in front of industrial implementation. To underline this, a figure of major equipment and their rough cost estimates depicts the difference of the two routes (see Figure 10.11).

Figure 10.11 shows that, due to the fewer reaction steps in the direct route, less equipment is used compared to the conventional route and thus the total costs of the

10.5 Green flow process design for bulk chemicals — 301

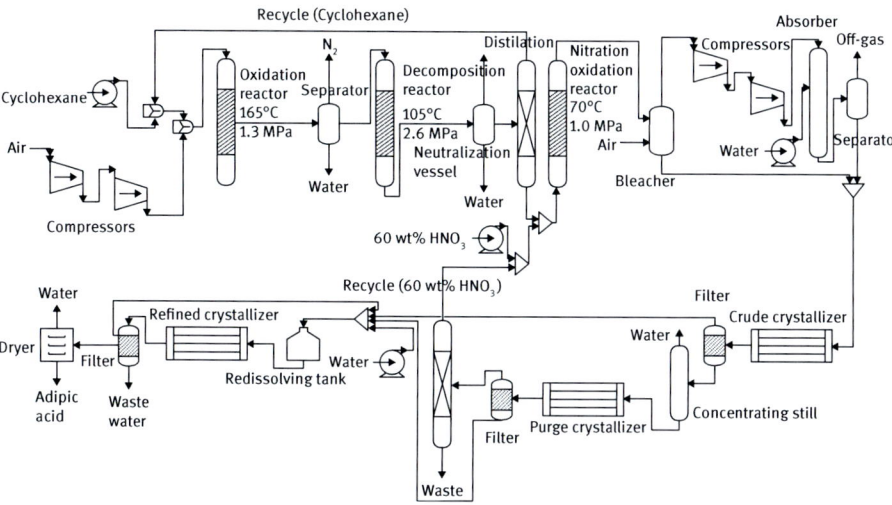

Fig. 10.10: Flow sheet for the two-step conventional synthesis of ADA (with permission from Elsevier (Wang, et al. 2013)).

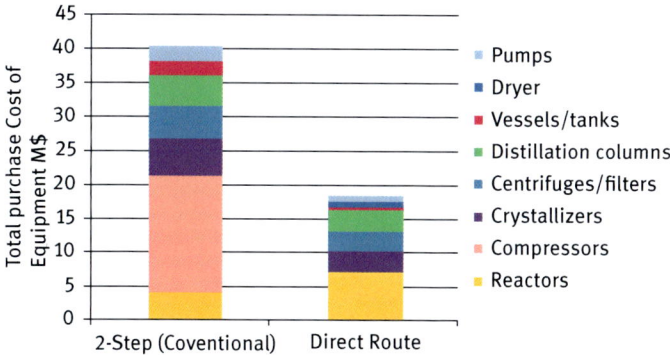

Fig. 10.11: Total purchase cost of all equipment for the conventional route and direct route (with kind permission of AIDIC Servizi S.r.l. (Vural-Gursel, et al. 2012)).

direct route benefit from this. It is evident that the continuous flow reactor cost constitutes about 40% of the total cost (see Figure 10.11). But the costs of continuous flow reactors will decrease with the development of manufacturing techniques, the numbers of plates/reactors manufactured, and new functionally advanced and constructional simplified designs. In the sense of decreasing the material consuming as well as process integration according to the principles of "Green Engineering", the continuous ADA synthesis process has the advantages compared to the conventional process.

10.5.2 LCA for continuous flow synthesis of ADA

The environmental profile for the direct micro-flow route of ADA is detailed for 9 impacts which contains: acidification potential (AP), average European, global warm potential: climate change in 20 years (GWP 20 a), eutrophication potential (EP), average European, freshwater aquatic ecotoxicity in 20 years (FAETP 20 a), human toxicity in 20 years (HTP 20 a), marine aquatic ecotoxicity in 20 years (MAETP 20 a), photochemical oxidant creation potential (POCP), depletion of abiotic resources, terrestrial ecotoxicity in 20 years (TAETP 20 a) (see Figure 10.12). Even at a glance it is obvious that the production of H_2O_2 dominates most of these impact categories. This is because H_2O_2 production is an energy and pollution intensive process. 95% of all H_2O_2 is produced on an industrial scale by the Anthraquinone Oxidation (AO) process, which was developed by BASF (www.BASF.com; Genti and Perathoner, 2009). It involves the sequential hydrogenation and oxidation of an anthraquinone precursor dissolved in a mixture of organic solvents followed by liquid-liquid extraction to recover H_2O_2 (Campos-Martin, *et al.* 2006). The high energy input and generated waste have negative effects on environmental impacts and production costs (Genti and Perathoner, 2009; Campos-Martin, *et al.* 2006). The transport, storage, and handling of bulk H_2O_2 involve hazards and escalating expenses (Campos-Martin, *et al.* 2006). Moreover, H_2 as raw material for producing H_2O_2 is also made in an energy consuming process, mostly by the steam reforming of methane to generate H_2.

The H_2O_2 impact is much dominant (> 50%) for all toxicity-related impacts (FAETP 20 a, HTP 20 a, MAETP 20 a and TAETP 20 a). Even when considering process scenarios based on experimental and superficial conversions (40, 50, 98%), the contribution of H_2O_2 to HTP is almost exclusive for all these scenarios, being > 90%. The

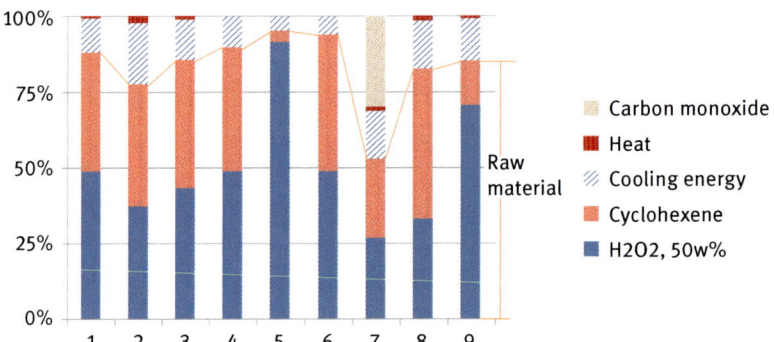

Fig. 10.12: Environmental profile of the direct micro-flow synthesis of ADA (functional unit: 1 kg ADA, conversion of cyclohexene is 50%, yield of ADA is 46.8%) 1. AP 2. GWP 20 a 3. EP 4. FAETP 20 a 5. HTP 20 a 6. MAETP 20 a 7. POCP 8. Depletion of abiotic resources 9. TAETP 20 a. The line drawn separates raw material- from process technology contributions. (with permission from Elsevier (Wang, *et al.* 2013)).

analysis, however, is somewhat different when referring to AP, GWP 20a, EP, POCP and Depletion of abiotic resources. Here the H_2O_2 is "only" 25–50%. It shows that H_2O_2 is not so "green" as claimed by chemists (Noyori, *et al.* 2003; Grigoropoulo, *et al.* 2003; Usui and Sato, 2003; Podgorsek, *et al.* 2009; Edwards, *et al.* 2005; Tse, et al. 2005).

10.5.3 LCA for two-step conventional synthesis of ADA

For the two-step conventional synthesis of ADA, the dominating contribution for most of the impact categories also comes from raw materials (see Figure 10.13). Nitrogen oxide is the dominating parameter for AP and EP, while waste treatment is the dominating parameter for POCP.

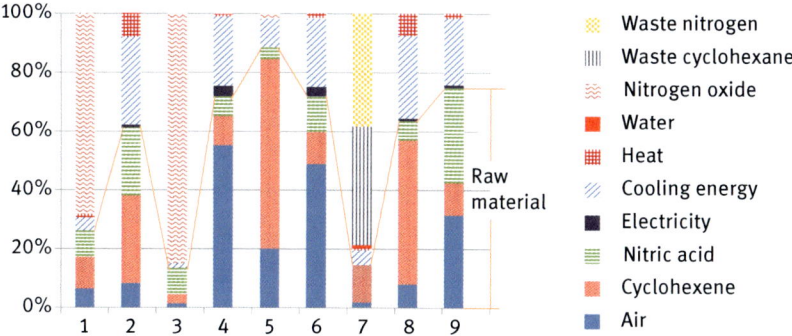

Fig. 10.13: Environmental profile of the two-step conventional synthesis of ADA (functional unit: 1 kg ADA, conversion of cyclohexene = 4.6%, yield of ADA = 4.16%, considering the recycle of unreacted cyclohexane and nitric acid). 1. AP 2. GWP 20 a 3. EP 4. FAETP 20 a 5. HTP 20 a 6. MAETP 20 a 7. POCP 8. Depletion of abiotic resources 10. TAETP 20 a (with permission from Elsevier (Wang, *et al.* 2013)).

10.5.4 Complete LCA picture

The quantitative results for the impact categories are listed in Table 10.5, giving the complete LCA picture. A sensitivity analysis with different conversions of cyclohexene for the direct micro-flow process of ADA is carried out in this study. This paper uses experimental data as important reference for the process simulation used in the LCA given here, for example, taking 40% and 50% conversion of cyclohexene. For sensitivity analysis, a superficial process scenario assuming 4.6% and 98% conversion of cyclohexene were chosen, which respectively match the practice for the first step in the conventional process and the best performance of the reaction in batch reactor. The

Table 10.5: Quantitative results for the ADA synthesis by the direct micro-flow route and two-step conventional route (functional unit: 1 kg ADA from raw material) (with permission from Elsevier (Wang, et al. 2013)).

Impact category	DR $X^* = 4.6\%$	DR $X = 40\%$	DR $X = 50\%$	DR $X = 98\%$	CTR $X = 4.6\%$
AP ($\times 10^{-3}$ kg SO_2)	18.7	13.8	13.8	13.9	48.8
POCP ($\times 10^{-3}$ kg ethylene)	2.7	1.8	1.8	1.8	10.1
EP ($\times 10^{-3}$ kg NO_x)	11.0	7.5	7.5	7.4	94.7
FAETP 20 a (kg 1,4-DCB)	1.9	1.5	1.5	1.3	1.0
MAETP 20 a (kg 1,4-DCB)	1.2	0.9	0.9	0.8	0.6
Depletion of abiotic resources ($\times 10^{-3}$ kg antimony)	83.7	53.2	53.0	51.0	46.7
TAETP 20 a ($\times 10^{-3}$ kg 1,4-DCB)	0.3	0.2	0.2	0.2	0.2
HTP	8.7	7.7	7.8	8.6	5.6
GWP 20 a	11.0	6.3	6.3	5.9	16.7

*X denotes the conversion of cyclohexene/cyclohexane

recycling of unreacted raw material was considered in the direct route with conversion of cyclohexene at 4.6%, 40%, 50% as well as in the two-step conventional route. At this, a crystallization step for ADA separation and direct reuse of the two-phasic reaction system was assumed. The recycling was not considered with higher conversion like 98% in direct route because of the low ratio of unreacted cyclohexene.

Two main conclusions arise. One impact-group of this LCA sometimes is in (strong) favor for the direct micro-flow process, another impact-group for the two-step conventional process. For example, the absolute value of GWP for the direct micro-flow route with different cyclohexene conversion (4.6%, 40%, 50% and 98%) is lower than that for the two-step conventional route. The absolute value of HTP for the direct route is higher than that for the conventional route. Considering the fact that the contribution of H_2O_2 to HTP is larger than 90%, it shows that H_2O_2 is not so "green" as claimed by chemists because of the high contribution in HTP profile. It is green in the waste context (only water emission), but not in a more holistic view "from cradle to factory gate".

On a full-scale process ("Green Engineering") level, there is no "greener process" as often claimed in literature focusing only on the chemical innovation (Noyori, et al. 2003; Grigoropoulo, et al. 2003; Usui and Sato, 2003; Podgorsek, et al. 2009; Edwards, et al. 2005; Tse, et al. 2005). It is a typical multi-criteria decision. It already supplied some evidence from the findings given above; yet is spotlighted in Table 10.5 by showing result of more impact categories. The second conclusion is that each impact category is improved because of the process improvements (such as increasing conversion from 4.6 to 98%). But the overall environmental profile as shown in Figure 10.12 will hardly be changed from the viewpoint of raw material domination, since the unreacted cyclohexene and H_2O_2 were ideally assumed according to the flow sheet shown in Figure 10.9. It will not increase any energy consumption for the phase separator with low conversion of cyclohexene, but will add working load on the crystallizers. Then the influence of cooling energy in the environmental profile as shown in Figure 10.13 will increase with the decrease of the conversion.

10.5.5 Enlightment

Also, due to widening of process view via the inclusion of superficial process scenarios, some generic conclusions can be drawn as well, besides the details of the analysis given above. This may shed a first light on the future use of micro- and flow reactor technology and respective processes from the standpoint of environmental benefits and disadvantages.

- *Green chemicals do not necessarily make green processes.* H_2O_2 is a green oxidant from the chemist point of view since it produces only water as waste within the synthesis. On the other side, when considered from cradle to process gate, the Anthraquinone process for H_2O_2 production is an energy and pollution intensive process. Only by replacing the latter through an environmental sustainable process, which can be the direct H_2O_2 synthesis out of the elements, the H_2O_2-based ADA making will become "green" on a process level.
- *Materials largely determine the environmental profile for large-scale chemical production.* The consumption of raw materials dominates most of the impact categories. This is commonly not seen in many statements and roadmaps of process intensification. Favored processes need first of all smart synthetic paths with "green" chemicals. Process technology then has the enabling function to bring innovation from reaction to process level.

From the environmental profile of the direct micro-flow synthesis in Figure 10.12 it becomes clear that the three dominant influential parameters on the environmental impacts are H_2O_2, cyclohexene, and cooling energy. From this, the following recommendations for further process optimization can be given.

- Greener production routes for H_2O_2 than given by the Anthraquinone process have to be found in the future or "green" H_2 as raw material should be used (the byproduct H_2 from electrolysis is much "greener" than H_2 from SMR process); for example by the direct route from the elements H_2 and O_2 (without carrier), that is, creating a double-direct route to ADA.
- If not possible, try to substitute H_2O_2 by a greener oxidant (for example, oxygen).

Note: Part of the content in Section 10.5 is reprinted from "Life cycle assessment for the direct synthesis of adipic acid in microreactors and benchmarking to the commercial process. Wang, Q., Vural-Gursel, I., Shang, M., Hessel, V., Chemical Engineering Journal, 2013, 234, 300–311, Copyright (2013), with permission from Elsevier."

Further readings
Wang, Q., Vural-Gursel, I., Shang, M., Hessel, V. Chemical Engineering Journal, 2013, 234, 300–311.

10.6 Outlook for green flow process design

The examples discussed before have shown that a broad range of green flow processes have already been developed.

Nevertheless, many open issues need to be solved for future green flow process design in order to make use of its full potential:
- Flow systems and compact modular plant concepts have to approach each other.
- Today's flow plants are either retrofitted conventional plants or flow plants with process design oriented on conventional process design. The reactor innovation degree being high, the systemic innovation degree is low. There is much space for process-design intensification.
- Today's flow plants center around reaction. Yet, catalyst manufacture, preconditioning, purification and even formulation need to be integrated.
- The thinking about suited process control specific for (fast) flow reactions has just been started.
- Flow plants are ideal for distributed manufacturing. This allows for new distribution schemes, new supply chains, and finally for new products. Such a groundbreaking idea, however, has not been realized yet.

On top of that all, one central issue is and remains – process integration. Here, the inspiration is mimicking nature's stunning ability to assemble molecules with unprecedented control (regio-/stereo-specificity) in complex chemical environments using multistep one-pot strategies: "This efficiency and orthogonality allows nature to develop the necessary toolbox of life" [81]. Indeed, some examples of sequential multistep flow synthesis have been described, even including complex natural product synthesis [52]. However, an intrinsic initial disadvantage of flow processing is presented here. The main bottleneck is the need for direct combination ('one stream') of process units for reaction control, work-up and purification (see Figure 10.13, whereas the same units can be run separately in batch mode, giving more degrees of freedom. The latter is characterized by the term "orthogonality", which is a term originating from statistical analysis [82] and taxonomy [83]). Concerning process chemistry, this means keeping the variables of the process protocol as independent as possible [84]. Yet, ex-

Current flow multi-step process scheme, restricted to laboratory

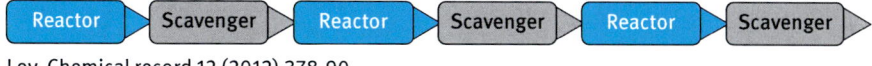

Ley, Chemical record 12 (2012) 378-90

Current industrial vision on flow multi-step process scheme

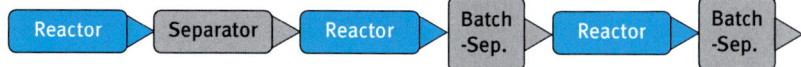

Kockmann, Gottsponer, Roberge, Chem. Eng. J. 167 (2011) 718-726

Vision on full-flow multi-step process scheme, with separation

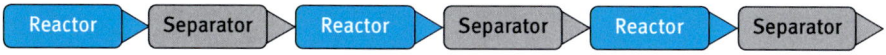

Hessel, Kralisch, Kockmann, Noël, ChemSusChem 6, 5 (2013) 746-789

Vision on full-flow multi-step process scheme, without separation

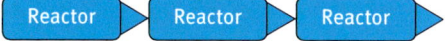

Benson, Ponton, Trans. IChemE 1993, 71 (A2) 160; Rinard, IMRET 2 Proc. at AIChE, New Orleans, 1998, 299

Vision on telescoped 'one-pot' multi-step process scheme, (with separation)

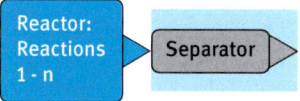

Hessel, Kralisch, Kockmann, Noël, Wang, ChemSusChem 6, 5 (2013)746-789

Fig. 10.14: Multistep micro-flow process schemes, reality and visions.

actly this, is much reduced in the current flow plants due to the interconnection of the reactor and separation units, that is, predefining the input stream of one reactor by the output stream of another reactor. This inevitably couples flow rates, concentrations, solvents, additives, and more. Part of it can be solved by flow separation. Yet, flow separation is underdeveloped and current concepts are not orthogonal. Thus, this also does not help. Rather, and now following nature again, the solution might be integrate processing and/or process functional units (like nature's organelles) in one system (see Figure 10.13). This might be termed *horizontal compartmentalization* in analogy to the *vertical compartmentalization*, practiced today in the flow plants.

An elongated way for horizontal compartmentalization is provided by flow reaction cascades which yield highly valuable products in one step (see last bar in Figure 10.13). Orthogonality is believed to be the key factor that allows to perform multistep reactions and "necessitates the development of new chemical reactions and strategies" [81]. Merrifield's orthogonal protection strategy has been extended to the development of orthogonal activation (e.g., used in Click Chemistry) for streamlining organic synthesis [84] and supramolecular chemistry [85]. There is large potential for bioorthogonal conjugation and soft functional material preparation via robust, efficient and orthogonal (REO) approaches [85] (see [86] for an example).

Study questions

10.1. Review the 12 Principles of Green Chemistry and 12 Principles of Green Engineering. Which of them can be integrated in your own experimental work?

10.2. What are the "green features" for continuous flow chemistry in microreactors? Please discuss the features according to the 12 Principles of Green Chemistry.

10.3. What makes flow processes essentially different from batch processes? Discuss this for the different scales – the reaction scale, process scale, company scale, and environment/society scale.

10.4. Take a chemical reaction which you know well and, at best, you also know the industrial processing well. Propose for this reaction, a flow processing and derive its advantages on a reactor scale. Imagine advantages on a process scale. Connect this chemistry- and chemical engineering advantages with benefits in "greenness". Use the roles of Green Chemistry and Green Engineering here.

10.5. Using a flow plant, can you realize greenness in process design without any such innovation in chemistry and transfer phenomena? Support your argument with how you think flow processing can change process design.

10.6. Which information is needed to compare the greenness of two synthesis strategies?

10.7. Make a list of pros and cons for the evaluation of green flow process design by means of a) a mass based metric and b) a full LCA.

10.8. Why were different evaluation approaches chosen to answer the question about the greenness of a chemical or pharmaceutical process? Discuss in the context of the case examples given above.

10.9. Rethink your own possibilities of contributing to the interdisciplinary task of green process design.

10.10. Please list the possibilities to decrease the environmental impacts for the continuous flow process of direct adipic acid synthesis.

10.11. Try to discuss the possible improvement considering the principles of "Green Chemistry" and "Green Engineering" by applying a greener process for H_2O_2 synthesis rather than the Anthraquinone process.

Bibliography

[1] P. T. Anastas, J. C. Warner, Green Chemistry: Theory and Practice. Oxford University Press: New York, 1998, p. 30.

[2] P. T. Anastas, J. B. Zimmerman, "Design through the Twelve Principles of Green Engineering", Env. Sci. Tech. 2003, 37(5), 94A-101A.

[3] S. J. Haswell, P. Watts Green Chemistry: synthesis in micro reactors. Green Chemistry, 2003, 5, 240–249.

[4] *For reviews about process intensification:* a) A. I. Stankiewicz, J. A. Moulijn, *Chem. Eng. Progress*, 2000, **1**, 22–34. b) J. R. Burns, J. N. Jamil, C. Ramshaw, *Chem. Eng. Sci.*, 2000, **55**, 2401–2415. c) A. I. Stankiewicz, *Chem. Eng. Proc.*, 2003, **42**, 137–144. d) D. Reay, C. Ramshaw, A. Harvey, *Process Intensification – Engineering for Efficiency, Sustainability and Flexibility*, Elsevier, Oxford, 2008. e) S. Becht, R. Franke, A. Geisselmann, H. Hahn, *Chem. Eng. Proc.*, 2009, **48**, 329–332. f) C. Ramshaw, *Green Chem.*, 1999, **2**, G15-G17. f) V. Kumar, K. D. P. Nigam, *Green Proc. Synth.*, 2012, **1**, 79–107.
[5] A. Gorak, A. I. Stankiewicz, *Annu. Rev. Chem. Bio. Eng.*, 2011, **2**, 431–451.
[6] V. Hessel, D. Kralisch, N. Kockmann, T. Noel, Q. Wang, *ChemSusChem*, 2013, **6** (5) 746–789.
[7] V. Hessel, I. Vural Gursel, Q. Wang, T. Noel, J. Lang, *Chem. Eng. Tech.*, 2012, **35**, 1184–1204.
[8] C. Jimenez-Gonzalez, P. Poechlauer, Q. B. Broxterman, B.-S. Yang, D. am Ende, J. Baird, C. Bertsch, R. E. Hannah, P. Dell'Orco, H. Noorman, S. Yee, R. Reintjens, A. Wells, V. Massonneau, J. Manley, *Org. Process. Res. Dev.*, 2011, **15**, 900–911.
[9] V. Hessel, T. Noël, *Micro Process Technology, 2. Processing*, Ullmann's Encyclopedia of Industrial Chemistry, 2013, DOI: 10.1002/14356007.b16_b37.pub2.
[10] For reviews about Novel Process Windows see: (a) V. Hessel, B. Cortese, M. H. J. M. de Croon, *Chem. Eng. Sci.* 2011, **66**, 1426–1448; (b) V. Hessel, P. Loeb, H. Loewe, *Curr. Org. Chem.*, 2005, **9**, 765–787 (c) V. Hessel, I. Vural-Guersel, Q. Wang, T. Noel, J. Lang, *Chem. Eng. Technol.*, 2012, **35**, 1184–1204; (d) V. Hessel, Q. Wang, *Chim. Oggi*, 2011, **29**, 81–84; (e) V. Hessel, Q. Wang, *Chim. Oggi*, 2011, **29**, 54–57; (f) T. Razzaq, C. O. Kappe, *Chem. Asian J.*, 2010, **5**, 1274–1289; (g) T. Illg, P. Loeb, V. Hessel, *Bioorg. Med. Chem.*, 2010, **18**, 3707–3719; (h) V. Hessel, *Chem. Eng. Technol.*, 2009, **32**, 1655–1681; (i) V. Hessel, D. Kralisch, U. Krtschil, *Energy Environ. Sci.*, 2008, **1**, 467–478; (j) T. Razzaq, T. N. Glasnov, C. O. Kappe, *Chem. Eng. Technol.*, 2009, **32**, 1702–1716.
[11] D. Kralisch, I. Streckmann, D. Ott, U. Krtschil, E. Santacesaria, M. Di Serio, V. Russo, L. De Carlo, W. Linhart, E. Christian, B. Cortese, M. H. J. M. de Croon, V. Hessel, *ChemSusChem*, 2012, **5**, 300–311.
[12] *For a selection of reviews which deal with micro process technology and flow chemistry:* (a) T. Noel, S. L. Buchwald, *Chem. Soc. Rev.*, 2011, **40**, 5010–5029; (b) R. L. Hartman, J. P. McMullen, K. F. Jensen, *Angew. Chem. Int. Edn.* 2011, **50**, 7502–7519; c) J. Wegner, S. Ceylan, A. Kirschning, *Chem. Commun.* **2011**, 47, 4583–4592; (d) T. N. Glasnov, C. O. Kappe, J. *Heterocycl. Chem.*, 2011, **48**, 11–30; (e) D. Webb, T. F. Jamision, Chem. Sci. **2010**, 1, 675–680; (f) A. Cukalovic, J.-C. M. R. Monbaliu, C. V. Stevens, *Top. Heterocycl. Chem.*, 2010, **23**, 161–198; (g) C. G. Frost, L. Mutton, Green Chem. 2010, 12, 1687–1703; (h) R. L. Hartman, K. F. Jensen, *Lab Chip*, 2009, **9**, 2495–2507; (i) K. Geyer, T. Gustafsson, P. H. Seeberger, *Synlett*, 2009, 2382–2391; (j) T. Fukuyama, M. T. Rahman, M. Sato, I. Ryu, Synlett 2008, **2**, 151–163; (k) S. V. Ley, I. R. Baxendale, *Chimia*, 2008, **62**, 162–168; (l) C. Wiles, P. Watts, *Eur. J. Org. Chem.*, 2008, 1655–1671; (m) B. P. Mason, K. E. Price, J. L. Steinbacher, A. R. Bogdan, D. T. McQuade, *Chem. Rev.*, 2007, **107**, 2300–2318; (n) K. Geyer, J. D. C. Codee, P. H. Seeberger, *Chem.-Eur. J.*, 2006, **12**, 8434–8442; (o) A. Kirschning, W. Solodenko, K. Mennecke, *Chem.-Eur. J.* 2006, **12**, 5972–5990; (p) K. Jaehnisch, V. Hessel, H. Loewe, M. Baerns, *Angew. Chem. Int. Edn.* 2004, 43, 406–446; (q) V. Hessel, H. Loewe, Chem. Eng. Technol. 2003, **26**, 13–24; (r) V. Hessel, H. Loewe, Chem. Eng. Technol. 2003, **26**, 391–408; (s) V. Hessel, H. Loewe, *Chem. Eng. Technol.*, 2003, **26**, 531–544; (t) K. F. Jensen, *Chem. Eng. Sci.*, 2001, **56**, 293–303; (u) K. F. Jensen, *AIChE J.* 1999, **45**, 2051–2054.
[13] *For some recent books pertaining flow chemistry:* (a) P. Watts, C. Wiles, *Microreactor Technology in Organic Synthesis* CRC Press, 2011; (b) T. Wirth, *Microreactors in Organic Synthesis and Catalysis*, Wiley-VCH, Weinheim, 2008; (c) J.-i. Yoshida, F*lash Chemistry: Fast Organic*

synthesis in Microsystems, Wiley-Blackwell, Hoboken, 2008; (d) P. H. Seeberger and T. Blume, *New Avenues to Efficient Chemical Synthesis – Emerging Technologies*, Springer-Verlag, Berlin, 2007.

[14] *For books about micro process technology:* (a) W. Ehrfeld, V. Hessel, H. Loewe, *Microreactors: New Technology for Modern Chemistry* Wiley-VCH, Weinheim, 2000; (b) V. Hessel, S. Hardt, H. Loewe, *Chemical Micro Process Engineering – Fundamentals, Modelling and Reactions* Wiley-VCH, Weinheim, 2004; (c) V. Hessel, H. Loewe, A. Mueller, G. Kolb, *Chemical Micro Process Engineering – Processing and Plants* Wiley-VCH, Weinheim, 2005; (d) V. Hessel, A. Renken, J. C. Schouten, J.-i. Yoshida, *Handbook of Micro Process Engineering* Wiley-VCH, Weinheim, 2009.

[15] C. Cremers, A. Pelz, U. Stimming, K. Haas-Santo, O. Goerke, P. Pfeifer, K. Schubert, *Fuel Cells*, 2007, **7**, 91–98.

[16] A. Nagaki, K. Imai, H. Kim, J.-i. Yoshida, *RSC Adv.*, 2011, **1**, 758–760.

[17] I. Vural-Gürsel, Q. Wang, T. Noël, V. Hessel, J. T. Tinge, *Ind. Eng. Chem. Res.*, 2013, **52** (23), 7827–7835.

[18] I. Vural Gursel, V. Hessel, Q. Wang, T. Noel, J. Lang, *Green Proc. Synth.*, 2012, **1** (4), 315–336.

[19] P. Anastas and J. Warner, *Green Chemistry: Theory and Practice*, Oxford Univ Press, 1998.

[20] P. T. Anastas and J. B. Zimmerman, *Environ. Sci. Technol.*, 2003, **37**, 95A-101A. Q

[21] S. Huebschmann, D. Kralisch, V. Hessel, U. Krtschil and C. Kompter, *Chem. Eng. Technol.*, 2009, **32**, 1757–1765.

[22] R. A. Sheldon, *CHEMTECH*, 1994, **24**, 38–47.

[23] B. M. Trost, *Science (Washington, D. C., 1883-)*, 1991, **254**, 1471–1477.

[24] A. D. Curzons, D. J. C. Constable, D. N. Mortimer and V. L. Cunningham, *Green Chemistry*, 2001, **3**, 1–6.

[25] C. Jimenez-Gonzalez, C. S. Ponder, Q. B. Broxterman and J. B. Manley, *Organic Process Research & Development*, 2011, **15**, 912–917.

[26] M. J. Gronnow, R. J. White, J. H. Clark and D. J. Macquarrie, *Organic Process Research & Development*, 2005, **9**, 516–518.

[27] D. Kralisch and G. Kreisel, *Chem. Ing. Tech.*, 2005, **77**, 784–791.

[28] D. Kralisch and G. Kreisel, *Chem. Eng. Sci.*, 2007, **62**, 1094–1100.

[29] International Organization for. Standardization, Geneva, 2006, vol. EN ISO 14040.

[30] International Organization for. Standardization, Geneva, 2006, vol. EN ISO 14044.

[31] *The eco-indicator 99, A damage oriented method for life cycle impact assessment*, PRe'Consultant, 2001.

[32] J. B. G. Guinée, M.; Heijungs, R.; Huppes, G.; Kleijn, R.; de Koning, A.; van Oers, L.; Wegener Sleeswijk, A.; Suh, S.; Udo de Haes, H. A.; de Bruijn, H.; van Duin, R.; Huijbregts, M. A. J., *Handbook on life cycle assessment. Operational guide to the ISO standards. I: LCA in perspective. IIa: Guide. IIb: Operational annex. III: Scientific background*, Dordrecht, 2002.

[33] M. J. H. Goedkoop, R; Huijbregts, M.; De Schryver, A.; Struijs, J.; van Zelm, R;, *ReCiPe 2008, A life cycle impact assessment method which comprises harmonised category indicators at the midpoint and the endpoint level*, 2009.

[34] J. Dewulf, M. E. Boesch, M. B. De, d. V. G. Van, L. H. Van, S. Hellweg and M. A. J. Huijbregts, *Environ. Sci. Technol.*, 2007, **41**, 8477–8483.

[35] G. Van der Vorst, J. Dewulf, W. Aelterman, W. B. De and L. H. Van, *Environ. Sci. Technol.*, 2011, **45**, 3040–3046.

[36] G. Van der Vorst, W. Aelterman, W. B. De, B. Heirman, L. H. Van and J. Dewulf, *Green Chem.*, 2013, **15**, 744–748.

[37] G. Geisler, S. Hellweg, T. B. Hofstetter and K. Hungerbuehler, *Environ. Sci. Technol.*, 2005, **39**, 2406–2413.

[38] O. G. Griffiths, J. P. O'Byrne, L. Torrente-Murciano, M. D. Jones, D. Mattia and M. C. McManus, *J. Cleaner Prod.*, 2013, **42**, 180–189.
[39] S. Huebner, S. Kressirer, D. Kralisch, C. Bludzuweit-Philipp, K. Lukow, I. Janich, A. Schilling, H. Hieronymus, C. Liebner and K. Jahnisch, *ChemSusChem*, 2012, **5**, 279–288.
[40] A. Azapagic and R. Clift, *Int. J. Life Cycle Assess.*, 1999, **4**, 133–142.
[41] A. Banimostafa, S. Papadokonstantakis and K. Hungerbuhler, *Comput.-Aided Chem. Eng.*, 2012, **31**, 1120–1124.
[42] J. Dufour, D. P. Serrano, J. L. Galvez, J. Moreno and C. Garcia, *Int. J. Hydrogen Energy*, 2009, **34**, 1370–1376.
[43] D. Kralisch, in *Green Chemistry Metrics*, John Wiley & Sons, Ltd, 2009, pp. 248–271.
[44] M. Pehnt, *Renewable energy*, 2006, **31**, 55–71.
[45] G. Finnveden, M. Z. Hauschild, T. Ekvall, J. Guinee, R. Heijungs, S. Hellweg, A. Koehler, D. Pennington and S. Suh, *Journal of Environmental Management*, 2009, **91**, 1–21.
[46] J. A. Assies, *Journal of hazardous materials*, 1998, **61**, 23–29.
[47] C. Hendrickson, A. Horvath, S. Joshi and L. Lave, *Environmental Science & Technology*, 1998, **32**, 184A-191A.
[48] S. Suh, M. Lenzen, G. J. Treloar, H. Hondo, A. Horvath, G. Huppes, O. Jolliet, U. Klann, W. Krewitt, Y. Moriguchi, J. Munksgaard and G. Norris, *Environmental Science & Technology*, 2003, **38**, 657–664.
[49] *Simplifying LCA: just a cut? – Final report of the SETAC-Europe Screening and Streamlining Working-Group*, Society of Environmental Chemistry and Toxicology (SETAC), Brussel, 1997.
[50] D. Kralisch, C. Staffel, D. Ott, S. Bensaid, G. Saracco, P. Bellantoni and P. Loeb, *Green Chem*, 2013, **15**, 463–477.
[51] D. Kralisch, I. Streckmann, D. Ott, U. Krtschil, E. Santacesaria, M. Di Serio, V. Russo, L. De Carlo, W. Linhart, E. Christian, B. Cortese, M. H. J. M. de Croon and V. Hessel, *ChemSusChem*, 2012, **5**, 300–311.
[52] J. F. Jenck, F. Agterberg and M. J. Droescher, *Green Chemistry*, 2004, **6**, 544–556.
[53] C. Jimenez-Gonzalez, P. Poechlauer, Q. B. Broxterman, B.-S. Yang, D. A. Ende, J. Baird, C. Bertsch, R. E. Hannah, P. Dell'Orco, H. Noorrnan, S. Yee, R. Reintjens, A. Wells, V. Massonneau and J. Manley, *Organic Process Research & Development*, 2011, **15**, 900–911.
[54] W. J. W. Watson, *Green Chem*, 2012, **14**, 251–259.
[55] C. Jiménez-González, A. D. Curzons, D. J. C. Constable, M. R. Overcash and V. L. Cunningham, *Clean Products and Processes*, 2001, **3**, 35–41.
[56] S. Huebschmann, D. Kralisch, H. Loewe, D. Breuch, J. H. Petersen, T. Dietrich and R. Scholz, *Green Chem*, 2011, **13**, 1694–1707.
[57] D. Kralisch, A. Stark, S. Koersten, G. Kreisel and B. Ondruschka, *Green Chem.*, 2005, **7**, 301–309.
[58] D. Reinhardt, F. Ilgen, D. Kralisch, B. Koenig and G. Kreisel, *Green Chem.*, 2008, **10**, 1170–1181.
[59] G. Van der Vorst, J. Dewulf, W. Aelterman, B. De Witte and H. Van Langenhove, *Industrial & Engineering Chemistry Research*, 2009, **48**, 5344–5350.
[60] G. Van der Vorst, W. Aelterman, B. De Witte, B. Heirman, H. Van Langenhove and J. Dewulf, *Green Chem.*, 2013, **15**, 744–748.
[61] J. Yoshida, H. Kim, A. Nagaki, Green and sustainable chemical synthesis using flow microreactors. ChemSusChem, 2011, 4, 331–340.
[62] A. Cybulski, J. A. Moilijn, M. M. Sharma, R. A. Sheldon. Fine Chemicals Manufacture: Technology and Engineering, 1st edn., Elsevier Science B. V.: Amsterdam, 2001.
[63] D. M. Roberge, L. Ducry, N. Bieler, P. Cretton, B. Zimmermann, Chem. Eng. Tech. 28, 3 (2005) 318–323.

[64] B. Herzog, M. I. Kohan, S. A. Mestemacher, R. U. Pagilagan, K. Redmond, 2013. Polyamides. Ullmann's Encyclopedia of Industrial Chemistry. Wiley-VCH, Weinheim
[65] J. W. M. Steeman, S. Kaarsemaker, Hoftyzer, P. J. Chemical Engineering Science, 1961, 14, 139–149.
[66] A. Castellan, J. C. J. Bart, S. Cavallaro, Catalysis Today, 1991, 9, 237–254.
[67] K. Sato, M. Aoki, R. Noyori, Science, 1998, 281, 1646–1647.
[68] M. G. Buonomenna, G. Golemme, De M. P. Santo, E. Drioli, Organic Process Research & Development, 2010, 14, 252–258.
[69] Y. Deng, Z. Ma, K. J. Wang, Chen. Green Chemistry, 1999, 1, 275–276.
[70] I. Vural-Gursel, Q. Wang, Noël, T., V. Hessel, Chemical Engineering Transactions, 2012, 29, 565–570.
[71] G. Centi, S. Perathoner, Catalysis Today, 2009, 143, 145–150.
[72] J. M. Campos-Martin, G. Blanco-Brieva, J. L. G. Fierro, Angewandte Chemie International Edition, 2006, 42, 6962–6984.
[73] R. Noyori, M. Aoki, K. Sato, Chemical Communications, 2003, 1977–1986.
[74] G. Grigoropoulo, J. H. Clark, J. A. Elings, Green Chemistry, 2003, 5, 1–7.
[75] Y. Usui, K. Sato, Green Chemistry, 2003, 5, 373–37
[76] A. Podgorsek, M. Zupan, J. Iskra, Angewandte Chemie International Edition, 2009, 48, 8424–8450.
[77] J. K. Edwards, B. E. Solsona, P. Landon, A. F. Carley, A. Herzing, C. J. Kiely, G. J. Hutchings, Journal of Catalysis, 2005, 236, 69–79.
[78] M. K. Tse, M. Klawonn, S. Bhor, C. Dobler, G. Anilkumar, H. Hugl, W. Magerlein, M. Beller, Organic Letters, 2005, 7, 987–990.
[79] G. N. Kulsrestha, U. Shankar, J. S. Sharma, J. Singh, Journal of Chemical Technology & Biotechnology, 1991, 50, 57–65.
[80] Q. Wang, I. Vural-Gursel, M. Shang, V. Hessel, Chemical Engineering Journal, 2013, 234, 300–311.
[81] P. Lundberg, C. J. Hawker, A. Hult, M. Malkoch, *Macromol. Rapid Commun.*, **2008**, 29, 998–1015.
[82] A. Papoulis, S. Unnikrishna, McGraw-Hill, Probability, *Random Variables and Stochastic Processes*, **2002**, 211.
[83] http://en.wikipedia.org/wiki/Orthogonality.
[84] C.-H.Wong, S. C. Zimmerman, *Chem. Comm.*, **2013**, 1679–95.
[85] R. K. Iha, K. L. Wooley, A. M. Nyström, D. J. Burke, Kade, C. J. Hawker, *Chem. Rev.*, **2009**, 109, 5620–5686.
[86] (a) M. D. Yilmaz and J. Huskens, *Soft Matt.*, **2012**, 8, 11768–11780; (b) D. Wasserberg, C. Nicosia, E. E. Tromp, V. Subramaniam, J. Huskens, P. Jonkheijm, J. Am. Chem. Soc., **2013**, 135, 3104–3111; (c) A. González-Campo, M. Brasch, D. A. Uhlenheuer, A. Gomez-Casado, L. Yang, L. Brunsveld, J. Huskens and P. Jonkheijm, *Langmuir*, **2012**, 28, 16364–16371; (d) A. González-Campo, B. Eker, H. J. G. E. Gardeniers, J. Huskensand, P. Jonkheijm, *Small*, **2012**, 8, 3531–3537.

Answers to the study questions

Chapter 1: Catalysis in flow

1.1. In an industry where each process usually generates more waste than product, catalytic reactions open the door to environmentally more benign and cost effective processes. The flow aspect enables the exploitation of parallel or consecutive reaction kinetics or thermodynamic equilibria by either stopping the reaction at an opportune moment in time, or changing reaction conditions as the reaction progresses to force the yield in favor of the desired product. A continuous reactor minimizes the reacting inventory, hence the hazard potential is much reduced in comparison with a batch reactor, allowing even highly energetic reactions to be carried out safely.

1.2. On the one side, the heterogeneous catalyst needs to interact intimately with the reaction mixture, on the other it needs to be easily separable from it. This effectively requires the reactor to perform the dual role of a mixer as well as a separator, which is usually achieved by trapping the catalyst through a system of filters. Smaller particle sizes facilitate better mass transfer but can lead to high pump energy requirements or operational issues such as line blockages. The influence of catalyst particle size on the reaction outcome should definitely be investigated.

1.3. The success of any multiphase reaction depends on the intimate contact between the two phases and thus on the amount of interfacial area. If the available interfacial area is large, it is likely that the potential of the chemistry is maximized. Conversely, if the available interfacial area is small, mass transfer effects will severely constrain the chemical reaction.

1.4. Olefin ring closing reactions to form C=C bonds using Ruthenium complex catalysts and cross-coupling reactions with Palladium catalysts.

1.5. The reactions generate volatile leaving groups, such as ethylene, which must be removed from the reaction mixture, and selectivity issues mandate the requirement for dilute reactant solutions. To achieve high levels of conversion under these conditions, a continuous reactor in re-circulation mode is highly suitable.

1.6. Palladium catalysts are susceptible to leaching, that is, the loss of the active metal. This causes loss of activity of the catalyst as well an impurity problem in the reaction mixture. Microencapsulation, functionalized solid supports, and silica-supported phosphine ligands are some approaches that can be used to minimize this effect.

1.7. Enzymes can easily be immobilized and as the resulting molecules are much larger than the substrates, membrane reactors can separate the catalyst from the reaction mixture, fulfilling the dual role of reactor and separator. The continuous removal of product or by-products may also be advantageous for reversible reactions from a chemical equilibrium perspective.

1.8. Any hydrogenation reactor using hydrogen gas will have to be pressurized to facilitate mass transfer of hydrogen from the gas into the liquid phase. A batch reactor has a larger pressurized volume than a continuous reactor and has hence a much higher hazard potential.

1.9. The most promising approach to carry out these reductions in continuous flow reactors is by immobilizing the organometallic catalyst on, for example, an inert carrier material (such as alumina) or making the metal-phosphine ligand system self-supporting.

1.10. The rate can be accelerated by increasing the mass transfer of oxygen into the system. This can be done by, for example, increasing oxygen pressure or the mixing efficiency between gas and liquid.

Chapter 2: Catalytic engineering aspects of flow chemistry

2.1. In monophasis (liquid) reaction and providing that the operating conditions are the same (concentrations, temperature, pressure), the contact time in the tube plug flow reactor can be directly used for comparison with the batch results (conversion, selectivity). In a multiphase reaction, generally with a solid catalyst, the proportion of the catalyst used in the Flow reactor, often a fixed bed of particles, is much higher than in Batch tests. Thus, some type of "normalized time" must be used. The concept of Hourly Space Velocity (HSV is very useful and the time in Flow reactor is defined as the reverse, i.e. 1/HSV). More discussion can be found in the chapter devoted to chemical engineering aspect of Flow Chemistry.

2.2.

Take, for the sake of conciseness, a first order rate law:

$$r = kC_S C_{cat}$$

The same S/C ratio of 100 could be obtained taking different concentrations in two different experiments but the results in terms of conversion or TOF or conversion versus time profile (see Figure 2.6) will be very different ...

	C_S	C_{cat}	C_S/C_{cat}	k	React. time	X_{300}	TOF_X
Exp 1	1	0.01	100	1	300	95%	333
Exp 2	0.05	0.0005				14%	17

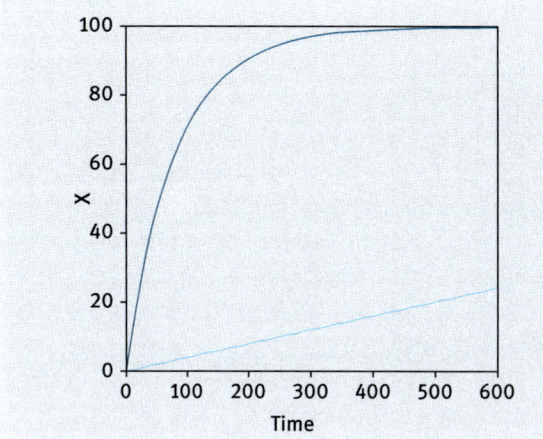

2.3. Both experimental and theoretical methods may be used. Without experiments but with some data collected during preliminary studies of the targeted reaction, several characteristic times for different processes can be estimated. For example, the reaction time, mixing time, liquid-solid mass transfer time, mass transfer (diffusion) time inside the catalyst porous network. Then these times may be compared using different dimensionless numbers (ratio of the characteristic times) and the general rule is that the time for reaction must be much longer than any other times. The intrinsic chemical reaction is thus the limiting process in the reactor. When further experiments can be performed, it is advised to check the influence of agitation, size of catalyst particles and the value of the apparent activation energy. More discussion can be found in the chapter devoted to chemical engineering aspect of Flow Chemistry.

2.4. Simple proportion rules may be used only when both the laboratory reactor and the production reactor are working under the chemical regime, that is, when the intrinsic chemical reaction is the limiting phenomenon. Indeed, several physical processes such as diffusion, convection and so on may falsify the true chemical performances. All these physical processes are length dependent. Thus it is obvious that scaling-up data that totally or partially result from physical phenomena will drive to bad quality prediction of the production reactor. Different characteristic lengths thus different impacts of those physical phenomena will play compared to the lab reactor.

Chapter 3: Continuous-flow photochemistry in microstructured environment

3.1. Upon interaction with electromagnetic radiation of appropriate wavelength, the molecule is converted into its electronically excited state. An electron is promoted to a higher energy level which results in a different reactivity and geometry of the molecule. In reverse, different deactivation pathways are possible: nonradiative deactivation, internal conversion, fluorescence, intersystem crossing to the triplet state, phosphorescence.

For further reading, see reference [6].

3.2. The absorption of light is based on the electronic properties of the molecules (number of π-electrons, molecular structure and alignment of π-electrons in molecule,...). As discussed, benzene with 6 π-electrons has another absorption range than its larger homologue anthracene with 14 π-electrons. In general, the higher the number of conjugated π-electrons, the higher the absorption wavelength. Electron donating (e.g., NH_2 or OH) or electron accepting groups (NO_2 or CN), which are in conjugation with the π-electron system can have a strong impact on the light absorption behavior of molecules.

For further reading, see Zollinger H. Color chemistry: syntheses, properties, and applications of organic dyes and pigments. 3rd edn, Zürich, Switzerland, Helvetica Chimica Acta, 2003.

3.3. The solvent must be compatible with the photochemical reaction. Besides being the transport medium for the reactants, the solvent must not absorb the incident light. The cut-off wavelength of the solvent defines, in general, its spectral window and in consequence its applicability for a photochemical process. Interactions must be circumvented between solvent molecules and reactants which lead to a quenching of the excited states generated during the irradiation. The solvent must act as inert medium regarding the photochemical processes. Except in gas/liquid reactions the solvent should be degassed. Especially, dissolved oxygen can lead to the formation of undesired by-products upon irradiation with appropriate wavelength. Contrary to the photophysical processes, sometimes the solvent must interact with a reactant to allow the progress of a chemical reaction, for example, the uptake or release of a proton.

For further reading, see references [7, 10].

3.4. The quantum yield Φ_λ expresses the ratio between the number of molecules produced during the reaction and the number of absorbed photons at a given wavelength. The ideal case is given when every absorbed photon yields one molecule of product. In reality, quantum yields are below 1 due to other deactivation pathways. Values far greater than 1 result from chain reactions as consequence of light-triggered start reactions.

For further reading, see reference [9].

3.5. Metal vapor and gas discharge lamps are light sources with a broad emission band, in case of mercury vapor lamps with spectral fine structure (polychromatic, nonthermal emitter), in case of xenon discharge lamps a more continuum spectrum (panchromatic, thermal emitter). In general, the light source should be compatible with photochemical reaction. In consequence, the light source should provide only light with appropriate wavelength for the desired photochemical reaction. Any irradiation which leads to decomposition, by-products or just heating of the reaction components reduces both the conversion or selectivity and the energy efficiency of the reaction.

3.6. Light emitting diodes are nearly monochromatic emitters circumventing the necessity of filter equipment. Their energy consumption is very low; the energy management is quite easy. Their life time can exceed 100,000 hours. LEDs do not need such an elaborated heat management compared to, for example, mercury vapor lamps. Therefore LEDs can be placed closer to the microreactor yielding a more efficient irradiation. LEDs are single point emitters. Their individual arrangement in arrays of different size and shape enables the adaption to many reactor concepts.

For further reading, see reference [18].

3.7. Although solar irradiation is panchromatic, the spectral window for photochemical reactions with solar light of adequate energy is limited to the terrestrial visible and infrared region. The largest part of high energy UV radiation is filtered by the atmosphere of our planet. Therefore, sensitizer molecules are necessary acting as energy collectors and transmitters to the real substrate which cannot be excited by visible light in the most cases.

For further reading, see König B. ed. Chemical photocatalysis. Berlin, Germany, De Gruyter, 2013.

3.8. Microchannels or micrometer-sized capillaries provide very thin liquid films or streams. Under this condition, a complete illumination of the reaction solution is possible. In microreactors, the reaction solution is pumped through a defined volume given by the length and diameter of the microchannel or the capillary. This allows an exact control of the irradiation time by fine-tuning the flow rate. Overexposure leading to by-products is minimized by this way. Beside this advantage, the optimized illumination of thin liquid films and streams allows the use of high photocatalyst concentrations, or light-absorbing molecules with very high extinction coefficients.

3.9. Numbering-up means to multiply, for example, the number of channels with the same architecture. In the ideal case, a chip reactor with ten channels produces the ten-fold amount of product compared to a single channel chip reactor. In consequence, 10 chip reactors with 10 channels each reactor unit can produce the 100-fold amount of product compared to one single channel chip reactor. This is a very elegant approach for photochemical conversions. On the contrary, the scale-up approach can lead to undesired results here since you change the architecture of the channel, for example, by enlarging the diameter of the channel or its length. The first change might lead to an incomplete illumination of the reaction solution while the length extension leads to a longer irradiation time with the possibility for by-product formation.

3.10. A falling film microreactor should be used. First, this reactor concept is excellently suitable for contacting a liquid phase with a gas phase due to its high surface-to-volume ratio. Second, both tetraphenylprophyrine and Rose Bengal are chromophores with very high extinction coefficients leading to a strong light absorption within a light path length as short as some tens of micrometers. This condition must be considered for the reactor concept applied. The falling film microreactor is able to produce liquid thin films of such small dimension due to the optimized architecture of the reaction plate.

3.11. Titanium dioxide can be immobilized on microchannels by a sol-gel process. This allows the formation of solid thin films on the channel walls. With this approach, heterogeneous photocatalysis is possible eliminating the separation of solid catalyst materials from the reaction solution. In addition, the roughness of the TiO_2 thin film enhances the available surface for irradiation and induces turbulences for better mixing and contacting of the liquid phase with the solid thin film.

3.12. The light absorption of TiO_2 is limited to the UV region ending at approx. 390 nm. In order to extend the absorption range of this semiconductor, a combination with other light-absorbing molecules is feasible. By dye-sensitizing of the TiO_2 surface with, for example, ruthenium polypyridyl complexes, the absorption of visible light is also possible. With this approach, a heterogeneous photocatalyst can be made available which might also be incorporated into microchannels provided that the dye is strongly bound to its inorganic substrate.

3.13. Visible light is of lower energy than UV light. This fact results in easier reaction control and minimizes by-product formation in the most cases. The photocatalysts used are carbon-based compounds which can be synthetically fine-tuned in order to fit as well as possible to the desired photochemical reaction parameters. The mild reaction conditions also allow the combination with sensitive reaction classes like organocatalysis. Beside solar light as the most environmentally friendly light source, household fluorescent bulbs or LED arrays can be used as well. As a consequence, both the energy and heat management is far easier to handle.

For further reading, see König B. ed. Chemical photocatalysis. Berlin, Germany, De Gruyter, 2013.

Chapter 4: Electrochemistry in flow

4.1. Electrochemical processes are useful for storage and generation of energy, for example in rechargeable batteries (MeH, LiPo, NiCd) and fuel cells. Electrochemical refinery (Cu, Al, Mg...), electroplating (Ni, Cu, Ag, Au...), surface treatment (oxidation) and electrolysis (NaOH) are also common industrial processes.

4.2. An electrochemical cell consists of two electrodes connected to an external DC power supply (current, voltage). Both electrodes are immersed in an electrolyte which contains chargeable anions and cations. By applying an electrical current, an oxidation reaction takes place on one electrode surface (anode) and a reduction on the opposite one (cathode).

Overall reaction: $2H_2O \rightarrow 2H_2 + O_2$
Anode reaction: $6H_2O \rightarrow O_2 + 4H_3O^+ + 4e^-$ (acidic solution)
$4OH^- \rightarrow O_2 + 2H_2O + 4e^-$ (alkaline solution)
Cathode reaction: $2H_3O^+ + 2e^- \rightarrow H_2 + 2H_2O$ (acidic solution)
$2H_2O + 2e^- \rightarrow H_2 + 2OH^-$ (alkaline solution)

4.3. It is necessary to add a conducting salt, which dissociates into anions and cations, to ensure the electrical transfer of charges (current) through the electrolyte. Without adding a conducting salt, the ohmic resistances between the electrodes raises and the electrode potential increases. Therefore, the electrochemical reaction becomes uncontrolled and the selectivity decreases.

A close proximity between anode and cathode, which is the case in a micro or thin-gap cell setup, reduces the ohmic resistance a priori and the concentration of the conducting salt can be remarkably decreased.

4.4. The *Wagner* number (W_A) is a measure for the distribution of the current density distribution along an electrode surface. It is defined as the quotient of the charge transfer resistance close to the electrode surface (R_S) and the overall resistance of the electrolyte (R_E):

$$W_A \approx \frac{R_S}{R_E}$$

For $W_A \rightarrow 0$, the charge transfer is equal over the whole electrode surface and the reaction is strictly controlled by diffusion.

4.5. The KFT is mostly used for the determination of the water content in organic solvents. To perform the KFT, a water-free solution of methanol, which contains iodine, SO_2 and pyridine or imidazole as buffer substance, is used as titrating solution. This solution appears as a brownish colorized liquid. By dropping this solution to an appropriate volume of the organic solvent, the iodine is reduced to the colorless I^- ion in the presence of water. Once all the water is consumed by the titration reaction, the solution remains colorized. With the used volume of the titrating solution the water content of the organic solvent can be calculated.

4.6. Too high current densities can cause high electrode potential due to the ohmic law, that is, it exceeds the activation potential of the targeted electrochemical reaction. As a result, the reaction becomes uncontrolled and side-reactions take place, mainly the organic solvent undergoes unwanted oxidation or reduction processes.

4.7. Performing an electrochemical reaction in a potentiostatic mode (electrode potential) instead of a galvanostatic one (current/current density) allow to select the chemical reaction by the applied potential and to suppress unwanted side-reactions. The potential defines the current and not vice versa. In this case, the current decreases proportional (at least to 0) by the progress of the electrochemical reaction.

Chapter 5: Synthesis of materials in flow – principles and practice

5.1. Low Reynolds number (below 2000) implies a laminar flow profile for a fluid flowing through a tube. When two fluids with low Reynolds number are combined they will co-flow and if miscible, will mix via diffusion. When Reynolds numbers are high the solutions will mix.

5.2. The major advantages of performing materials synthesis in flow are the excellent control over mass transfer, heat transfer, pressure attributes, fluid structure and behavior.

5.3. Droplets can be prepared using a T-junction microfluidic device, or a coaxial flow focusing device. Satellite droplets can be observed in each of these strategies (depending on flow rates and properties of liquids involved).

5.4. O/W/O emulsions can be generated using a coaxial flow focusing device. Emulsions of this type are used to prepare responsive capsules.

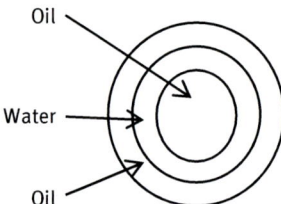

5.5. Janus-faced particles are anisotropic particles with two "faces" or regions, each having physical and (or) chemical properties that differ from the other. Low Reynolds conditions allow for these particles to be prepared in flow. The laminar flow profile enables two immiscible fluids to be co-flowed yielding Janus droplets that are subsequently converted to Janus particles.

Chapter 6: Flow chemistry for nanotechnology

6.1. Nanomaterials are chemical substances or materials having a diameter in the range of 1–100 nm that are manufactured and used at the nanometer scale.

6.2. Flow-chemistry based approaches offer inexpensive, scalable, reproducible and safe routes for the development and production of uniform nanosized materials. Flow-chemistry based approaches have numerous advantages such as good mixing, improved heat and mass transfer and high space-time-yields due to short diffusion lengths and high surface-to-volume ratios of the reactors, green and sustainable production (increased safety (toxic materials) and reduced costs (expensive materials)), very homogeneous conditions (reduced turbulence, low Reynolds numbers) and fast and precise adjustment and modification of process parameters (T, p, x) over conventional batch methods.

6.3. The particle formation and growth can be described by the Classical Nucleation Theory. The theory describes the nucleation process in terms of the change in Gibbs free energy of the system upon transfer of molecules from the liquid phase to a solid cluster with the following equation:

$$\Delta G = -\frac{4}{V}\pi r^3 k_B T \ln(S) + 4\pi r^2 \gamma$$

where V is the molecular volume of the precipitated species; r is the radius of the nuclei; k_B is the Boltzmann constant; T is the absolute temperature; S is the saturation ratio and γ is the surface free energy per surface area.

The particles can only be formed if $S > 1$, the system is supersaturated, and the free energy term is negative, favoring generation of solid particles and their growth.

The maximum value of ΔG corresponds to the nucleus with critical radius (r^*). A thermodynamically stable nucleus exists when the radius of the nucleus reaches to r^*. Therefore, the slope of ΔG at critical radius of nucleus will be zero:

$$\frac{d\Delta G}{dr} = 0.$$

In this situation, the critical radius (r^*) of the spherical nucleus can be obtained by:

$$r^* = \frac{2V\gamma}{3k_B T \ln(S)}.$$

The behavior of formed solid particles in supersaturation solution depends on their size. Particles with the radius smaller than r^* will dissolve because this is the only way that leads to the reduction of the particle's free energy. Similarly, if $r > r^*$ particle growth will occur (see Figure 6.8).

The colloid stability is the balance of various interaction forces such as van der Waals attraction, double-layer repulsion and steric interaction. These interaction forces have been described at a fundamental level by Deryaguin and Landau and Verwey and Overbeek (DLVO theory). In this theory, the van der Waals attraction is combined with the double-layer repulsion and an energy–distance curve can be established to describe the conditions of stability/instability. The electrical forces increase exponentially as particles approach one another and the attractive forces increase as an inverse power of separation. As a consequence, these additive forces may be expressed as a potential energy versus separation curve. A positive resultant corresponds to an energy barrier and repulsion, while a negative resultant corresponds to attraction and hence aggregation (Figure 6.7 (a) and (b)). It is generally considered that the basic theory and its subsequent modifications provide a sound basis for understanding colloid stability.

6.4. There are two main approaches to make nanomaterials: "top-down" and "bottom-up" technologies. Top-down approach basically relies on mechanical attrition to render large components into nanosized substances, for example nanomilling. The bottom-up approach relies on the arrangement of smaller component into more complex assemblies at molecular level, for example continuos flow metal nanocrystal production in microreactors.

6.5. Gold, silver, platinum, palladium, iron/gold, iron/copper, cobalt/platinum and so on.

6.6. Organic nanoparticles consist of active pharmaceutical, agrochemical ingredients, nutraceuticals or food supplements, active ingredients of skin care products, and so on. For examples, drugs encapsulated in liposomes or dentrimers, nanocrystals of pesticides, fungicide, encapsulated vitamines.

Chapter 7: Continuous-flow synthesis of carbon-11 radiotracers on a microfluidic chip

7.1. Radiolabeling is a synthetic process that involves incorporating a radioisotope into a chemical compound.
Positron Emission Tomography (PET) is an imaging technique in various areas of medicine such as oncology, cardiology, neurology and pharmacology that involves that use of radiolabeled compounds. Radioisotopes used in this imaging technique must be positron ($_1\beta^+$) emitters.

7.2. One half-life will result in 50% of the original amount, and so the radioactivity loss is 50%. Two half-lives will result in the 50% amount to become 25%, and so the loss is 75%. Three half-lives will make 25% become 12.5%, resulting in a loss of 87.5%.

7.3. Some advantages the investigation of [^{11}C]raclopride synthesis on a microfluidic chip explored:
   – Reduce reaction times.
   – Can carry out multiple reactions from the same batch while manipulating the reaction conditions.
   – Reduce work space and increase safety.
   – Reduce amount of reagents used and hence costs.

7.4. A passive micromixing loop design is incorporated in the microchannels in an attempt to increase mixing and reagent interactions and hence increase product generation.

7.5. A slower flow rate will increase the amount of [^{11}C]raclopride produced because more time is allowed for the reaction to proceed.
7.6. Since the radiochemical yield (now based on the limiting reagent of [^{11}C]methyl iodide) was not optimized, the relative activity of [^{11}C]raclopride was adopted for comparison purpose.
7.7. PDMS is a porous polymer, and therefore volatile [^{11}C]methyl iodide will be lost through evaporation.

Chapter 8: Lab environment: in-line separation, analytics, automation & self optimization

8.1. In-line: IR, UV-VIS, Fluorescent spectroscopy; On-line: MS, FID-GC
8.2. 35–40 °C
8.3. Signal intensity of starting material would be constantly 0, while the signal intensity of the product would stay constantly at 100%.
8.4. The parameter space is not used optimally; most of the space was not sampled. See graph below.

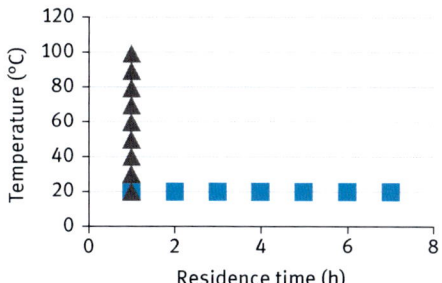

Badly investigated parameter space.

8.5. Benzaldehyde will be adsorbed by the silica, while benzylalcohol will pass through the gel with the eluent. So the extract will contain the benzaldehyde, and the raffinate will contain the benzylalcohol.

Chapter 9: Safety aspects related to microreactor technology

9.1. Safety (or loss prevention): the prevention of accidents through the use of appropriate technologies to identify the hazards of a chemical plant and eliminate them before an accident occurs.

Hazard: a chemical or physical condition that has the potential to cause damage to people, property, or the environment.

Risk: a quantitative or qualitative measure of human injury, environmental damage, or economic loss in terms of both the incident likelihood and the magnitude of the loss or injury. The risk is usually defined as $R = P \times S$ (P = probability and S = Severity).

9.2. The high surface-to-volume ratio ensures improved control over the operating conditions and reduces the risk of thermal runaway during the process itself or downstream processing steps. The ability to efficiently remove heat allows for exothermic reactions to be carried out at higher temperatures and concentrations, which has a beneficial impact on the global efficiency of the process. Unstable materials are immediately processed rather than being stored or shipped. Small volumes significantly reduce the hazards associated with the handling of toxic/reactive materials or gases under high temperatures and pressures.

9.3. The removal of reaction heat proceeds mainly through the surface of the reactor, that is, the heat exchanger, where thermofluids are usually circulated to keep the reactor isothermal. The ratio between the surface of the reactor and its volume decreases with an increase in reactor size, and strongly affects the efficiency of heat removal. Thus, in practice, smaller volumes and larger surfaces make it easier to manage heat removal.

9.4. Processes Involving Exothermic Reactions and Highly Reactive, Potentially Explosive Materials; Processes Involving Toxic Materials; Processes Involving Harsh Conditions (elevated temperatures and pressures).

9.5. See Sections 9.2.2.1 (Nitrations) and 9.2.2.1 (Azides).

Chapter 10: From green chemistry principles in flow chemistry towards green flow process design in the holistic viewpoint

10.1. See Table 10.1. Open question.

10.2. Open question. Examples below.

12 principles of Green Chemistry	Green features for continuous flow chemistry in microreactor
1. Prevention: prevent waste but not treat or clean up waste afterwards.	Less waste or no waste because of high conversion and selectivity
3. Less Hazardous Chemical Syntheses: Design synthetic methods to use and generate substances that minimize toxicity to human health and the environment.	More safe to operate because of the possibility to skip intermediate product treatment in flow chemistry
5. Safer Solvents and Auxiliaries: Minimize the use of auxiliary substances wherever possible, make them innocuous when used	Less solvent or greener solvent in micro-flow synthesis, because of the possibility to operate under high concentration

10.3. Many issues, basically to be grouped hierarchically from reaction- to process scale and finally to the environment/-society scale. The major ones are – on reaction scale: short residence times – narrow residence time distribution; faster mixing of miscible phases; higher dispersion of immiscible phases; improved heat transport; improved safety. On a process scale: increased selectivity, increased reactivity & selectivity, increased enantioselectivity – changed isomer ratios. On a company scale, this refers to lowering CAPEX and OPEX costs, the net present value (NPV) in cash flow, faster process development (50% idea), and gives rise to new business chances/products/markets ('windows of opportunity'). On an environment/scociety scale, maximal use of starting materials (reactants) and minimize waste, minimize solvent load, have maximal process safety, minimize toxicity (to human), and maximize process integration. There is still more, but goes beyond a simple answer.

10.4. Obviously, this is a creative question and has not one answer. Nitrations are of major industrial interest. Flow processing has at an early stage considered nitrations. For reason of enhanced mass transport – segmented flow with immiscible phases. For reason of increased heat transport – nitrations are highly exothermic and hot spots cause side-product formation (with higher activation energy). Thus cooling of the segmented flow is demanded, possibly by microchannels. Several green advantages are possible. Lower solvent load due to higher workable concentration. Less waste due to higher selectivity and less side-products. Less energy consumption despite higher cooling efforts due to higher productivity. Less toxicity due to encased processing.

10.5. Yes, one can; although the scope now seems limited and an exact picture remains to be answered by future research. For example, the tighter connection of heat-releasing and heat-receiving units allows for efficient energy use (e.g., endothermic/exothermic). The microeffects are then on the heat integration side and not on the reaction side. The compactness of flow units allows to integrate several units in close proximity which can give rise to synergistic effects. In a vision, to have reaction cascades using intermediates as reactants.

10.6. (a) The mass balance of (i) minimum all substances included in the reaction mixtures and (ii) at best all material flows induced by these syntheses.
 (b) The energy demand (i) minimum for the syntheses itself and (ii) at best all energy flows induced by these syntheses.

10.7.

	Metrics	Full LCA
Pro	– Fast – Low evaluation effort – Data directly available – Results easy comprehensible	– High validity – Holistic view – Analysis of environmental impacts instead of mass balances – Improved comparability
Con	– Holistic view is missing – May result in problem shifting to up-stream or down-stream processes – Effects on the environment only indirectly considered	– Time demanding – Quality and validity strongly depends on data availability – Difficult to perform in early reaction and process design stages

10.8. The choice depended on:
    (a) Goal and scope of the evaluation
    (b) Stage of development and data availability

10.9. Open question (recommendation: have a look on the 12 principles of green chemistry and green engineering (Table 10.1) as basis for your ideas)

10.10.(a) Increase the yield of adipic acid from cyclohexene
    (b) Decrease the decomposition of H_2O_2
    (c) Find a green route for H_2O_2 synthesis
    (d) Substitute H_2O_2 by another greener oxidant
    (e) Apply high efficient heat exchanger (micro heat exchanger) to decrease the cool energy utilization

10.11. Open question.

Index

A
absorbance 64, 68
Adipodinitrile 116
anatas 81, 82
artemisinin 77, 95
asymmetric 86, 87, 91, 92, 97
automated optimization v
automation v

B
Barton reaction 79
Batch Reactors
– Electrochemistry 100
Biological Process Windows 63
black light 80

C
capillary microreactor 89
capillary reactor 76, 86
catalysis v
Cathodic Reaction 110
chemistry v, vi
chip reactor 73
chlorine 75, 78
clogging 77
conduction band 70, 81
Continuous-flow 189, 193
countercurrent flow 75, 85
cut-off wavelength 66

D
deactivation 65, 66
diazonium salt 89, 90
Duran glass 70
dye-sensitized 69
dye-sensitized solar cell 81

E
Electrochemistry 99
– Activation of Chemicals 119
– Cation Flow Method 120
– Cation Pool Method 120
– Direct Synthesis 116
– Electrolyte Free Synthesis 117
– Ionic Liquids 122
– Organic Synthesis 100
– Process Parameters 103, 105
– Segmented Thin-gap Cell 115
– Wagner Number 107
enantiomeric excess 83, 86
enantioselective 86
Eosin Y 84
excimer 74
excitation 65, 69, 70
excited state 64, 66, 69, 72, 84

F
falling film microreactor 75, 79, 85, 97
falling film reactor 74
FEP 77, 86
flow v, vi
Flow Chemistry
– Electrochemistry 100
flow chemistry v, vi
fluorescence 66, 69, 86
fluorinated ethylenepropylene 76
fluorine 88

G
Grätzel cell 81
green chemistry v

H
heterogeneous v, 73, 80, 84
hydrogen peroxide 66

I
immersion well 73, 76, 77
Ionic Liquids 122

L
Lambert–Beer law 64, 65, 67
light emitting diode 70, 75, 76, 80, 85, 86, 88, 89
light intensity 64, 68
light transmission 64

M
mass transfer v
mercury vapor lamp 69, 76, 78

methylene blue 82
Microreactor
– Double Channel Thin-gap Cell 119
– Electrochemical 103
– ELMI 111
– Segmented Thin-gap Cell 115
– Thin-gap Cell 104
molar decadic extinction coefficient 64, 68

N

nanotechnology v
nitrite photolysis 79, 80
nonthermal emitter 69, 70
Number of Transfer Units 106, 107
numbering-up 74, 76, 83

O

Ohmic Drop 113, 115
organic light emitting diode 71
organocatalysis 86, 87, 91
organocatalyst 86, 87
ozone 71, 75

P

panchromatic 69
PET 190–192
pharmaceuticals v
photocatalysis 64, 80, 81, 84, 86, 91, 92
photocatalyst 72, 77, 81, 85, 87, 92, 97
photochemistry v, 72, 73, 75, 78, 87, 91, 92
photochlorination 78
photodegradation 82
photon 63, 66, 69
photon flux 63
photooxidation 66, 74
photooxygenation 66, 74, 85, 91
polychromatic 69
Pyrex glass 70, 80, 82

Q

quantum yield 65
quartz glass 70, 73, 80

R

Radiolabeling 189–191, 194
– [11C]methyl iodide 192, 193, 196, 197
– [11C]Raclopride 198, 199, 205–207
– [11C]raclopride 192–194, 196–205
– abacus 193–199, 203, 207

– automation 190
– Carbon-11 191
– desmethyl racloipride 204
– Desmethyl raclopride 196, 197
– desmethyl raclopride 193, 195, 196, 198, 199, 205, 206
– full loop 193, 194, 198, 199, 203, 204, 207
– methyl iodide 194–199, 204–207
– methylation 191, 192, 197, 198, 204
– no loop 193–196, 203, 204, 206
– nonradioactive 195
– **nonradioactive synthesis of raclopride** 194
– raclopride 189, 192–196, 198, 199, 202, 204, 206, 207
– **raclopride** yield 206
– radioactive 192, 196, 201, 202
radiolabeling
– full loop 206
riboflavin tetraacetate 64, 65, 84
Rose Bengal 64, 84, 85

S

safety v
scalability v
scaling-up 76
semiconductor 70, 81, 92
semiconductors 71
single electron transfer 84, 86
singlet oxygen 77, 85, 91
soda lime glass 80
sodium vapor lamps 69
solar irradiation 68, 85, 92
sol-gel process 73, 82
steady state v
Stirred Tank Reactor 101
supercritical v
surface-to-volume ratio 75, 81

T

terrestrial sunlight 71
tetraphenylporphyrine 65
thermal emitter 69
TiO_2 64, 73, 81–83, 96
titanium dioxide 81
toxic v
trifluoroethylation 88
triplet oxygen 85
triplet sensitizer 85

V
valence band 70, 81
visible light 72, 77, 85, 91, 92, 95–98

W
Wagner Number 107
wastewater treatment 81

X
Xenon arc lamp 69

Z
Ziegler–Stadler reaction 89, 90

Made in the USA
Middletown, DE
14 March 2020